Pharmaceutical Biotechnology

Pharmaceutical Biotechnology

Dr. Pankaj Goyal

RANDOM

RANDOM PUBLICATIONS

NEW DELHI (INDIA)

Pharmaceutical Biotechnology

ISBN 978-93-5111-211-2

Published in 2014 in India by

RANDOM PUBLICATIONS

4376-A/4B, Gali Murari Lal, Ansari Road
New Delhi-110 002
Phone : +91-11-43580356, +91-11-23289044
e-mail: randomexports@gmail.com, sales@randompublications.com, info@randompublications.com

Reprinted 2026

Type Setting by : Keystoneprintads, Delhi-110051
Digitally Printed at: Replika Press Pvt. Ltd.

Preface

Pharmaceutical Biotechnology is an increasingly important area of science and technology, and contributes to design and delivery of new therapeutic drugs, the development of diagnostic agents for medical tests, and the beginnings of gene therapy for correcting the medical symptoms of hereditary diseases.

The societal implications of Pharmaceutical Biotechnology are similarly widespread, ranging from many ethical issues of identifying and treating various hereditary diseases, to changes in healthcare practices and a significant contribution to national economic development.

Pharmacogenomics is the study of genes and the gene products (proteins) essential for pharmacological or toxicological responses to pharmaceutical agents. Oligonucleotide-based DNA chip technology or cDNA microarray can be used to analyse gene expression profiles that are induced or repressed by xenobiotics.

An attempt has been made to develop microarray genotyping system for multiplex analysis of a panel of single nucleotide polymorphisms (SNPs) in genes encoding proteins involved in blood pressure regulation, and to apply this system in a pilot study demonstrating its feasibility in the pharmacogenetics of anti-hypertensive drug response.

All the antibodies the immune system normally produces in response to a specific antigen are capable of marking (binding to) that antigen, but these antibodies—termed "polyclonal"—are varied, not identical. Monoclonal antibodies (MoAbs) that share a specific antigenic target are identical and are more sensitive to that target than are polyclonal antibodies for the same antigen. MoAbs are the products of hybridomas—cells that result from the biotech fusion of bone-marrow tumor cells and B lymphocytes. Hybridomas can be geared to produce specific MoAbs continuously.

This has been designed for undergraduate, postgraduate, medical, biochemistry, biotechnology and those students opting pharmacology biotechnology as a new subject.

I thank all members of my team who have helped in the preparation of the book. My special thanks go to "Random Publications" who have published the book.

– Dr. Pankaj Goyal

Contents

Preface .... *v-vi*

1. **Structure of Antifungal Drugs** **1**
   - Antifungal Drug Therapy 4
   - Characterization of Thermosensitive Gels 17
   - Characterization of Inclusion Complex 20
2. **Pharmaceutical Engineering** **23**
   - Buffering 23
   - Variety of Alkaloids 27
   - Role of the Kidneys in Acid-Base Balance 34
   - Pharmaceutical Nomenclature 36
   - Routes of Drug Administration 38
   - Pharmaceutical Dosage Forms 44
   - Adsorption in General Pharmaceutical Processing 51
3. **Antibiotics** **60**
   - History of Antibiotics 60
   - Specific Type of Drug Resistance 62
   - Chemical Substances of Antibiotic 71
   - Oncology Antibiotics Drugs 81
   - Classifications of Antibiotic 88
4. **Pharmacogenomics and DNA Microarrays** **92**
   - The DNA Microarray 93
   - Pharmacognosy and DNA Microarray 96
   - The Pharmaceutical Industry 98
   - DNA Microarrays in Pharmacodynamics 100
5. **Absorption Mechanisms of Drug** **106**
   - Objectives of Drug Absorption 106
   - Desiccant for Stabilizing Moisture-Sensitive Drug 107

Parenteral Routes of Drug 109
Absorption Mechanisms 118
Dosage Routes and Drug Absorption 118
Oral Dosing of Absorption Enhancers 127
The Cell Membrane 131

6. Pharmacodynamics 135

Methods of Measurement 135
Antimicrobial Pharmacodynamics 140

7. Biotransformation of Drugs 143

Pathways of Biotransformation 143
Drug Biotransformation Reactions 145
Need for Drug Biotransformation 155
Chemical Conversion of Drugs 155
Methods for the Study of Drug Biotransformation 156
Stereoselective and Stereospecific Biotransformation 158

8. Vitamins 164

Organic Compounds of the Vitamins 164
Vitamins and Minerals 165
Fat-Soluble Vitamins A, D, E, and K 183

9. Structures of Protein 190

Complex Protein Structures 190
Forces Controlling Protein Structure 191
Primary Structure of Proteins 193
Protein Folding and Tertiary Structure 196
Amino Acid Sequence in Protein Structure 199

10. Role of Enzymes 200

Functional Role of Coenzymes 200
Enzymes in the Diagnosis of Pathology 200
Regulation of Enzyme Activity 202
Enzymes as Biological Catalysts 205
Enzyme Relative to Substrate Type 212

11. Animal Models in Drug Discovery 216

Proper Care of Laboratory Animals 219
Principles of Animals in Drug Safety Testing 220
Sustainable Sources of Plant and Animal Drugs 233

12. Classification of Nutraceuticals 238

Effectiveness and Safety 240
Processed Foods 241
Nutritious Meal in Mental Agility 245
Functional Beverage of Drink Products 248

13. Nutrient Utilization **254**

Carbohydrate Utilization During Exercise 254
Factors Influencing Energy Metabolism 261
Factors Influencing Digestion and Absorption of Nutrients 264
Factors Influencing Food Consumption 267
Nutrition Tactics 267
Nutrition Tactics for Power Athletes 278

Index **284**

1

Structure of Antifungal Drugs

Genital tract infections, which are among the most frequent gynecological diseases, are treated commonly with imidazole-derivative antifungal agents such as clotrimazole since these drugs are locally active with no major side effects. Current vaginal delivery systems include creams, foams, gels, tablets, pessaries, and irrigations, which are limited because of residence time at the genitourinary tract: they are removed rather rapidly by the self-cleansing action of the vaginal tract.

Moreover, the physiological conditions imposed by the protective mechanism of the genital tract, limiting the residence time and thus impairing the therapeutic efficacy of the drug, make multiple and frequent administration necessary for treatment. Patient compliance when administering the dosage forms and following the repeated-dose therapeutic regimen is an important challenge in vaginal drug delivery. Patients are generally reported to tolerate gels better than other dosage forms, but it is believed that vaginal therapy can be significantly improved if a delivery system can keep the drug at the site of administration for longer than the conventional dosage forms can.

Mucoadhesive vaginal gels containing polycarbophil, Carbopol, hydroxypropylcellulose, and polyvinylpyrrolidone as polymers enhancing the adhesion of the administered dosage form to the mucosal tissue have been added to in situ–gelling thermosensitive gels prepared with Pluronic F127. Different techniques are adopted in drug delivery with thermosensitive gels, including dispersing the drug in the gel with a concentration higher than its solubility value and dispersing drug-loaded nanoparticles, liposomes, and drug:cyclodextrin complexes.

Cyclodextrins are used in the pharmaceutical field to form inclusion complexes with drug molecules to increase their aqueous solubility, to enhance their aqueous stability or photostability, to mask unwanted characteristics, or to reduce side effects. Cyclodextrins are also reported to convey controlled-release properties to certain active ingredients. The objective of this study was to design a vaginal gel formulation with thermosensitive and mucoadhesive properties to ensure longer residence at the infection site, providing a

Favourable release profile for the antifungal drug clotrimazole with the help of β-cyclodextrin complexation. The effect of β-cyclodextrin on drug release was believed to arise from its solubility-enhancing properties.

Clotrimazole (molecular weight: 344 g/mol, aqueous solubility: 5.5 μmol/L) was purchased from Sigma Chemicals, and β-cyclodextrin (Kleptose) was supplied by Roquette Freres. Pluronic F127 was purchased from Sigma-Aldrich, Carbopol 934PH was of pharmaceutical grade and purchased from BF Goodrich, and ydroxypropylmethylcellulose (HPMC) USP was purchased as Metolose 90SH 15000 from Seppic.

Polyethylene glycol 400 (PEG400), used to solubilize the drug in the gel formulation, was purchased from Fluka BioChemika. All other reagents were of analytical grade and used without further purification.

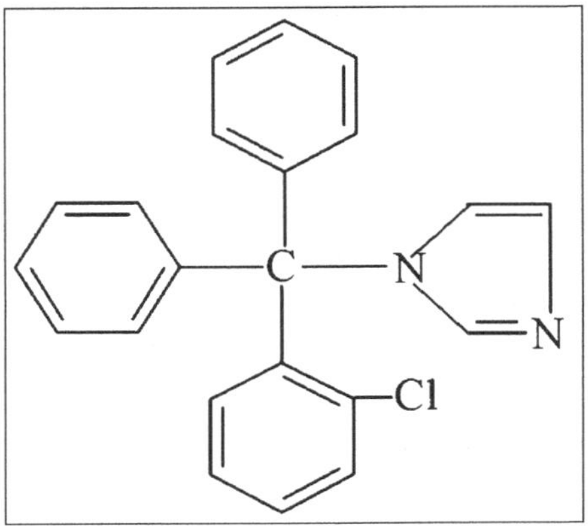

Fig. Structure of Antifungal Drug Clotrimazole.

Preparation of Clotrimazole:β-cyclodextrin Complex

The inclusion complex of clotrimazole with β-cyclodextrin in a 1:1 molar ratio was prepared according to the colyophilization technique. According to this technique, a fixed amount of clotrimazole was weighed and transferred to a round-bottomed flask. β-cyclodextrin solution in distilled water was added onto clotrimazole, and the powder was dispersed evenly. After ensuring that all clotrimazole powder was wetted by the β-cyclodextrin aqueous solution, the dispersion was left to equilibrate for 7 days at room temperature under constant stirring.

At the end of equilibrium time, the dispersion was filtered and filtrate containing soluble clotrimazole:β-cyclodextrin complex (1:1) was lyophilized.

Characterization of Clotrimazole:β-cyclodextrin Complex

Protein nuclear magnetic resonance (^1H NMR) spectra of clotrimazole, β-cyclodextrin, and clotrimazole:β-cyclodextrin complex were taken by a Bruker DPX 400 digital Fourier transform-nuclear magnetic resonance (FT-

NMR) spectrophotometer at 400 MHz. Chemical shifts were given to external tetramethylsilane at 0 ppm with calibration using solvent signals (dimethylsulfoxide at 2.5 ppm, deuterated water at 4.75 ppm, and chloroform $CHCL_3$ at 7.25 ppm).

Fourier transform infrared (FT IR) spectra of clotrimazole, β-cyclodextrin, and clotrimazole:β-cyclodextrin complex were taken with a Nicolet 520 FT IR spectrophotometer (Thermo Electron Corp, Waltham, MA) using discs of each sample and previously prepared potassium bromide containing 0.01 g of sample in 0.1 g of potassium bromide between the wavelengths of 400 and 4000 cm^{-1}. Differential scanning calorimetry (DSC) analyses were performed on lyophilized clotrimazole, lyophilized β-cyclodextrin, and clotrimazole:β-cyclodextrin complex samples with a DuPont 910 differential scanning calorimeter. Each sample weighing ~3 mg was heated in hermetically sealed aluminum pans at a rate of 10ºC/min up to 200ºC in a dynamic nitrogen atmosphere.

Scanning electron microscopy (SEM) was performed to elucidate the crystal structure of lyophilized clotrimazole, β-cyclodextrin, and clotrimazole:β-cyclodextrin complex. The samples were mounted on metal stubs with a double-sided adhesive band and then sputtered with a 100 Å thick layer of gold. They were examined with a JEOL scanning electron microscope at an acceleration voltage of 80 kV. Phase solubility studies were performed as follows: 10 mg of clotrimazole was mixed with 10 mL of aqueous solutions containing varying amounts of β-cyclodextrin within the range of 0 to 16 mM. The suspensions were shaken until equilibrium was reached (7 days). The suspension was filtered through a 0.22-μm membrane filter. The filtrate was Analysed spectrophotometrically for the total amount of solubilized clotrimazole.

Preparation of Clotrimazole Gel Formulations

Thermosensitive and mucoadhesive clotrimazole gels were prepared according to the cold method. Briefly, mucoadhesive polymer (Carbopol 934 or HPMC) (0.2% wt/vol) was slowly added to citrate phosphate buffer (0.1M, pH 4.0) at 4ºC with gentle mixing. Pluronic F127 (20% wt/vol) was then added to Carbopol or HPMC solution and allowed to dissolve overnight at 4ºC. Clotrimazole in free form (1%) was initially dissolved in a mixture of PEG400 and ethanol (5:3) and added to cold Pluronic F127 solution containing 0.2% Carbopol 934 or HPMC. On the other hand, for formulations containing drug:cyclodextrin complex, clotrimazole:β-cyclodextrin was dissolved directly in the cold Pluronic F127 solution containing 0.2% Carbopol 934 or HPMC with gentle mixing.

Characterization of Gels

The rheological Behaviour of gel formulations was determined. The flow curves of thermosensitive and mucoadhesive gels containing either free clotrimazole or clotrimazole:β-cyclodextrin complex were determined using a Brookfield Model DV-II Viscometer at controlled-rate mode. Measurements were made in a cone-and-plate geometry with a diameter of 40 mm (cone angle 2º).

The shear rates ranged from 0.5 to 100 seconds^{-1}. Samples were applied to the lower plate using a spatula to ensure that formulation shearing and air bubbles did not occur. Measurements were made at 22ºC. Each point is the mean of at least triplicate analysis. Error bars have been omitted to retain clarity; however, in all cases, the coefficient of variation of replicate analyses was less than 5%. The gelation temperature of the formulations was determined as follows: a 20-mL transparent vial containing a magnetic bar in 5 mL of Pluronic F127 gel was placed in a water bath. A thermometer connected to a thermistor was immersed in the gel, which was heated at a rate of 2ºC/min with constant (150 rpm) stirring. When the magnetic bar stopped moving because of gelation, the temperature displayed on the thermistor was recorded as the gelation temperature.

The in vitro release of clotrimazole in free form and in complex with β-cyclodextrin was determined from different vaginal gel formulations using a dialysis bag placed in a sealed glass vial under constant magnetic stirring. The gel formulations (2.5 g) were packed into the dialysis bags (Spectra/Por Cellulose Ester Membrane MWCO: 100 000 Da, Spectrum Labs, Rancho Dominguez, CA) sealed with closures of 50 mm. The release medium was 100 mL 0.1M citrate-phosphate buffer (pH 5.5) containing 1% Tween 80, providing sink conditions for clotrimazole. The medium was maintained at 37ºC and stirred at 100 rpm. At various time intervals, 5 mL of dissolution fluid was collected. Levels of clotrimazole in the samples were Analysed with a Shimadzu UV-VIS 160A spectrophotometer at ë = 210 nm. The exact amount of clotrimazole released from the formulation was calculated with a calibration curve with an analytically validated method (r^2 = 0.9994, repeatability coefficient of variation (CV) = 0.18%, reproducibility CV = 2.7%).

ANTIFUNGAL DRUG THERAPY

Fungi, like mammalian cells but unlike bacteria, are eukaryotic and possess nuclei, mitochondria and cell membranes. Their membranes are unusual in containing distinctive sterols. The similarity between fungal and mammalian cells mitigates against selective toxicity and antifungal drugs are in general more toxic than antibacterial agents. The very success of antibacterial therapy has created ecological situations in which opportunistic fungal infections can flourish. In addition, potent immunosuppressive and cytotoxic therapies have produced patients with seriously impaired immune Defences, in whom fungi that are non-pathogenic to healthy individuals become pathogenic and cause disease.

AMPHOTERICIN B

Amphotericin is uniquely valuable in treating life-threatening systemic fungal infectins, but has considerable toxicity. It is given as an intravenous infusion freshly prepared in 5% dextrose over 4-6 hours or topically as lozenges (10 mg 3 hourly) or suspension for oral or esophageal and gastrointestinal moniliasis, respectively. There is some evidence that effective therapy may be achieved by reduced doses and therefore lower toxicity if amphotericin is combined with 5-fluorocytosine A test dose of 1 mg is given at least 6 hours before starting treatment. The initial dose is 5 mg daily in 500 ml of 5% dextrose which is increased daily by 5 mg to a dose of 0,5-1 mg/kg daily and continued for 6-12 weeks. Treatment on alternate days at 1-1,5 mg/kg may reduce toxicity. The antifungal spectrum amphotericin B is broad.

Mechanism of Action

Amphotericin binds to a sterol in fungal cell membranes and increases their permeability, allowing leakage and loss of small molecules such as glucose and potassium ions.

Adverse Effects

- Fever, chills, headache, nausea and vomiting, and hypotension during intravenous infusion. Pulse and temperature should be monitored every 30 min and the infusion can be halted if necessary.
- Nephrotoxicity is almost invariable and results from vasoconstriction, tubular damage resulting in renal tubular acidosis, and acute renal failure. Fortunately most of these effects are reversible if detected early and the drug discontinued or the dose reduced.
- Hypokalemia.
- Normochromic normocytic anemia due to temporary marrow suppression is common.

Amphotericin B should not be withheld in serious progressive infection caused by a sensitive fungus despite toxic effects. Dose reduction may be appropriate.

Pharmacokinetics

Amphotericin *is poorly absorbed* following oral administration, therefore for systemic mycoses it must be given *by intravenous infusion. Liposomal delivery systems* deliver adequate plasma concentrations, with a lower incidence of toxicity. Given intravenously it distributes very unevenly throughout the body: concentrations in the CSF are only one-fortieth of the plasma concentration. It is concentrated in the reticuloendothelial system.

NYSTATIN

Nystatin is another polyene antifungal antibiotic isolated from

Streptomyces with an identifical mode of action to amphotericin B, but its *greater toxicity precludes systemic use*. Nystatin has *a broad antifungal spectrum*. Its indications are limited to *cutaneous and mucocutaneous* infections, especially those caused by *Candida* spp. which do not gain resistance to nystatin during therapy. Preparations of nystatin include tablets, pastilles, lozenges or suspension, given in doses of 100 000-500 000 units three times daily for oral or intestinal Candida infections. Patients often prefer amphotericin B because nystatin has an intensely bitter taste. Cutaneous infections are treated with ointment, vaginitis by suppositories and aerosol has been used for bronchopulmonary fungal colonization.

Adverse Effects

- Adverse effects seldom result from the topical use of nystatin.
- Large oral doses cause *nausea and diarrhea.*

Very little nystatin is absorbed from the gastrointestinal tract.

Griseofulvin

Griseofulvin was isolated from Penicillium griseofulvium. It is *systemically active,* but unlike amphotericin is useful only for mild infections since its spectrum is limited to *dermatophytes* (ringworm fungi). Distance is not a problem. Treatment is given *orally* (0,5-1 g daily in two divided doses) with *meals.* Treatment should be for 6 weeks in skin infections and up to 12 months for nail infections.

Mechanism of Action

Griseofulvin is *actively taken up by fungi*. Its mode of action is obscure, but it binds to the *microtubules* that form the mitotic spindle and *blocks polymerization* of the microtubule. It also interferes with fungal DNA replication.

Adverse Effects

- Headaches and mental dullness or inattention (uncommon).
- Diarrhea or nausea (uncommon).
- Rashes

Pharmacokinetics

Griseofulvin is *nearly insoluble in water* and is formulated as *micronized particles*. Its absorption is *facilitated by a fatty meal*. It has a slow onset of action because it must first be taken up into slow *growing keratinized structures to reach the site of infection.*

Cell turnover time thus determines efficacy of treatment so that palmar and *plantar skin requires at least 8 weeks´ treatment, fingernails 6 months and toenails up to 1 year for eradicative treatment.*

Drug Interactions

Griseofulvin induces hepatic cytochrome P_{450} enzyme activity and consequently interacts with warfarin reducing its anticoagulant effect.

Flucytosine (5-fluorocytosine)

Flucytosine is used for systemic candidiasis and cryptococcosis, providing the strain is sensitive. The optimal oral dose is 200 mg/kg/day in 6-hourly divided doses. For very ill patients an intravenous preparation is available. *It is used in combination therapy with amphotericin.* Topical use is unacceptable because of the danger of widespread emergence of resistant Candida spp. Its spectrum is relatively restricted to Cryptococcus neoformans, Candida albicans and some other Candida spp., Torulopsis spp. and Cladosporum spp. There are big differences in sensitivity between strains: 5-15% have innate resistance, and resistance is relatively easily acquired during therapy. Filamentous fungi, especially Aspergillus, are resistant.

Mechanism of Action

Flucytosine enters the fungus by active transport. It is deaminated to *5-fluorouracil,* a known antimetabolite that inhibits thymidylate synthetase thereby depressing DNA synthesis. Its relative specificity is due to the presence of cytosine deaminase in fungi but not in mammalian cells.

Adverse Effects

Flucytosine is less toxic than amphotericin B.

- Gastrointestinal upsets.
- Leukopenia.
- Hepatitis may occur so liver function tests should be monitored.

At plasma concentrations below 100 mg/ml there is little danger of toxicity. Depression of bone marrow and hepatotoxicity are associated with higher concentrations. Plasma drug assays are useful in patients with impaired renal function.

Pharmacokinetics

Flucytosine is well absorbed from the gut, peak concentrations of 75-90 mg/ml being attained on a dose of 50 mg/kg 6 hourly. It penetrates adequately into CSF, in contrast to amphotericin B, and is consequently particularly useful when combined with amphotericin in treating cryptococcal meningitis.

Drug Interactions

Amphotericin acts additively or synergistically with flucytosine, and flucytosine should always be used with amphotericin.

This combination is useful because amphotericin is more effective but more toxic than *flucytosine which also penetrates the blood-brain barrier* better than amphotericin. *Concurrent use of amphotericin reduces the likelihood of resistance to flucytosine emerging during therapy.*

Imidazoles

Imidazole antifungal drugs are *fungistatic* at low concentrations and *fungicidal* at higher concentrations.

They are *used topically* and are active against both *dermatophytes* and *yeasts* such as Candida. Some imidazoles are also used systemically although they have limited efficacy and significant toxicity, limiting systemic use. They act similarly to one another by inhibiting *fungal ergosterol synthesis*. Ergosterol is an important constituent of fungal membranes. Imidazoles inhibit lanosterol 14a-demethylase (which is a fungal cytochrome P_{450} enzyme) and have considerable specificity for fungal cytochromes. Membrane leakage and dysfunction of membrane adenosine triphosphatase (ATPase) ensue.

KETOCONAZOLE

Ketoconazole is used *as oral or topical* therapy for dermatophytic infections and some phycomycetes. It is active against systemic infection with Candida, Blastomyces, Histoplasma capsulatum and Cryptococcus neoformans. Aspergillus and Mucor spp. are resistant. It is given *orally* (200-400 mg once daily). Its systemic use has waned because of the high incidence of hepatic and endocrine side effects.

Adverse effects

- *Nausea and vomiting,* reduced by giving the drug with food.
- *Transient liver function* abnormalities in 5-10% of patients, fulminant hepatic damage, jaundice and fever are rare.
- *Gynecomastia* (by blocking testosterone synthesis).
- Impotence and azoospermia.
- *Adrenal insufficency* (by inhibiting cortisol biosynthesis).

Pharmacokinetics

Ketoconazole is given *orally* and achieves maximum plasma concentrations in 1-2 hours.

Drug interactions

- Antacids and H_2 -blockers reduce ketoconazole absorption
- Ketoconazole should not be used with amphotericin B because it reduces its effectiveness.

CLOTRIMAZOLE

This imidazole is only used for topical treatment of Candida or dermatophyte infections as a 1% cream, solution of powder. It is *poorly absorbed* from the gastroinstestinal tract and *induces its own metabolism, therefore it is not used systemically.*

MICONAZOLE

Miconazole (2% cream or powder applied twice daily) is usually used topically to treat cutaneous Candida, ringworm and pityriasis rosea.

Triazoles

This group of drugs (e.g. fluconazole) is derived from the imidazoles. They are nitroimidazoles and have a *wider antifungal spectrum*. Their mechanism of action *is idential* to that of other imidazoles (e.g. ketoconazole), by inhibition of lanosterol a-demethylase.

FLUCONAZOLE

Fluconazole is a *potent and broad-spectrum* antifungal agent. It may be *given orally or intravenously as a once daily dose.* For superficial infections it is given as 50-100 mg/day. In systemic or meningitic infections 200-400 mg *intravenously daily* is required for 4-6 weeks, followed by a daily maintenance dose. For prophylaxis in cytotoxic immunosuppressed patients 50-100 mg is adequate.

Adverse Effects

- Gastroinstestinal upsets with nausea, abdominal distension, diarrhea and flatulence.
- Skin rashes - erythema multiforme.
- Hepatitis - raised liver enzymes.

Contraindications

Fluconazole is *contraindicated in pregnancy* because of fetal defects in rodents. Breast milk concentrations are similar to those in plasma, and fluconazole should not be used in *nursing* mothers.

Pharmacokinetics

Fluconazole is absorbed rapidly after oral administration with maximum plasma concentrations achieved within 1-2 hours. Absorption is virtually *complete* and is unaffected by food or gastric pH. Fluconazole *is widely distributed throughout the body penetrating the CSF well.*

ITRACONAZOLE

Itraconazole *is similar to fluconazole in its antifungal spectrum and mechanism of action. It is* given orally once daily as a 100 mg dose to treat *dermatophyte infections, superficial oropharyngeal candidiasis and pityriasis versiColour*. Its major side effects are gastrointestinal disturbances. It causes several adverse drug interactions, increasing cyclosporin and warfarin plasma concentrations.

TIACONAZOLE

Tiaconazole is a *broad-spectrum antifungal triazole,* marketed only as a solution for topical administration *to treat nail infections with dermatophytes and yeasts.* No other topical imidazole preparations are effective in fungal nail infections. The recommended duration of therapy is 6-12 months.

Terbinafine

Terbinafine is fungicidal. It can be given orally and is used to treat ringworm (tinea pedis, cruris or corporis) or dermatophyte infections of the nails if oral therapy is considered appropriate. It is given 250 mg once daily for 2-6 weeks or longer in infections of the nailbed as an alternative to griseofulvin. It acts by inhibiting sterol synthesis by the fungal enzyme squalene epoxidase. It *is well absorbed,* strongly bound to plasma proteins and *concentrated* in *the stratum corneum.* Its major side effects *are nausea, abdominal discomfort, anorexia, diarrhea and rashes (including urticaria).*

ANTIVIRAL DRUG THERAPY

Many viral illnesses are mild and/or self-limiting, but a few are inherently deadly (witness the now extinct smallpox, and the all too prevalent AIDS). Patients who are immunocompromised, especially by AIDS, are also at risk of serious illness from viruses that are seldom serious in otherwise healthy hosts. Antiviral drug therapy is therefore increasingly important. Antiviral chemotherapy is intrisically more difficult than antibacterial therapy because:

- Viral replication is essentially intracellular so that drugs must penetrate cells to be effective.
- Viral replication, although under the control of the viral genome, involves metabolic processes of the host cells.
- Although viral replication begins almost immediately the host cell is penetrated, the clinical signs and symptoms of infection often appear after peak replication is over. Clinical manifestations often result from inflammatory and other processes mounted by the host in response to viral damage to tissue.
- Few viral infections can be routinely rapidly and specifically diagnosed with certainy using present techniques.

There are some events in viral colonization which might be susceptible to drug action:

- When the virus is outside cells it is susceptible to antibody attack, but it has proved difficult to find drugs that are non-toxic yet can destroy viruses in this situation.
- Viral attachment to the cell surface probably involves a specific chemical reaction between the virus coat and the cell surface. Neurominidase, an enzyme that destroys myxovirus receptors, including those on cell surfaces, has an effect on experimental infections in animals.

- Penetration of the cell membrane can be prevented (e.g. by amantadine for influenza A).
- Uncoating of the virus with release of viral nucleic acid intracellularly.
- Viral nucleic acid acts as a template for new strands of nucleic acid that in turn direct the production of new viral components utilizing the host cell′s synthetic mechanisms. Most antiviral drugs in current use act at this stage of viral grouwth.
- Release of new viral particles.

ACYCLOVIR

Acyclovir is a potent and selective inhibitor *of herpes viruses* (particularly herpes simplex). It is also effective against *herpes zoster* but much less active against *cytomegalovirus* (which also belongs to the group of herpes viruses).

- Three per cent acyclovir ointment applied five times daily for up to 14 days accelerates healing in herpetic keratitis.
- The efficacy of local application of acyclovir in genital and labial herpes simplex has been unimpressive. In spite of its low bioavailability, 200 mg five times daily orally accelerates healing in genital herpes. It is much less effective in secondary than primary infection.
- Treatment of shingles (herpes zoster) should be started within 72 hours of the onset, and is useful for patients with severe pain, although it shortens the illness by only a short time and is expensive. Acyclovir, 800 mg, five times a day is given for 7 days.
- In generalized herpes simplex or herpetic meningoencephalitis acyclovir can be given intravenously in doses of 5 mg/kg infused intravenously over 1 hour three times daily for 5 days. The acyclovir dose in immunocompromised hosts with herpes simplex, shingles or herpes simplex encephalitis is 10 mg/kg intravenously three times daily for at least 10 days.

Mechanism of Action

Acyclovir is a cyclic analog of the purine guanoside. Its selective action results from metabolic activation *to its monophospate,* solely in infected cells, via a specific thymidine kinase coded for by the virus but not by the host genome. *Acyclovir monophosphate is converted to the di- and triphosphates and the triphosphate inhibits viral DNA synthesis.*

Adverse Effects

- Reversible rise in plasma urea and creatinine.
- Neurological disturbances.
- Rashes.

- Nausea and vomiting.
- Increase in liver enzymes.

Acyclovir is *contraindicated in pregnancy* as it is an analog of guanosine and so potentially *teratogenic.*

Pharmacokinetics

Acyclovir *crosses the blood-brain barrier to give a CSF concentration of approximately* 50% of the plasma concentration.

AMANTADINE

Amantadine has a prophylactic action in preventing *the spread of influenza* A, and also has an unrelated action in *Parkinson's disease*. Its usefulness as an antiviral agent is limited to influenza A: it is inactivate on influenza B and only weakly active on inluenza C and rubella. These are all RNA viruses. Its mode of action is unknown. *Prophylaxis with amantadine* (200 mg daily) has an advantage over immunization in that the latter can be ineffective when a new antigenic variant arises in the community and spreads too rapidly for a killed virus vaccine to be prepared and administered. Prophylaxis by amantadine during an epidemic should be considered for persons at special risk, for example with severe lung or cardiac disease. Amantadine is less effective during periods of antigenic variation than during periods of relative antigenic stability.The effectiveness of amantadine as treatment rather than prophylaxis when given during the first 48 hours of illness is slight.

Adverse Effects

Dizziness, nervousness and headaches with short-term use.

FOSCARNET

Foscarnet was originally found to have activity against *the influenza virus* and more recently has been found to be active against several other important viruses, *notably HIV-1* and *al human herpes viruses* inluding *acyclovir-resistant herpes viruses* and *cytomegalovirus (CMV)*. It is used to treat CMV infections (retinitis, pneumonitis, colitis and esophagitis) and acyclovir-resistant herpes simplex virus (HSV) infections in *immunocompetent and immunosuppressed* hosts.

Foscarnet is given as a loading dose of 30 mg/kg intravenously over half an hour followed by 60-90 mg/kg as a 1 hour infusion 8 hourly. After 23 weeks treatment doses are reduced to 20-30 mg/kg/day. Dose reduction is required in patients with renal failure.

Mechanism of Action

Foscarnet is a nucleotide analog *that inhibits viral DNA synthesis*. It acts as a non-competitive inhibitor of viral *DNA polymerase* and *inhibits reverse transcriptase* from several *retroviruses*.

Adverse Effects

- *Nephrotoxicity* is reduced by adequate hydration and dose reduction if the creatinine rises. Monitoring of renal function is mandatory.
- *Central nervous system effects* include irritability, anxiety and fits.
- *Nausea, vomiting and headache.*
- *Thrombophlebitis.*
- *Hypocalcemia and hypomagnesemia.*

Drug Interactions

The nephrotoxicity of foscarnet is potentiated by other nephrotoxins, including gentamicin, cyclosporin and amphotericin B.

GANCICLOVIR

Ganciclovir is the first antiviral drug to be licensed for sight or life-threatening *CMV infections* (retinitis, pneumonitis, colitis and esophagitis) *in immunocompromised hosts*. It also has potent activity against *herpes virus 1 and 2*. It is an analog of guanine. It is administered at an induction dose of 5 mg/kg every 12 hours for 14-21 days as an intravenous infusion over 1 hour, followed by a maintenance dose of 6 mg/kg for 5 days every week or 5 mg/kg for 7 days a week, continuously.

Mechanism of Action

Ganciclovir *inhibits viral replication*. It is metabolized intracellularly in herpes-infected cells by virally encoded thymidine kinase to its monophosphate. It undergoes further phosphorylation by host kinases to the triphosphate, which competitively *inhibits the DNA polymerase of CMV* and, if incorporated into nascent viral DNA causes chain termination. Ganciclovir is concentrated ten times in infected cells compared to uninfected cells.

Adverse Effects

- *Neutropenia and bone marrow suppression* (thrombocytopenia and less often anemia), cell counts usually return to normal within 2-5 days of discontinuing the drug.
- Temporary or possibly permanent *inhibition of spermatogenesis or oogenesis.*
- *Phlebitis and* pain at the infusion site.
- Rashes and fever.
- *Gastrointestinal upsets.*
- *Transient increases in liver enzymes and creatinine in underhydrated patients.*

Contraindications

Ganciclovir is *contraindicated in pregnancy (it is teratogenic in animals) and in breastfeeding women.*

Pharmacokinetics

Absorption of ganciclovir is poor following oral administration, necessitating intravenous administration.

Drug Interactions

Antineoplastic drugs, co-trimoxazole and amphotericin B increase its toxic effects on rapidly dividing tissues including bone marrow, skin and gut epithelium. *Zidovudine* should not be given concomitantly with ganciclovir because of the potentiation of bone marrow suppression.

IDOXURIDINE

Idoxuridine is *an analog of the DNA nucleotide thymidine* and *inhibits replication* of some DNA viruses, particularly *herpes viruses*. Probable mechanisms of action include *incorporation* into DNA and/or competitive *inhibition of enzymes involved in nucleic acid synthesis.* It is only relatively selective for viral nucleic acid synthesis and causes toxic effects on mammalian *bone marrow resulting in leukopenia and thrombocytopenia.*

Viral resistance to idoxuridine develops readily and to be effective the drug must be started early after the onset of infection. It is too toxic for systemic use, but has several indications for topical use:

- Herpetic keratitis Hourly drops of 0,1% idoxuridine in saline encourage resolution of corneal ulcers providing a deep stromal reaction has not occurred. Prolonged use however may cause blepharitis, conjunctivitis and punctate lesions of the corneal epithelium.
- Herpes labialis Recurrent lesions may be aborted by treating the affected area with 5,0% idoxuridine solution at the first sign of recurrence (usually a burning sensation in the affected skin precedes vesicle formation).
- Herpes zoster Applied topically as a 5% idoxuridine solution in dimethyl sulfoxide idoxuridine hastens healing of shingles and may reduce the occurrence of postherpetic neuralgia. Dimethyl sulfoxide enhances skin penetration but is an irritant and must never be applied to the eye. It is important to begin treatment as early as possible, preferably as lesions are developing, and is especially valuable in the elderly and those with immunological impairment (e.g. steroid or cytotoxic treatment).

TRIBAVIRIN (RIBAVIRIN)

Tribavirin is active *against a number of RNA and DNA (HSV-1 and -2, influenza) viruses*. Its major indication is for the treatment of *bronchiolitis secondary to respiratory syncytial virus* infection in infants and children. It is also effective, *given systemically, in Lassa fever*. Administration for bronchiolitis *is via aerosol inhalation or nebulizer,* of a solution of 20 mg/ml for 12-18 hours (dose - 6 g/day) for at least 3 days and up to a maximum of 7 days.

Mechanism of Action

Tribavirin *is taken up into cells* and *phosphorylated* to tribavirin *5´-monophosphate* by adenosine kinase and then rapidly phosphorylated to *the di- and triphosphates* by other cellular kinases.

Tribavirin 5´-monophosphate is *a powerful inhibitor of inosine monophosphate dehydrogenase* which affects the pool size of *intracellular ribonucleotides* and *deoxynucleotides*. What effect this has on viral replication is presently unclear. *Tribavirin 5´-triphosphate inhibits the guanylation reaction in formation of the 5´ cap of mRNA and inhibits viral RNA methyltransferase..*

Adverse Effects

No systemic adverse effects of tribavirin have been reported following aerosol or nebulizer administration.

- Worsening respiration and bacterial pneumonia.
- Pneumothorax.

Pharmacokinetics

Following nebulized administration negligible amounts of tribavirin are absorbed systemically.

IMMUNOGLOBULINS

Some pooled human immunoglobulins have high titers against specific viruses such as *cytomegalovirus, hepatitis A, hepatitis B, varicella and measles.* Immunoglobulin is given *intramuscularly* as near as possible to the time of exposure and may be used *prophylactically. Parenterally administered immunoglobulins reach peak serum concentrations in 4-6 days then decline with a half-life of 20-30 days.*

INTERFERONS

Interferons are cytokines (mediators of cell growth and function). They are glycoproteins secreted by cells infected with viruses or foreign double-stranded DNA. They are non-antigenic and are active against a wide range of viruses, but unfortunately are relatively species specific.

Thus it is necessary to produce human interferon to act on human cells. Interferon production is triggered not only by viruses but also by tumor cells or previously encountered foreign antigens. Some interferons are important in immune cell regulation.

Four main types of interferon with different structures, derivation and properties are recognized:

1. Interferon-α known previously *as leukocyte or lymphoblastoid interferon.*

Subspecies of the human a gene produce variants designated by the addition of a number, e.g. interferon-a_2, or in the case of a mixture of proteins, by N1, N2 etc. Two methods of commercial production have been developed and these are indicated by: rbe - produced from bacteria (Escherichia coli) genetically modified by recombinant DNA technology, and 1ns - produced from cultured lymphoblasts stimulated by Sendai virus.
Interferon-α_2 may also differ in the amino acids at positions by the addition of a letter. Thus α_2a has lys-his at these sites whilst α_2b has arg-his. At present it is not certain whether these different molecules have different therapeutic properties.
2. Interferon-β from fibroblasts.
3. Interferon-ω 60% homology to interferon-a and not yet clinically available.
4. Interferon-γ formerly called „immune" interferon as it is produced by lymphocytes in response to antigens and mitogens.

Commercial production of pure interferon by cloning of interferon genes into bacterial and yeast plasmids is now available to facilitate large scale production.

Interferon-*α is of benefit in chronic hepatitis B and chronic hepatitis C infection,* and is given *thrice weekly by subcutaneous injections of 1-5 million units, for 6-12 months*. All three interferons are used to treat hairy cell leukemia. Interferon-α_{2a} and - α_{2b} are used to treat Kaposi's sarcoma in AIDS patients, and interferon- α_{N1} is licensed for recurrent or metastatic renal cell carcinoma.

Recombinant interferon-g_{1B} has been used for the treatment of chronic granulomatous disease and is under investigation for the treatment of mesothelioma and carcinoma of the ovary. Interferon therapy is beneficial in chronic myelogenous leukemia, multiple myeloma and refractory lymphoma, and is being further investigated in rheumatoid arthritis and mycosis fungoides.

Mechanism of Action

Interferons bind to receptors on the cell membrane. Interferon-*α*, -β and -ω bind to a common cell membrane receptor, but interferon-g binds to its own receptor. Following receptor binding inteferons induce adenosine triphosphate (ATP)- dependent, tyrosine kinase-stimulated phosphorylation of three proteins - interferon-stimulated gene factors (ISGF).

These then bind together to form ISGF 3α, which translocates to the nucleus and binds to DNA regulating RNA transcription of certain proteins. Consequently a number of enzymes with antiviral activity are increased, namely, 2′, 5′ *oligoadenylate synthetase* (which activates ribonuclease L which preferentially cuts viral RNA), and *protein kinase activity*. The onset of these effects takes several *hours but may persist for days even after plasma interferon concentrations* have been completely cleared. Interferon also increases presentation of viral antigens in infected cells and *up-regulates macrophage activation and T and natural killer cell cytotoxicity, thereby increasing viral elimination.*

Adverse Effects

- Fever, malaise, chills - an influenza-like syndrome, and neuropsychiatric symptoms similar to a postviral syndrome.
- Lymphocytopenia and thrombocytopenia are reversible and tolerance may occur after a week or so.
- Anorexia and weight loss.
- Alopecia.
- Confusion, tremor and fits.
- Transient hypotension or cardiac arrhythmias.
- Hypothyroidism.

Pharmacokinetics

Most clinical experience is with interferon-a, which has been given intravenously, intramuscularly, subcutaneously and topically (nasal, vaginal, mucosal and eye). More than 80% of an intramuscular dose is absorbed. After subcutaneous administration, peak plasma concentrations occur after 4-8 hours and decline over 1-2 days.

Cerebrospinal fluid concentrations are 1-5% of those in the plasma. Mean half-life is 3-5 hours. Elimination of interferons is complex, inactivation occurring in liver, lung, kidney, heart and striated muscle.

CHARACTERIZATION OF THERMOSENSITIVE GELS

Different formulations of Pluronic F127 gel with or without the mucoadhesive polymers Carbopol 934 and HPMC using free clotrimazole and clotrimazole:β-cyclodextrin complex were prepared using the cold method. Formulation components influenced gelling properties significantly. Carbopol 934, a mucoadhesive polymer known to interact strongly with low-molecular-weight molecules, polymers, and drugs, displayed the same phenomenon with both free drug and the complex, causing precipitation within a few hours for gels containing Carbopol 934.

It is believed that Carbopol interacts with the free drug, which has a considerably lower molecular weight, causing the drug's precipitation. Carbopol-containing formulations were not used for further studies, including viscosity measurements and in vitro release studies, because of this precipitation effect. Another important factor influencing gel characteristics was whether the drug was employed in free form or in complex. The significant differences in gelation temperature and in vitro release profiles of different formulations. Gelation temperature is dependent largely on polymer content; when clotrimazole is administered in free form, additional use of PEG400 as a solubilizing agent is required.

Together with the high percentage of Pluronic (20%), known to lower gelation temperature, an excess amount of polymer content with the presence of PEG causes the formation of gels at even 4°C. On the other hand, mucoadhesive polymer type seems to have no significant effect on gelation temperature. The presence of cyclodextrins such as HP-β-cyclodextrin was previously reported to lower gel viscosity. This effect was attributed to the fact that the binding force (hydrogen bonding) of cross-linked Pluronic gel became weaker when HP-β-cyclodextrin was replaced in the gel matrix. The phenomenon was reported to be similar to the decrease in gel strength of sodium salicylate.

Table. Gelation Temperatures (ºC) of Different Formulations of 20% Pluronic F127 Vaginal Gel (n = 3)*

Formulation	Gelation Temperature (ºC) ±SD
CLOT/PEG400-EtOH/PF127	10 ± 2
CLOT/PEG400-EtOH/HPMC/PF127	12 ± 1.5
CLOT:β-CD/PF127	32 ± 0.3
CLOT:β-CD/HPMC/PF127	33 ± 2

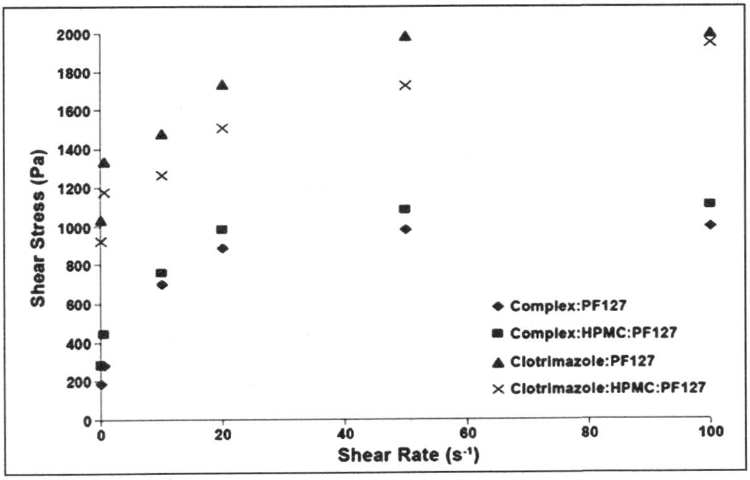

Fig. Flow Curves of Different Clotrimazole Vaginal Gel Formulations at Room Temperature (22°C) (n = 3).

Flow plots of gels containing free and complexed clotrimazole with and without the mucoadhesive polymer HPMC. Steady shear properties of Pluronic formulations were affected by temperature and overall polymer content. Rheological properties of gels formulated with free clotrimazole are very much different from that of clotrimazole:β-cyclodextrin complex. Gels containing free clotrimazole show significantly higher viscosity that was not affected by temperature. Gels containing clotrimazole complexed to β-cyclodextrin, however, show a temperature-sensitive rheological Behaviour. In vitro release experiments were realized on formulations that did not display precipitation. Thus, gels formulated with or without HPMC containing either free clotrimazole or clotrimazole:β-cyclodextrin complex were used. The in vitro release profiles of clotrimazole thermosensitive gel formulations.

Gels containing free drug liberated most of the drug within 4 days. However, formulations containing complexed clotrimazole released only ~40% of the drug during this period. This fact could be attributed to 2 different phenomena. Complexation to β-cyclodextrin was performed because of the poor aqueous solubility of clotrimazole. Complexation with cyclodextrins is known to enhance clotrimazole aqueous solubility considerably. On the other hand, this solubilization effect has a different impact on release Behaviour of the gels. In extensive studies performed on the release Behaviour of thermosensitive hydrogels, it was reported that although the rate of dissolution of Pluronic gel is actually the controlling factor in drug release, it is not the only parameter affecting drug release.

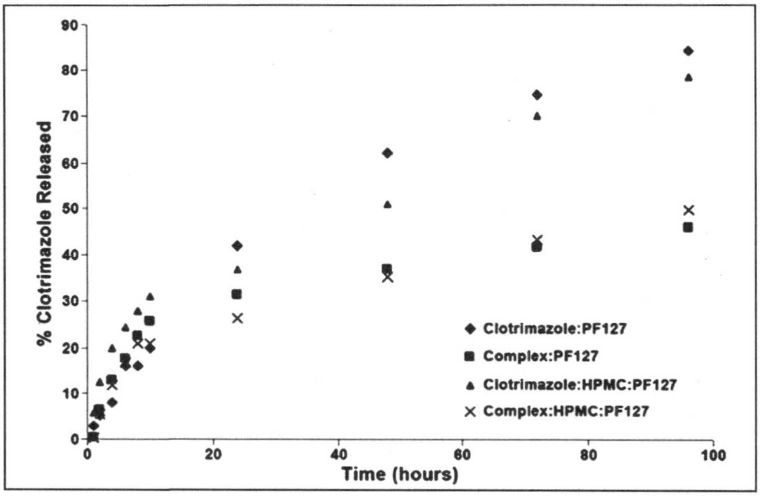

Fig. In vitro Release Profiles of Clotrimazole Vaginal Gels at 37°C in pH 5.5 Citrate Buffer.

Because the interface is eroding at a constant rate, the surface concentration of the drug is held relatively constant and the drug diffuses out at a constant rate. More concentrated gels were reported to dissolve at a slower rate than less concentrated ones because of the decreased water diffusion coefficient for the rate of water diffusing into the gel. Release from the complex is, on the other hand, a competitive process controlled by diffusion of the drug upon dilution and competitive displacement of the drug by components in the dissolution media. These 2 factors may have contributed to the prolonged release of clotrimazole from the thermosensitive gel in the complexed form. During the 92-hour release study, released amounts were significantly above the minimum inhibitory concentration (MIC) value for clotrimazole determined previously. This release period could be helpful in predicting the gel's Behaviour in the vagina. Most probably, a faster release of clotrimazole would be achieved in vivo because of the presence of salts and proteins. Even if the vaginal formulation's residence time was much less than 96 hours, this could be sufficient to obtain a high local dose of the drug for therapeutic efficacy.

CHARACTERIZATION OF INCLUSION COMPLEX

First, formation of a genuine inclusion complex of 1:1 molar ratio was evaluated using different techniques, including ^1H NMR spectroscopy, FT IR spectrophotometry, DSC, and SEM analysis. NMR spectroscopy has been used to elucidate the structure of drug:cyclodextrin complexes. NMR allows a clear distinction between inclusion and any other possible external interaction, with large effects observed on the protons located inside the hydrophobic cavity (H-3 and H-5, with H-5 more pronouncedly affected) clearly proving inclusion.

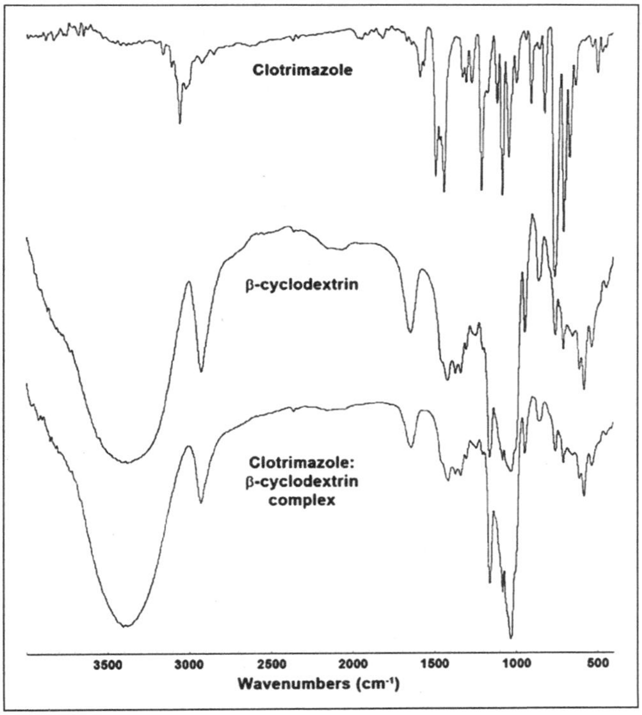

Fig. FT IR Spectra of Clotrimazole, b-cyclodextrin, and Clotrimazole:b-cyclodextrin 1:1 complex.

The chemical environment of the internal protons changes, resulting in changes in the chemical shifts of the protons because of shielding or deshielding effects. On the other hand, protons located on the outside experience very little or no effect details the shifts observed for the internal proton H-5, suggesting that the drug has entered the cyclodextrin cavity to form an inclusion complex. There are no shifts for external protons, indicating that the complex formed can be classified as an inclusion complex.

Another indicator of the formation of a drug:cyclodextrin complex is the changes observed for the FT IR spectra of the drug in complex form. It can be clearly observed that the FT IR spectrum for the complex is almost identical to that of β-cyclodextrin alone. On the other hand, clotrimazole's aromatic region peaks at around 600 to 800 cm^{-1} have disappeared, which suggests that these groups of the drug are included in the cavity.

For the demonstration of a genuine inclusion complex, results should be confirmed with different and independent techniques. The DSC thermogram clearly indicates that clotrimazole's sharp melting endotherm at 148ºC and β-cyclodextrin's dehydration observed between 50 and 120ºC have completely disappeared for the thermogram of the complex, with a very slight water loss around 100ºC. Clotrimazole's DSC thermogram was taken after freeze drying in order not to overlook the effects of the lyophilization process on the crystalline structure of the drug. This might have affected the thermogram of the drug if a transformation from the crystalline state to the amorphous state had taken place, but this did not happen in this study.

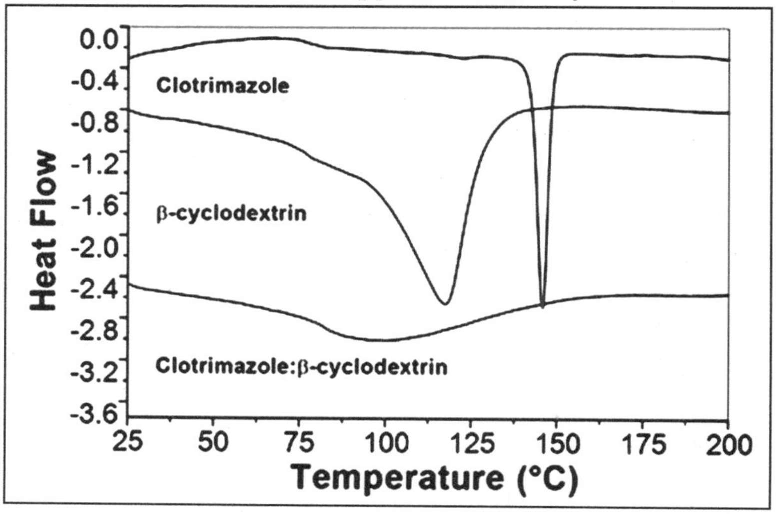

Fig. Differential Scanning Calorimetry Data of Lyophilized Clotrimazole, β-cyclodextrin, and Clotrimazole:β-cyclodextrin (1:1) Complex.

SEM photomicrographs of clotrimazole, β-cyclodextrin, and the complex were taken with ×2500 magnification represents the crystalline structure of clotrimazole, which is close to cubic after lyophilization.

On the other hand, β-cyclodextrin is more of a planar structure with large crystals. The complex exhibits a totally different crystalline structure than the drug or the β-cyclodextrin, with needle- and flake-type crystals suggesting the formation of a complex.

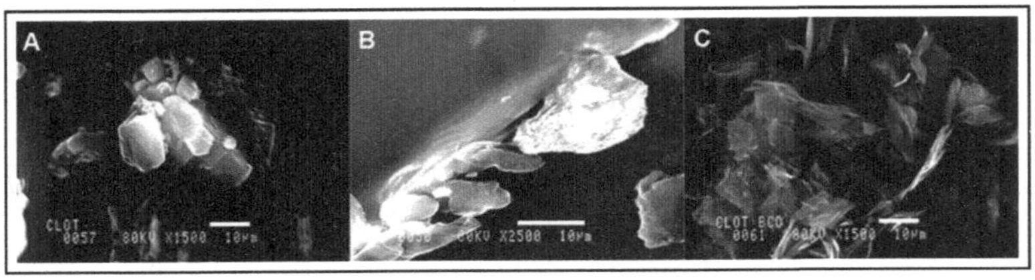

Fig. Scanning Electron Microscopy Photomicrographs of Clotrimazole (A), β-cyclodextrin (B), and Complex (C) (magnification ×2500).

The phase solubility diagram of clotrimazole with β-cyclodextrin can be observed. As can be seen, this diagram is an AL-type curve with a linear relationship between solubilized clotrimazole and molar β-cyclodextrin with a negative curvature.

The initial linear ascending part of the solubility is generally ascribed to the formation of a 1:1 complex. The apparent stability constant ($K_{1:1}$) can be calculated from the solubility data using this formula:

K 1: 1 = S l o p e S o (1 — s l o p e)

where S_0 is the intrinsic solubility of clotrimazole.

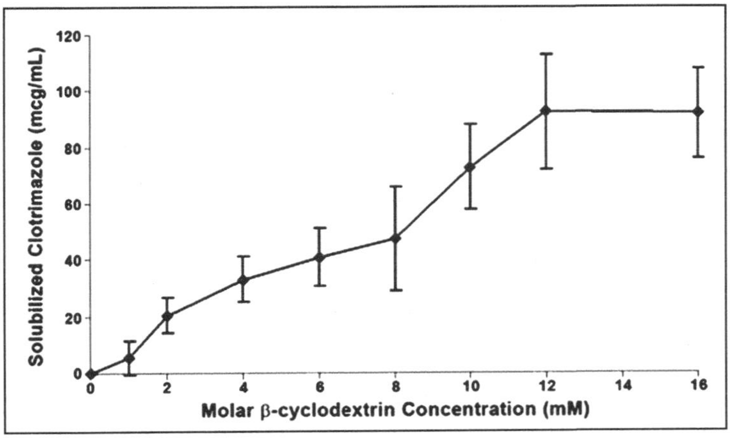

Fig. Phase Solubility Diagram of Clotrimazole with β-cyclodextrin in Pure Water (n = 3).

The apparent stability constant ($K_{1:1}$) for the clotrimazole:β-cyclodextrin complex was calculated from the solubility data and found to be $701M^{-1}$, which confirms previously reported values.

2

Pharmaceutical Engineering

BUFFERING

It should be noted that around the pK_a the pH of a solution does not change appreciably even when large amounts of acid or base are added. This phenomenon is known as buffering. In most biochemical studies it is important to perform experiments, that will consume H^+ or OH^- equivalents, in a solution of a buffering agent that has a pK_a near the pH optimum for the experiment.

CLINICAL SIGNIFICANCE OF BLOOD BUFFERING

The pH of blood is maintained in a narrow range around 7.4. Even relatively small changes in this value of blood pH can lead to severe metabolic consequences. Therefore, blood buffering is extremely important in order to maintain homeostasis. Although the blood contains numerous cations (e.g., Na^+, K^+, Ca^{2+} and Mg^{2+}) and anions (e.g., Cl^-, PO_4^{3-} and SO_4^{2-}) that can, as a whole, play a role in buffering, the primary buffers in blood are hemoglobin in erythrocytes and bicarbonate ion (HCO_3^-) in the plasma. Buffering by hemoglobin is accomplished by ionization of the imidazole ring of histidines in the protein. The formation of bicarbonate ion in blood from CO_2 and H_2O allows the transfer of relatively insoluble CO_2 from the tissues to the lungs, where it is expelled. The major source of CO_2 in the tissues comes from the oxidation of ingested carbon compounds.

Carbonic acid is formed from the reaction of dissolved CO_2 with H_2O. The relationship between carbonic acid and bicarbonate ion formation is shown in the two equations below:

$$CO_2 + H_2O \leftrightarrow H_2CO_3$$

$$H_2CO_3 \leftrightarrow H^+ + HCO_3^-$$

The reactions shown in equations 16 and 17 occur predominately in the erythrocytes, since nearly all of the CO_2 leaving tissues via the capillary endothelium is taken up by these cells. This reaction is catalyzed by carbonic anhydrase. Ionization of carbonic acid then occurs spontaneously, yielding bicarbonate ion. Carbonic acid is a relatively strong acid with a pK_a of 3.8.

However, carbonic acid is in equilibrium with dissolved CO_2. Therefore, the equilibrium equation for the sum of equations 16 and 17 requires a conversion factor, since CO_2 is a dissolved gas. This factor has been shown to be approximately 0.03 times the partial pressure of CO_2 (PCO_2). When this is entered into the Henderson-Hasselbalch equation:

$$pH = 6.1 + \log\left[\frac{[HCO_3^-]}{(0.03)(PCO_2)}\right]$$

where the apparent pK_a for bicarbonate formation, 6.1, has been introduced into the equation.

The PCO_2 in the peripheral tissues is approximately 50mm Hg, whereas in the blood entering the peripheral tissues it is approximately 40mm Hg. This difference results in the diffusion of CO_2 from the tissues into the blood in the capillaries of the periphery. When the CO_2 is converted to H_2CO_3 within the erythrocytes and then ionizes, the hydrogen ions (H^+) are buffered by hemoglobin.

The production of H^+ ions, within erythrocytes, and their subsequent buffering by hemoglobin results in a reduced affinity of hemoglobin for oxygen. This leads to a release of O_2 to the peripheral tissues, a phenomenon is termed the Bohr effect

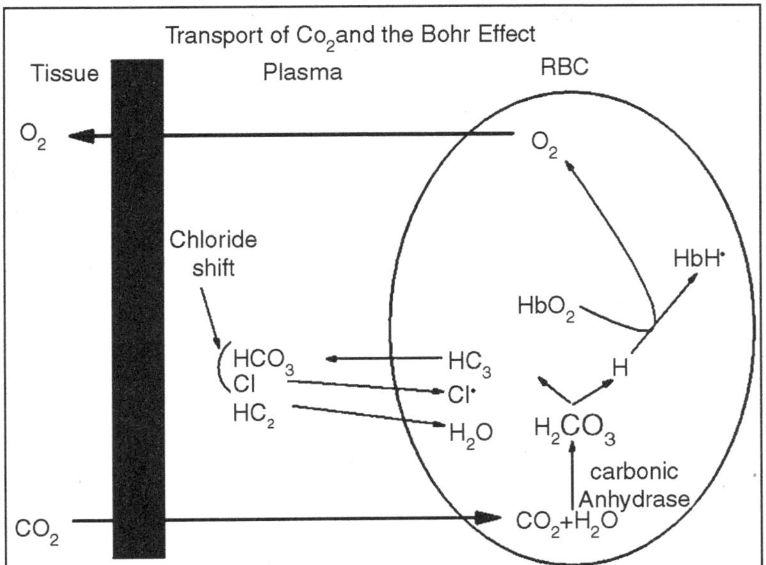

Representation of the transport of CO_2 from the tissues to the blood with delivery of O_2 to the tissues. The opposite process occurs when O_2 is taken up from the alveoli of the lungs and the CO_2 is expelled. All of the processes

of the transport of CO_2 and O_2 are not shown such as the formation and ionization of carbonic acid in the plasma.

The latter is a major mechanism for the transport of CO_2 to the lungs, i.e. in the plasma as HCO_3^-. The H^+ produced in the plasma by the ionization of carbonic acid is buffered by phosphate (HPO_4^{2-}) and by proteins. Additionally, some 15% of the CO_2 is transported from the tissues to the lungs as hemoglobin carbamate as shown in the above equation. As CO_2 passes from the tissues to the plasma a minor amount of carbonic acid takes form and ionizes. The H^+ ions are then buffered predominantly by proteins and phosphate ions in the plasma. As the concentration of bicarbonate ions rises in erythrocytes, an osmotic imbalance occurs.

The imbalance is relieved as bicarbonate ion leaves the erythrocytes in exchange for chloride ions from the plasma. This phenomenon is known as the chloride shift which is also shown in the diagram above. Therefore, the majority of the bicarbonate ion formed as CO_2 leaves the peripheral tissues is transported by the plasma to the lungs.

Around 15% of CO_2 transport from the tissues to the lungs occurs through a reversible combination with non-ionized amino groups ($-NH_2$) of hemoglobin forming hemoglobin carbamate.

$$\text{Hemoglobin-NH}_2 + CO_2 \leftrightarrow \text{Hemoglobin-NH-COO}^- + H^+$$

The formation of hemoglobin carbamate results in a reduced affinity of hemoglobin for O_2 thus favoring dissociation of bound oxygen in the tissues where the concentration of CO_2 is high. The process is reversed when the erythrocytes enter the lungs and the partial pressure of O_2 is elevated. The partial pressure of O_2 (PO_2) in the pulmonary alveoli is higher than the PO_2 of the entering erythrocytes that contain predominantly deoxygenated hemoglobin. This increased PO_2 leads to oxygenation of hemoglobin and release of H^+ ions from the hemoglobin. The released H^+ ions combine with the bicarbonate ions to form H_2CO_3.

Cellular carbonic anhydrase then catalyzes the reverse reaction, leading to release of CO_2 from erythrocytes. Owing to the PCO_2 gradient, the CO_2 diffuses from the blood to the alveoli where it is expelled. The great utility of bicarbonate as a physiological buffer stems from the fact that if excess acid is added to the blood the concentration of bicarbonate ion declines and the level of CO_2 increases. The CO_2 then passes from capillaries in the pulmonary alveoli and is expelled.

As a consequence, the H^+ ion concentration drives the reaction to the left and bicarbonate ion acts as a buffer until all of the hydrogen ion is consumed. Conversely, when excess base is added to the blood, CO_2 is consumed by carbonic acid and replaced by metabolic reactions within the body. If blood is not adequately buffered, the result may be metabolic acidosis or metabolic alkalosis. These physiological states can be reached if a metabolic defect results in the inappropriate accumulation or loss of acidic or basic compounds. These compounds may be ingested, or they may accumulate as metabolic by-products such as acetoacetic acid and lactic acid.

Both of these will ionize, thereby increasing the level of H^+ ions that will in turn remove bicarbonate ions from the blood and alter blood pH. The predominant defect in acid or base elimination arises when the excretory system of the kidneys is impaired. Alternatively, if the lungs fail to expel accumulated CO_2 adequately and CO_2 accumulates in the body, the result will be respiratory acidosis. If a decrease in PCO_2 within the lungs occurs, as during hyperventilation, the result will be respiratory alkalosis.

K_{EQ} K_W AND pH

As H_2O is the medium of biological systems one must consider the role of this molecule in the dissociation of ions from biological molecules. Water is essentially a neutral molecule but will ionize to a small degree. This can be described by a simple equilibrium equation:

$$H_2O \leftrightarrow H^+ + OH^-$$

This equilibrium can be calculated as for any reaction:

$$K_{eq} = \frac{[H^+][OH^-]}{[H_2O]}$$

Since the concentration of H_2O is very high (55.5M) relative to that of the $[H^+]$ and $[OH^-]$, consideration of it is generally removed from the equation by multiplying both sides by 55.5 yielding a new term, K_w:

$$K_w = [H^+][OH^-]$$

This term is referred to as the ion product. In pure water, to which no acids or bases have been added:

$$K_w = 1 \times 10^{-14}\ M^2$$

As K_w is constant, if one considers the case of pure water to which no acids or bases have been added:

$$[H^+] = [OH^-] = 1 \times 10^{-7}\ M$$

This term can be reduced to reflect the hydrogen ion concentration of any solution. This is termed the pH, where:

$$pH = -\log[H^+]$$

PK_A

Acids and bases can be classified as proton donors and proton acceptors, respectively. This means that the conjugate base of a given acid will carry a net charge that is more negative than the corresponding acid. In biologically relevant compounds various weak acids and bases are encountered, e.g. the acidic and basic amino acids, nucleotides, phospholipids etc.

Weak acids and bases in solution do not fully dissociate and, therefore, there is an equilibrium between the acid and its conjugate base. This equilibrium can be calculated and is termed the equilibrium constant = K_a.

This is also sometimes referred to as the dissociation constant as it pertains to the dissociation of protons from acids and bases.

In the reaction of a weak acid:

$$HA \leftrightarrow A^- + H^+$$

the equilibrium constant can be calculated from the following equation:

$$K_a = \frac{[H^+][A^-]}{[HA]}$$

As in the case of the ion product:

$$pK_a = -\log K_a$$

Therefore, in obtaining the –log of both sides of the equation describing the dissociation of a weak acid we arrive at the following equation:

$$-\log K_a = -\log \frac{[H^+][A^-]}{[HA]}$$

Since as indicated above $-\log K_a = pK_a$ and taking into account the laws of logrithms:

$$PK_a = -\log[H^+] - \log \frac{[A^-]}{[HA]}$$

$$PK_a = pH - \log \frac{[A^-]}{[HA]}$$

From this equation it can be seen that the smaller the pK_a value the stronger is the acid. This is due to the fact that the stronger an acid the more readily it will give up H^+ and, therefore, the value of [HA] in the above equation will be relatively small.

VARIETY OF ALKALOIDS

A variety of alkaloids have been used as pharmaceuticals and most of them are plant metabolites. The typical tropane alkaloids, atropine, hyoscyamine, scopolamine and cocaine, were widely used as blockers of the parasympathetic nervous system such as anodyne and antispasmodic. Research on production of these useful alkaloids by plant cell cultures has been carried out for more than 25 years, however, industrial production has not yet succeeded because of low producing ability of the cultured cells. Plants used for these studies are mainly Atropa belladonna, Hyoscyamus niger, Datura meteloids and others.

Alkaloid, any of a class of naturally occurring organic nitrogen-containing bases. Alkaloids have diverse and important physiological effects on humans and other animals. Well-known alkaloids include morphine, strychnine, quinine, ephedrine, and nicotine. Alkaloids are found primarily in plants and are especially common in certain families of flowering plants. More than 3,000 different types of alkaloids have been identified in a total of more than 4,000 plant species. In general, a given species contains only a few kinds of alkaloids, though both the opium poppy (Papaver somniferum) and the ergot fungus (Claviceps) each contain about 30 different types. Certain plant families are particularly rich in alkaloids; all plants of the poppy family (Papaveraceae) are thought to contain them, for example. The Ranunculaceae (buttercups), Solanaceae (nightshades), and Amaryllidaceae (amaryllis) are other prominent alkaloid-containing families. A few alkaloids have been found in animal species, such as the New World beaver (Castor canadensis) and poison-dart frogs (Phyllobates). Ergot and a few other fungi also produce them.

The function of alkaloids in plants is not yet understood. It has been suggested that they are simply waste products of plants' metabolic processes, but evidence suggests that they may serve specific biological functions. In some plants, the concentration of alkaloids increases just prior to seed formation and then drops off when the seed is ripe, suggesting that alkaloids may play a role in this process. Alkaloids may also protect some plants from destruction by certain insect species.

The chemical structures of alkaloids are extremely variable. Generally, an alkaloid contains at least one nitrogen atom in an amine-type structure—i.e., one derived from ammonia by replacing hydrogen atoms with hydrogen-carbon groups called hydrocarbons. This or another nitrogen atom can be active as a base in acid-base reactions. The name alkaloid ("alkali-like") was originally applied to the substances because, like the inorganic alkalis, they react with acids to form salts. Most alkaloids have one or more of their nitrogen atoms as part of a ring of atoms, frequently called a cyclic system. Alkaloid names generally end in the suffix -ine, a reference to their chemical classification as amines. In their pure form most alkaloids are colourless, nonvolatile, crystalline solids. They also tend to have a bitter taste.

Interest in the alkaloids stems from the wide variety of physiological effects (both wanted and unwanted) they produce in humans and other animals. Their use dates back to ancient civilizations, but scientific study of the chemicals had to await the growth of organic chemistry, for not until simple organic bases were understood could the intricate structure of the alkaloids be unraveled. The first alkaloid to be isolated and crystallized was the potent active constituent of the opium poppy, morphine, in about 1804.

Alkaloids are often classified on the basis of their chemical structure. For example, those alkaloids that contain a ring system called indole are known as indole alkaloids. On this basis, the principal classes of alkaloids are the pyrrolidines, pyridines, tropanes, pyrrolizidines, isoquinolines, indoles, quinolines, and the terpenoids and steroids. Alternatively, alkaloids can be classified according to the biological system in which they occur. For example, the opium alkaloids occur in the opium poppy (Papaver somniferum). This dual classification system actually produces little confusion because there is a rough correlation between the chemical types of alkaloids and their biological distribution.

The medicinal properties of alkaloids are quite diverse. Morphine is a powerful narcotic used for the relief of pain, though its addictive properties limit its usefulness. Codeine, the methyl ether derivative of morphine found in the opium poppy, is an excellent analgesic that is relatively nonaddictive. Certain alkaloids act as cardiac or respiratory stimulants. Quinidine, which is obtained from plants of the genus Cinchona, is used to treat arrhythmias, or irregular rhythms of the heartbeat. Many alkaloids affect respiration, but in a complicated manner such that severe respiratory depression may follow stimulation. The drug lobeline (from Lobelia inflata) is safer in this respect and is therefore clinically useful. Ergonovine (from the fungus Claviceps purpurea) and ephedrine (from Ephedra species) act as blood-vessel constrictors. Ergonovine is used to reduce uterine hemorrhage after childbirth, and ephedrine is used to relieve the discomfort of common colds, sinusitis, hay fever, and bronchial asthma.

Many alkaloids possess local anesthetic properties, though clinically they are seldom used for this purpose. Cocaine (from Erythroxylon coca) is a very potent local anesthetic. Quinine (from Cinchona species) is a powerful antimalarial agent that was formerly the drug of choice for treating that disease, though it has been largely replaced by less toxic and more effective synthetic drugs. The alkaloid tubocurarine is the active ingredient in the South American arrow poison, curare (obtained from Chondrodendron tomentosum), and is used as a muscle relaxant in surgery. Two alkaloids, vincristine and vinblastine (from Vinca rosea), are widely used as chemotherapeutic agents in the treatment of many types of cancer.

Nicotine obtained from the tobacco plant (Nicotiana tabacum) is the principal alkaloid and chief addictive ingredient of the tobacco smoked in cigarettes, cigars, and pipes. Some alkaloids are illicit drugs and poisons. These include the hallucinogenic drugs mescaline (from Anhalonium species) and psilocybin (from Psilocybe mexicana). Synthetic derivatives of the alkaloids morphine and lysergic acid (from C. purpurea) produce heroin and LSD, respectively. The alkaloid coniine is the active component of the poison hemlock (Conium maculatum). Strychnine (from Strychnos species) is another powerful poison.

Special methods have been developed for isolating commercially useful alkaloids. In most cases, plant tissue is processed to obtain aqueous solutions of the alkaloids. The alkaloids are then recovered from the solution by a process called extraction, which involves dissolving some components of the mixture with reagents. Different alkaloids must then be separated and purified from the mixture. Chromatography may be used to take advantage of the different degrees of adsorption of the various alkaloids on solid material such as alumina or silica. Alkaloids in crystalline form may be obtained using certain solvents.

MORPHINAN ALKALOIDS

Codeine is an analgesic and cough-suppressing drug and Papaver somniferum L.is a traditional commercial source of codeine, and morphine which can be converted to codeine. Mature capsules of P. bracteatum accumulates up to 3.5% of thebaine which also can be converted to codeine. Although many researchers have tried to produce codeine by undifferentiated cells of these plants, little success has been achieved.

It is indicated that morphogenetic differentiation from cultured cells of P. bracteatum was prerequisite for producing higher levels of thebaine. The researchers also showed the similar results as Kamimura's report using P. bracteatum and P. somniferum. For example, it is reported that cells of P. somniferum produced norsanguinarine, sanguinarine, cryptopine and other various alkaloids, but not codeine and morphine.

Concerning sanguinarine, shows that fungal mycelium of Botrytis sp. elicited production of sanguinarine by P. somniferum cells. The level of this alkaloid increased 26 times in the presence of the elicitor, 29% of the dry cell weight, compared to the medium without it. Since the alkaloid extracted from intact plants is added to toothpastes as an anti-plaque agent, the commercial production of sanguinarine by cell culture technology was intensively investigated by Kurz's group at the National Research Council of Canada and a company in the U.S. several years ago, but the process has not yet been commercialized.

The efficient production of thebaine and codeine using cell culture systems by de novo synthesis was not successful, therefore studied the biotransformation of codeinone to codeine using immobilized cells of P. somniferum. The conversion yield was 70.4% and about 88% of codeine converted was excreted into the medium.

BERBERINE

Berberine is an isoquinoline alkaloid which is distributed in roots of Coptis japonica and cortex of Phellondendron amurense. Berberine chloride is used for intestinal disorders in the Orient and it takes 5 to 6 years to produce Coptis roots as the raw material.

Mitsui Petrochemical has improved the productivity, and found that addition of 10^{-8} M gibberellic acid into the medium stimulated berberine productivity up to 1.66 g per L of the medium. Using a cell sorter and protoplasts of C. japonica, they selected many higher alkaloid-producing cell lines. Thus, the Mitsui group produces berberine in a large scale at a level of 1.4 g per L of their optimized medium within 2 weeks. Furthermore, they established a "high-density cell culture" process to produce berberine much more efficiently. In order to achieve a cell mass of 70 g/L on a dry weight basis, stirring without damaging the cells, supply of sufficient amounts of oxygen, and that of appropriate nutrients were optimize. As a result, the yields reached 70 g per L of cell mass and 0.45 g/day of berberine. The continuous culture with high cell density was also conducted successfully.

Addition of a polyamine, spermidine, was found to stimulate the production of berberine by Thalictrum minus cell suspension cultures by Hara *et al.* in Tabata's laboratory although other polyamines such as cadaverine, putrescine and spermine were ineffective. They indicated that spermidine effected an increase of ethylene generation which was associated with activation of berberine synthesis. The maximum stimulative effect was obtained by addition of 2 mM spermidine.

TROPANE ALKALOIDS

Scopolamine and hyoscyamine are being used commercially as anesthetic and antispasmodic drugs. These alkaloids occur in leaves of (Solanaceae) plants including D. myoporoides and D. leichhardtii. Scopolia, Atropa, Hyoscyamus and Datura also contain tropane alkaloids.

Studies on production of tropane alkaloids by plant tissue cultures have been actively carried out by many researchers since West *et al.* found tropane alkaloids in an Atropa belladonna callus more than 30 years ago. However, the concentrations of scopolamine and hyoscyamine in cultured cells are generally very low in spite of many efforts to increase the yield using various approaches. Therefore, the plant cell culture has not yet been employed to manufacture these alkaloids. For example, Tabata *et al.* added tropic acid into Scopolia japonica suspension cultures as a precursor and could increase the level of alkaloids up to 15 times. Mitsuno *et al.* (159) selected a high tropane alkaloid producing strain of Hyoscyamus niger which produced about 7 times more hyoscyamine (13.9×10^{-3}% fresh cells), than that of the parent strain. According to their results, there was no direct correlation between high producing ability and variation of chromosomal numbers.

Cardinolides

Cardiac glycosides or cardenolides are products of Digitalis species. Some of these compounds have been employed in treatment of heart diseases. Chemical structures of the major aglycone of cardenolides and various sugar residues link the 3-hydroxy group of aglycones. For commercial production Digitalis plants are being cultivated in the fields of several countries including the Netherlands, Hungary and Argentina.

Although there are a number of papers describing production of cardiac glycosides in Digitalis tissue cultures, generally the yield was very low, and moreover, during the successive transfers of the cultured cells the amount of cardenolides often decreased and disappeared completely. Instead, many researchers indicated that morphological differentiation caused an increase in productivity. For example, it shows that organ cultures of D. lanata leaves and roots produced cardenolides and during the cultivation the level of digoxin in the tissues rose with increasing age. Hirotani and Furuya also found that renewed organ differentiation from callus tissues of D. purpurea led to a new formation of cardinolides.

Nutritional factors including growth regulators, sugars, nitrogen sources, vitamins and so on in the medium affect on differentiation of shoots and other organs and the secondary metabolites such as cardiac glycosides are often synthesized. The differentiated tissue culture is prerequisite for secondary metabolites production in some cases, however in general, the method requires much longer culture time than the suspension cell culture and consequently it is not efficient.

Various secondary metabolites other than digitoxin and digoxin have been found in callus tissues of D. lanata and D. purpurea. These include cholesterol, campesterol, stigmasterol, β-sitosterol, 4-hydroxy-digitolutein and others. It is found that the levels of those secondary products tended to decrease over several passages in cultivation. In contrast to the de novo synthesis, the biotransformation process with Digitalis plant cells seems to be more promising from a commercial point of view. It is reported that D. lanata and D. purpurea callus cultures rapidly transformed progesterone to pregnane. Leaf and root cultures of D. lanata and shoot-forming callus tissues of D. purpurea accumulated an increased level of digoxin and/or digitoxin when progesterone was added to the cultures. Stohs and Staba studies on the biotransformation of cardenolides by Digitalis cells and recognized that the glycosylation reaction occurred.

Among many studies, the biotransformation from digitoxin to digoxin using Digitalis lanata cells investigated by Reinhard and Alfermann is the most interesting approach in terms of commercial application since digoxin has a higher demand as a drug for heart diseases than digitoxin. It is advantageous that Digitalis leaves contain a larger amount of digitoxin which can be used as a substrate. It is a hydroxylation reaction at the 12 β-position of digitoxin, and Reinhard *et al.* found that β-methyldigitoxin was the most suitable substrate in this biotransformation as methyldigoxin is the major product.

L-DOPA

L-DOPA, L-3,4-dihydroxyphenylalanine, is an important intermediate of secondary metabolism in higher plants and is known as an precursor of alkaloids, betalain, melanine, and others. It is also a precursor of cathecolamines in animals and is being used as a potent drug for Parkinson's disease.

It is found that the callus tissue of Mucana pruriens accumulated 25 mg/L DOPA in the medium containing very high concentration of 2,4-D. Wichers *et al.* immobilized cells of M. pruriens within alginate and found that the cells produced DOPA from tyrosine in up to 2% of dry cell weight. The DOPA synthesized was secreted mostly into the medium.

Teramoto and Komamine induced callus tissues of Stizolobium hassjoo (Mucuna hassjoo), M. pruriens and M. deeringiana, and optimized the culture conditions. The highest concentration of DOPA was obtained when S. hassjoo cells were cultivated in MS medium containing 0.025 mg/l 2,4-D and 10 mg/l kinetin. The level of DOPA in the cells was about 80 nmol/g-f.w.

VALEPOTRIATES

Plants in the Valerianaceae have been used as folk medicines. For example, Nardostachys jatamansi, Valeriana wallichii and V. officinalis L. var. angustifolia have been used in India, and N. chinensis has been employed in China for hundreds of years. Partrinia plants are also being used as sedative drugs in former Soviet Union. Although the active principles in these plants have not been identified, a group of compounds having biological activities such as sedative, tranquilization, cytotoxicity and antitumor activities were named "valepotriates". Thies synthesized a series of valepotriate derivatives and tested for biological activities.

It has investigated plant tissue cultures for producing valepotriates because of their limited supply and uncertain availability. They induced callus tissues of nine different species of Valerianaceae on MS media and found that Fedia cornucopiae and V. locusta cells produced higher levels of the compounds than the intact plants. Isolation of cell lines resistant to trifluoroleucine and to nystatin, treatment with colchicine, cultivation of the cells in two-phase culture media and addition of several bioregulators were carried out intensively.

As valepotriates have monoterpene skeletons, L-leucine was considered as a precursor. In an isolated cell lines of V. wallichii resistant to trifluoroleucine, a leucine analog, but the yield of valepotriates in the cells was not increased although the intracellular level of leucine was increased. It was reported that fungi resistant to nystatin, a polyene antibiotic, produced a high level of steroids.

One of the strains of V. wallichii to resistant to nystatin increased the amount of valepotriates produced up to 3 times, which was 88 mg/g-d.w. The same group treated a suspension culture of V. wallichii with colchicine which was expected to induce polyploid cells. As a result, they found that the colchicine treated cells produced higher amount valepotriates than the respective untreated cultures. A two-phase culture provided by addition of RP-8 into the culture medium was effective in inducing secretion of the lipophilic compounds from the cells, and the total yield of valepotriates was substantially increased. Since the valepotriate skeleton is of the iridoid nature, they added some plant bioregulators such as dimethyl-morpholinium-bromide, dimethyl-piperidinium-bromide, dimethyl-piperidinium-chloride as well as 2-(3,4-dichloro-phenoxy)-triethylamine and 2-(3,5-diisopropylphenoxy)-triethylamine to cell suspension cultures of F. cornucopiae and V. wallichii. When they were employed in concentrations of 0.01 to 0.04 mmol during the early exponential growth stages of the cells, the levels of valepotriates were increased significantly.

ROLE OF THE KIDNEYS IN ACID-BASE BALANCE

The kidneys function to filter the plasma that passes through the nephrons. Filtration of the plasmas occurs in the glomerular capillaries of the nephron.

These capillaries allow the passage of water and low molecular weight solutes (less than 70 kDa) into the capsular space. The filtrate then passes through the proximal and distal convoluted tubules where reabsorption of water and many solutes takes place. In the course of glomerular filtration and tubule reabsorption the composition of the plasma changes generating the typical composition of urine. From a biochemical standpoint the kidneys serve important roles in the regulation of plasma acid-base balance and the elimination of nitrogenous wastes.

SODIUM BICARBONATE REABSORPTION

Regulation of plasma acid-base balance is primarily effected within the kidney through control over HCO_3^- reabsorption and secretion of H^+. Secretion of H^+, in excess of its capacity to react with HCO_3^- in the tubular lumenal fluid, requires the presence of other buffers. The generation of HCO_3^- and H^+ occurs by dissociation of carbonic acid (H_2CO_3), formed in the tubule cells from H_2O and CO_2, through the action of carbonic anhydrase. Secretion of H^+ into the lumen of the tubule is accompanied by an exchange for Na^+. This reabsorption of Na+ occurs by an antiport mechanism during the exchange for H^+.

Reduction in the intracellular concentration of Na^+ occurs by an active transport process involving a Na^+/K^+-ATPase pump which pumps the excess Na^+ into the interstitial fluid. The intracellular HCO_3^- then diffuses from the tubule cell into the interstitial fluid. The capacity of the kidney to secrete H^+ is regulated by the maximal H^+ gradient that can form between the tubule and lumen and still allow transport mechanisms to operate.

This gradient is determined by the pH of the urine which in humans is near 4.5. The capacity to secrete H^+ would be rapidly reached if it were not for the presence of buffers within the interstitial fluid. The H^+ secreted into the tubular lumen can undergo three different fates depending upon the concentration of the three primary buffers of the interstitial fluid.

These buffers are,

HCO_3^-, HPO_4^{2-} and NH_3.

Reaction of H^+ with HCO_3^- forms H_2O and CO_2 which diffuse back into the tubule cell. The net result of this process is the regeneration of HCO_3^- within the tubule cell. This process is termed reabsorption of sodium bicarbonate. The reabsorption of sodium bicarbonate takes place primarily within the proximal convoluted tubules.

Excretion of Acid

As the concentration of HCO_3^- in the tubular lumen drops, the pH of the fluid drops due to an increasing concentration of H^+. The pH of the tubular fluid gradually approaches the pK_a for the dibasic/monobasic phosphate buffering system (pK_a = 6.8). The excess H^+ reacts with dibasic phosphate (HPO_4^{2-}) forming monobasic phosphate ($H_2PO_4^-$). The $H_2PO_4^-$ so formed is not reabsorbed and its excretion results in the net excretion of H^+. The greatest extent of $H_2PO_4^-$ formation occurs within the distal convoluted tubules and the collecting ducts.

Ammonia Secretion

Buffering of H^+ is also accomplished by reaction with ammonia, NH_3, to form ammonium ion, NH_4^+. Elimination of NH_4^+ is the major contributory factor in the ability of the body to excrete acid. Because the pK_a of NH_4^+ is 9.3, excretion of acid in this form can be accomplished without lowering the pH of the urine. Additionally important is the fact that excretion of acid in the form of NH_4^+ occurs without depleting Na^+ nor K^+.

Two principal reactions within tubule cells result in the generation of NH_3, conversion of glutamine to glutamate and conversion of glutamate to α-ketoglutarate.

These reactions are catalyzed by glutaminase and glutamate dehydrogenase, respectively as shown in the Figures below.

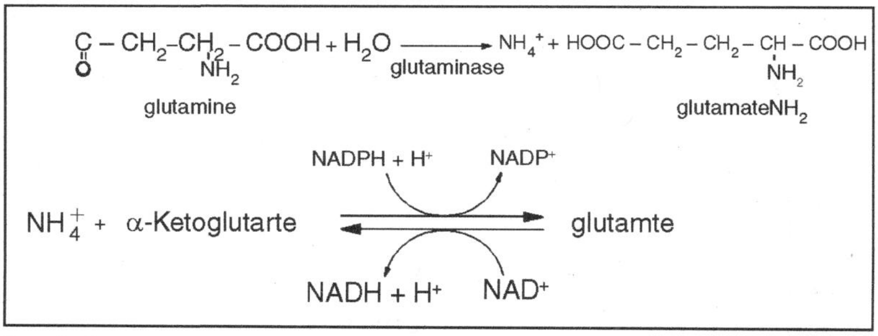

Both of these enzymes are abundant in tubule cells. Ammonia is lipid soluble and will diffuse down its concentration gradient out of the tubule cell into the tubular fluid. There it reacts with H^+ to yield NH_4^+ which is excreted in the urine.

Neurotoxicity of Ammonia

Excess ammonia is severely neurotoxic. Marked brain damage is seen in cases of failure to make urea via the urea cycle or to eliminate urea through the kidneys. The result of either of these events is a buildup of circulating levels of ammonium ion.

Aside from its effect on blood pH, ammonia readily traverses the brain blood barrier and in the brain is converted to glutamate via glutamate dehydrogenase, depleting the brain of α-ketoglutarate. As the α-ketoglutarate is depleted, oxaloacetate falls correspondingly, and ultimately TCA cycle activity comes to a halt. In the absence of aerobic oxidative phosphorylation and TCA cycle activity, irreparable cell damage and neural cell death ensue.

In addition, the increased glutamate leads to glutamine formation. This depletes glutamate stores which are needed in neural tissue since glutamate is both a neurotransmitter and a precursor for the synthesis of ã-aminobutyrate: GABA, another neurotransmitter. Therefore, reductions in brain glutamate affect energy production as well as neuro-transmission.

Additional untoward consequences are the result of elevations in neural glutamine concentration. Glial cell (astrocytes) volume is controlled by intracellular organic osmolyte metabolism. The organic osmolyte is glutamine. As glutamine levels rise in the brain the volume of fluid within glial cells increases resulting in the cerebral edema seen in infants with hyperammonemia caused by urea cycle defects.

PHARMACEUTICAL NOMENCLATURE

First of all substances have their chemical name, which gives the chemical composition of that drug. Since these names are generally very long and complicated they are also given a shorter, international name, a generic name. Finally, the manufacturer gives the product a trade name.

This naming process means that it is possible for a substance to be known under many different trade names. Trade names are always indicated by the sign R inside a circle, following the name.

PHARMACOKINETICS

This is the branch that is concerned with how a pharmacon acts after it is introduced into the body. This is firstly dependent on the way in which it was administered;

This can be:

- *Enteral*: By way of the intestine
- *Oral, per os*: By way of the mouth, swallowing sublingual: Beneath the tongue, sucking
- *Rectal*: By way of the anus (suppository) parenteral: by injection through the skin
- *Intracutaneous*: Within the skin

- *Subcutaneous*: Beneath the skin
- *Intramuscular*: Within the muscles intravenous: within a vein
- *Intraperitoneal*: Within the peritoneal cavity intracardiac: within the heart
- *Inhalation*: Drawing into the lungs (smoking)
- *Transcutaneous*: Absorption through the skin (bandage) The most important aspect of admini-stration per os is that when the substance has been absorbed into the blood of the gastrointestinal tract it goes on to the liver.

One important function of the liver is to break down foreign substances and/or prepare them for elimination by the kidneys by changing them chemically. This process is called biotransformation. Biotransformation reduces the effect of many substances when taken per os. This is not the case with smoking or shooting up, which means that in these cases the effect is much stronger.

Another way to avoid biotransformation in the liver is by rectal or sublingual administration (the nitrobate in angina pectoris) as the blood vessels from the mouth and rectum do not first lead to the liver.

Another factor influencing the strength of the effect IS that it is not so much the amount of the substance in the blood (concentration in the blood) that determines the severity of the effect, but rather how quickly that concentration rises. A substance administered per os is absorbed slowly by the blood; its concentration in the blood rises slowly. With shooting up or smoking, concentration in the blood rises rapidly to very rapidly, causing a much stronger effect.

Blood transports the substance administered to the place where it is to have its effect. Whether the substance is easily soluble in water, or in fat is also important. If a substance is easily soluble in water, it mixes well with the blood; if it is easily soluble in fat, it mixes less easily, but stores well in the adipose tissue for later use.

Substances are also broken down: Although the liver is often mentioned in this context, the breakdown of enzymes also takes place elsewhere in the body. An enzyme is a substance that causes a chemical reaction to occur without being changed itself. Outside the medical world this reaction is called catalysation; a substance that has such a function is called a catalyst. Of importance in this connection is that some substances (also medicines) can speed up or slow down the breakdown process. In cases of combined usage this can lead to a lowered or heightened effect. Examples of this are:

- Rifampicine (an antibiotic which is also used in tuberculosis cases): speeds up methadone breakdown, which reduces its effect; methaqualone (a sleepinducing drug): slows down the breakdown of many substances (including opiates), which heightens their effect. An important concept with this is halflife, which is the amount of time the body needs to eliminate half of the substance present as measured by its concentration in the blood.

The substances, whether biotransformed or not, are then excreted again, usually with the urine. Substances can also be excreted with the feces, sweat, saliva, tears and mother's milk.

If they are gasforming, they can be exhaled by the lungs:

- *Agonist antagonist*: Pharmaca are administered for their effect. They have that effect because they react with special molecules, called receptors. The pharmacon is then anagonist with regard to a specific receptor. Other substances can have the opposite effect; these latter are referred to as the antagonists. An example of this is the agonist morphine of which nalorphine is the antagonist that counteracts the effects of morphine. It often 'happens that the antagonists 'fight' with the agonists for an effect on the receptor; we then speak of competitive antagonists. This in contrast to situations in which the antagonist blocks the receptor just like that: we then speak of noncompetitive blocking. In the first case, the receptor blockade can be lifted again by a higher dosage of the agonist, in the second case it cannot.

ROUTES OF DRUG ADMINISTRATION

- Intravenous
- Buccal
- Rectal
- Transdermal
- Inhalational
- Oral
- Sublingual
- Intramuscular
- Subcutaneous
- Topical

Of all of these routes you are most likely to be asked about the transdermal, as it is fashionable. Otherwise, most other basic pharmacology questions tend to concern the pharmaco-logy of intravenous agents; that is what is discussed below.

FIRST ORDER KINETICS

A constant fraction of the drug in the body is eliminated per unit time. The rate of elimination is proportional to the amount of drug in the body. The majority of drugs are eliminated in this way.

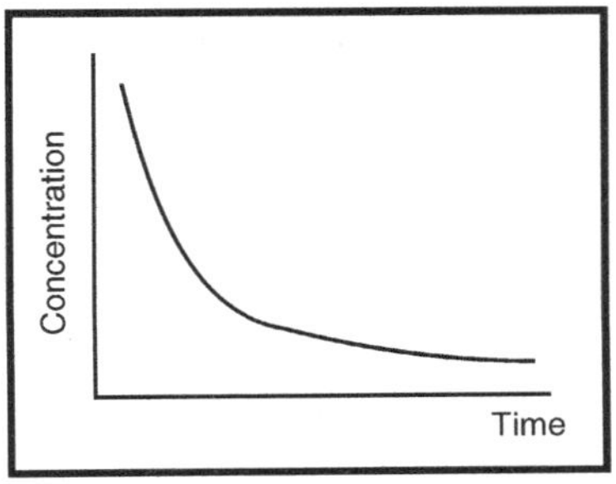

What follows concerns drugs which follow first order kinetics. The Volume of Distribution (Vd) is the amount of drug in the body divided by the concentration in the blood. Drugs that are highly lipid soluble, such as digoxin, have a very high volume of distribution (500 litres). Drugs which are lipid insoluble, such as neuromuscular blockers, remain in the blood, and have a low Vd.

The Clearance (Cl) of a drug is the volume of plasma from which the drug is completely removed per unit time. The amount eliminated is proportional to the concentration of the drug in the blood.

The fraction of the drug in the body eliminated per unit time is determined by the elimination constant (kel). This is represented by the slope of the line of the log plasma concentration versus time.

$$Cl = kel \times Vd$$

Rate of elimination = clearance × concentration in the blood.

Elimination half life (t1/2): The time taken for plasma concentration to reduce by 50%. After 4 half lives, elimination is 94% complete.

It can be shown that the kel = the log of 2 divided by the t1/2 = 0.693/t1/2.

Likewise, Cl = kel × Vd, so, Cl = 0.693Vd/t1/2.

And t1/2 = 0.693 × Vd/ cl

The rate of elimination is the clearance times the concentration in the plasma,

$$Roe = Cl \times Cp$$

Fraction of the total drug removed per unit time = Cl/Vd.

If the volume of distribution is increased, then the kel will decrease, the t1/2 will increase, but the clearance won't change.

Example

You have a 10ml container of orange squash. You put this into a litre (ok 990ml!) of water. The Vd of the orange squash is 1000ml. If, each minute, you empty 10ml of the orange liquid into the 10ml container, discard this, and replace it with 10ml of water. The clearance is 10 ml per minute. The elimination half life is: 70 minutes. The kel is Cl/Vd = 10/1000 = 0.01. Shown the other way, 0.693/50 = 0.01.

If the volume of the container is increased to 2000ml, then the clearance remains the same, but the Vd, and consequently the t1/2, increases (to 140 minutes).

Simple Example

What is described above is a single compartment model, what would occur if the bloodstream was the only compartment in the body (or if the Vd = the blood volume). But the human body is more complex than this: there are many compartments: muscle, fat, brain tissue etc. In order to describe this, we use multicompartment models.

MULTICOMPARTMENT MODELS

Why does a patient wake up after 5 minutes after an injection of thiopentone when we know that it takes several hours to eliminate this drug from the body? What happens is that, initially the drug is all in the blood and this blood goes to "vessel rich" organs; principally the brain. After a few minutes the drug starts to venture off into other tissues (fat, muscle etc) it redistributes, the concentration in the brain decreases and the patient wakes up! The drug thus redistributes into other compartments.

If you were to represent this phenomenon graphically, you would follow a picture of rapid fall in blood concentration, a plateau, and then a slower gradual fall. The first part is the rapid redistribution phase, the alpha phase, the plateau is the equilibrium phase (where blood concentration = tissue concentration), and the slower phase, the beta phase, is the elimination phase where blood and tissue concentrations fall in tandem. This is a simple two compartment model and is as much as you need to know.

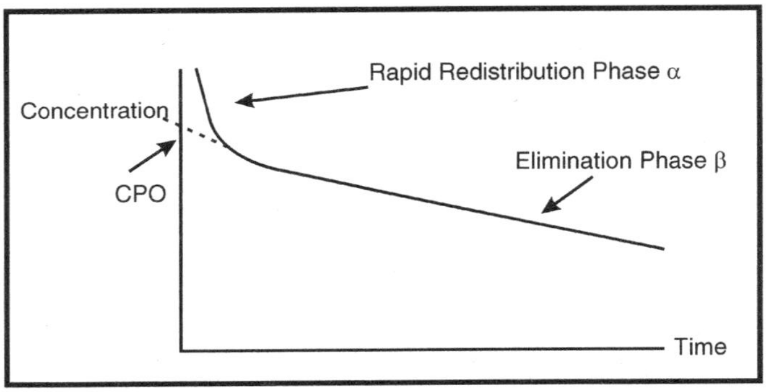

An couple of interesting pieces of information can be derived from the log concentration versus time graph. If you extrapolate back the elimination line to the y axis, then you get to a point called the CP0 - a theoretical point representing the concentration that would have existed at the start if the dose had been instantly distributed (dose/Vd). From this new straight line you can figure out how long it takes for the concentration to drop by 50%: the elimination half life. Likewise, a similar procedure can be performed on the α phase: the redistribution half life. While it is very important that you understand these concepts, the reality is that most drugs are infinitely more complicated that this, and computer calculations are required to derive this data.

BIOAVAILABILITY

This is the fraction of the administered dose that reaches the systemic circulation. Bioavailability is 100% for intravenous injection. It varies for other routes depending on incomplete absorption, first pass hepatic metabolism etc. Thus one plots plasma concentration against time, and the bioavailability is the area under the curve.

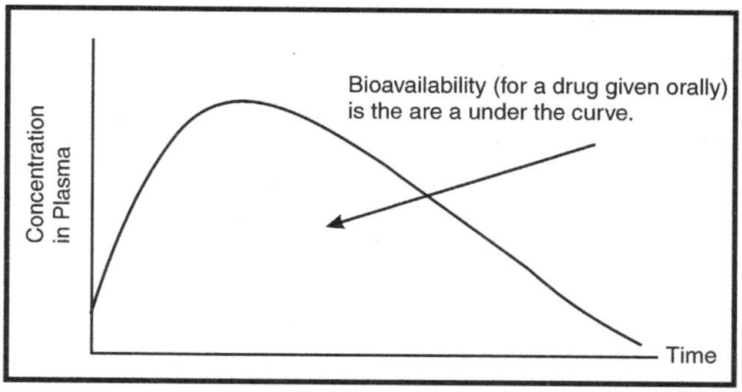

ZERO ORDER ELIMINATION

Why if I have 10 pints of beer before midnight will I fail a breathalyser test at 8 am the following morning? Either this is due to alcohol having a very long half life (which it does not) or that alcohol is cleared in a different way.

What happens is that the metabolic pathways responsible for alcohol metabolism are rapidly saturated and that clearance is determined by how fast these pathways can work. The metabolic pathways work to their limit. This is known as zero order kinetics: a constant amount of drug is eliminated per unit time.

This form of kinetics occours with several important drugs at high dosage concentrations: phenytoin, salicylates, theophylline, and thiopentone (at very large doses). Because high dose thio is very slow to clear, we no longer use it in infusion for status epilepticus.

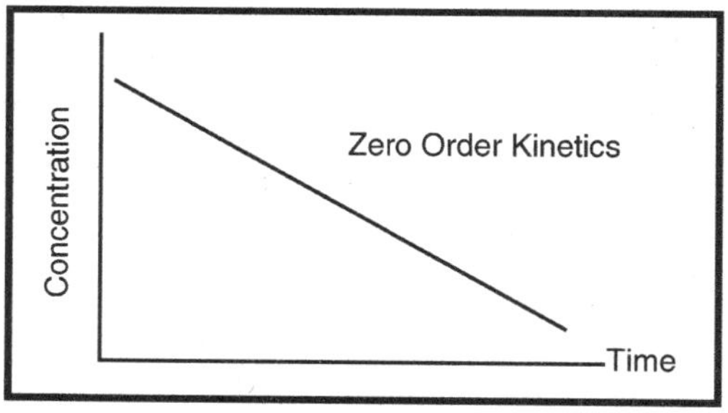

DOSAGE REGIMENS

The strategy for treating patients with drugs is to give sufficient amounts that the required theraputic effect arises, but not a toxic dose. The maintenance dose is equal to the rate of elimination at steady state (i.e.at steady state, rate of elimination = rate of administration):

Dosing rate = clearance × desired plasma concentration.

Drugs will accumulate within the body if the drug has not been fully eliminated before the next dose. Steady state concentration is thus arrived at after four half lives.

This is all very well if you are willing to wait 4 half lives for the drug to be fully effective, but what if you are not? What you may need to do is to "load" the volume of distribution with the drug to achieve target plasma concentrations rapidly: the loading dose.

HEPATIC DRUG CLEARANCE

Many drugs are extensively metabolised by the liver.

The rate of elimination depends on:

- The liver's inherent ability to metabolise the drug,
- The amount of drug presented to the liver for metabolism.

This is important because drugs administered orally are delivered from the gut to the portal vein to the liver: the liver gobbles up a varying chunk of the administered drug (pre-systemic elimination) and less is available to the body for theraputic effect. This is why you have to give a higher dose of morphine, for examole, orally, than intravenously.

Hepatic drug clearance (i.e. the amount of each drug gobbled up by the liver) depends on:

- The Intrinsic clearance (Cl int).
- Hepatic blood flow.

These two factors are independent of one another, and their combined effect is the proportion of drug gobbled up: the extraction ratio.

For drugs that have a low intrinsic clearance, this effect can be increased by giving a second agent that boosts the effect of the liver's enzyme system; these are enzyme inducers. Examples of such drugs are cigarrettes, antiepileptics (carbamazepine & phenytoin), rifampicin, griseofulvin, alcohol and spironolactone (CAR GAS) [also barbiturates].

Consequnetly if a drug addict is given rifampicin or tuberculosis, a higher dose of heroin is required for the same effect. Enzyme inhibitors have the opposite effect: examples are flagyl, allopurinol, cimetidine, erythromycin, dextropropoxyphene, imipramine, (the) pill (FACE DIP).

Likewise, if the blood flow increases, the liver has less chance to gobble up the drug, and the extraction ratio falls. This is particularly the case, as you would expect, of the intrinsic clearance is low.

Illustration: Think of factory workers picking bad apples out of a pile on a conveyor belt, if only one person (low intrinsic clearance) is doing the picking and the speed of the conveyor belt is increased, more bad apples get through.

If there are several pickers (high intrinsic clearance) then they are much more able to cope with an increase in the speed of the conveyor belt, but there will come a rate at which they will become overwhelmed, and bad apples will get through.

From this example you can take home this message: For drugs with a low extraction ratio, the kinetics (the body's ability to deal with the drug) depends on enzymatic activity (giving an enzyme inducer effectively gives the single picker 4 arms!). For high extraction ratio drugs, kinetics depends on liver blood flow - the slower the flow the higher the extraction, the higher the flow the lower the extraction ratio.

DRUG DISTRIBUTION

When a drug is introduced into the body, where it ends up depends on a number of factors:

- Blood flow, tissues with the highest blood flow receive the drug first,
- Protein binding, drugs stuck to plasma proteins are crippled, they can only go where the proteins go (and that's not very far!), 3) lipid solubility and the degree of ionisation, this describes the ability of drugs to enter tissues (highly lipid soluble/ unionised drugs can basically go anywhere).

PROTEIN BINDING

Most drugs bind to proteins, either albumin or alpha-1 acid glycoprotein (AAG), to a greater or lesser extent. Drugs prefer to be free, it is in this state that they can travel throughout the body, in and out of tissues and have their biological effect. The downside of this is that they are easy prey for metabolising enzymes. As you would expect, more highly bound drugs have a longer duration of action and a lower volume of distribution.

Generally high extraction ratio drugs' clearance is high because of low protein binding and, conversely, low extraction ratio drugs' clearance is strongly dependent on the amount of protein binding.

Why is this important? If a drug is highly protein bound, you need to give loads of it to get a theraputic effect; as so much is stuck to protein. But what happens if another agent comes along and starts to compete with the drug for the binding site on the protein? Yes, you guessed it, the amount of free drug is increased. This is really important for drugs that are highly protein bound: if a drug is 97% bound to albumin and there is a 3% reduction in binding (displaced by another drug), then the free drug concentration doubles; if a drug is 70% bound and there is a 3% reduction in binding, this will make little difference. The drugs that you really need to keep an eye on are: warfarin, diazepam, propranolol and phenytoin. For example, a patient on warfarin is admitted with seizures, you treat the patient with phenytoin, next thing you know - his INR is 10.

The amount of albumin does not appear to be hugely relavent. In disease states such as sepsis, the serum albumin drops drastically, but the free drug concentration does not appear to increase

DEGREE OF IONIZATION

This is really important with regard to local anaesthetics. The essential fact to know is that highly ionized drugs cannnot cross lipid membranes (basically they can't go anywhere) and unionised drugs can cross freely.

Morphine is highly ionised, fentanyl is the opposite. Consequently the latter has a faster onset of action. The degree of ionisation depends on the pKa of the drug and the pH of the local environment.

The pKa is the the pH at which the drug is 50% ionised. Most drugs are either weak acids or weak bases. Acids are most highly ionised at a high pH (i.e. in an alkaline environment). Bases are most highly ionised in an acidic environment (low pH). For a weak acid, the more acidic the environment, the less ionised the drug, and the more easily it crosses lipid membranes. If you take this acid, at pKa it is 50% ionised, if you add 2 pH points to this (more alkaline), it becomes 90% ionised, if you reduce the pH (more acidic) by two units, it becomes 10% ionised. Weak bases have the opposite effect. Local anaesthetics are weak bases: the closer the pKa of the local anaesthetic to the local tissue pH, the more unionised the drug is. That is why lignocaine(pKa 7.7) has a faster onset of action than bupivicaine (pKa 8.3). If the local tissues are alkalinised (e.g. by adding bicarbonate to the local anaesthetic), then the tisssue pH is brought closer to the pKa, and the onset of action is hastened.

PHARMACEUTICAL DOSAGE FORMS

Dosage forms are provided for most of the Pharmacopeial drug substances, but the processes for the preparation of many of them are, in general, beyond the scope of the Pharmacopeia. In addition to defining the dosage forms, this section presents the general principles involved in the manufacture of some of them, particularly on a small scale. Other information that is given bears on the use of the Pharmacopeial substances in extemporaneous compounding of dosage forms.

BIOAVAILABILITY

Bioavailability, or the extent to which the therapeutic constituent of a pharmaceutical dosage form intended for oral or topical use is available for absorption, is influenced by a variety of factors. Among the inherent factors known to affect absorption are the method of manufacture or method of compounding; the particle size and crystal form or polymorph of the drug substance; and the diluents and excipients used in formulating the dosage form, including fillers, binders, disintegrating agents, lubricants, coatings, solvents, suspending agents, and dyes. Lubricants and coatings are foremost among these. The maintenance of a demonstrably high degree of bioavailability requires particular attention to all aspects of production and quality control that may affect the nature of the finished dosage form.

TERMINOLOGY

Occasionally it is necessary to add solvent to the contents of a container just prior to use, usually because of instability of some drugs in the diluted form. Thus, a solid diluted to yield a suspension is called [DRUG] for Suspension; a solid dissolved and diluted to yield a solution is called [DRUG] for Solution; and a solution or suspension diluted to yield a more dilute form of the drug is called [DRUG] Oral Concentrate. After dilution, it is important that the drug be homogeneously dispersed before administration.

AEROSOLS

Pharmaceutical aerosols are products that are packaged under pressure and contain therapeutically active ingredients that are released upon activation of an appropriate valve system. They are intended for topical application to the skin as well as local application into the nose (nasal aerosols), mouth (lingual aerosols), or lungs (inhalation aerosols). These products may be fitted with valves enabling either continuous or metered-dose delivery; hence, the terms "[DRUG] Metered Topical Aerosols," "[DRUG] Metered Nasal Aerosols," etc.

The term "aerosol" refers to the fine mist of spray that results from most pressurized systems. However, the term has been broadly misapplied to all self-contained pressurized products, some of which deliver foams or semisolid fluids. In the case of Inhalation Aerosols, the particle size of the delivered medication must be carefully controlled, and the average size of the particles should be under 5 μm. These products are also known as metered-dose inhalers (MDIs). Other aerosol sprays may contain particles up to several hundred micrometers in diameter.

The basic components of an aerosol system are the container, the propellant, the concentrate containing the active ingredient(s), the valve, and the actuator. The nature of these components determines such characteristics as particle size distribution, uniformity of dose for metered valves, delivery rate, wetness and temperature of the spray, spray pattern and velocity or plume geometry, foam density, and fluid viscosity.

Types of Aerosols

Aerosols consist of two-phase (gas and liquid) or three-phase (gas, liquid, and solid or liquid) systems. The two-phase aerosol consists of a solution of active ingredients in liquefied propellant and the vaporized propellant. The solvent is composed of the propellant or a mixture of the propellant and cosolvents such as alcohol, propylene glycol, and polyethylene glycols, which are often used to enhance the solubility of the active ingredients.

Three-phase systems consist of a suspension or emulsion of the active ingredient(s) in addition to the vaporized propellants. A suspension consists of the active ingredient(s) that may be dispersed in the propellant system with the aid of suitable excipients such as wetting agents and/or solid carriers such as talc or colloidal silicas.

A foam aerosol is an emulsion containing one or more active ingredients, surfactants, aqueous or nonaqueous liquids, and the propellants. If the propellant is in the internal (discontinuous) phase (i.e., of the oil-in-water type), a stable foam is discharged; and if the propellant is in the external (continuous) phase (i.e., of the water-in-oil type), a spray or a quick-breaking foam is discharged.

Propellants

The propellant supplies the necessary pressure within an aerosol system to expel material from the container and, in combination with other components, to convert the material into the desired physical form. Propellants may be broadly classified as liquefied or compressed gases having vapor pressures generally exceeding atmospheric pressure. Propellants within this definition include various hydrocarbons, especially halogenated derivatives of methane, ethane, and propane, low molecular weight hydrocarbons such as the butanes and pentanes, and compressed gases such as carbon dioxide, nitrogen, and nitrous oxide. Mixtures of propellants are frequently used to obtain desirable pressure, delivery, and spray characteristics. A good propellant system should have the proper vapor pressure characteristics consistent with the other aerosol components.

Valves

The primary function of the valve is to regulate the flow of the therapeutic agent and propellant from the container. The spray characteristics of the aerosol are influenced by orifice dimension, number, and location. Most aerosol valves provide for continuous spray operation and are used on most topical products. However, pharmaceutical products for oral or nasal inhalation often utilize metered-dose valves that must deliver a uniform quantity of spray upon each valve activation. The accuracy and reproducibility of the doses delivered from metering valves are generally good, comparing favorably to the uniformity of solid dosage forms such as tablets and capsules. However, when aerosol packages are stored improperly, or when they have not been used for long periods of time, valves must be primed before use. Materials used for the manufacture of valves should be inert to the formulations used. Plastic, rubber, aluminum, and stainless steel valve components are commonly used. Metered-dose valves must deliver an accurate dose within specified tolerances.

Actuators

An actuator is the fitting attached to an aerosol valve stem, which when depressed or moved, opens the valve, and directs the spray containing the drug preparation to the desired area. The actuator usually indicates the direction in which the preparation is dispensed and protects the hand or finger from the refrigerant effects of the propellant. Actuators incorporate an orifice that may vary widely in size and shape. The size of this orifice, the expansion chamber design, and the nature of the propellant and formulation influence the delivered dose as well as the physical characteristics of the spray, foam, or stream of solid particles dispensed. For inhalation aerosols, an actuator capable of delivering the medication in the proper particle size range and with the appropriate spray pattern and plume geometry is utilized.

Containers

Aerosol containers usually are made of glass, plastic, or metal, or a combination of these materials. Glass containers must be precisely engineered to provide the maximum in pressure safety and impact resistance. Plastics may be employed to coat glass containers for improved safety characteristics, or to coat metal containers to improve corrosion resistance and enhance stability of the formulation. Suitable metals include stainless steel, aluminum, and tin-plated steel. Extractables or leachables (e.g., drawing oils, cleaning agents, etc.) and particulates on the internal surfaces of containers should be controlled.

Manufacture

Aerosols are usually prepared by one of two general processes. In the "cold-fill" process, the concentrate (generally cooled to a temperature below 0) and the refrigerated propellant are measured into open containers (usually chilled). The valve-actuator assembly is then crimped onto the container to form a pressure-tight seal. During the interval between propellant addition and crimping, sufficient volatilization of propellant occurs to displace air from the container. In the "pressure-fill" method, the concentrate is placed in the container, and either the propellant is forced under pressure through the valve orifice after the valve is sealed, or the propellant is allowed to flow under the valve cap and then the valve assembly is sealed ("under-the-cap" filling). In both cases of the "pressure-fill" method, provision must be made for evacuation of air by means of vacuum or displacement with a small amount of propellant vapor. Manufacturing process controls usually include monitoring of proper formulation and propellant fill weight and pressure testing, leak testing, and valve function testing of the finished aerosol. Microbiological attributes should also be controlled.

Extractable Substances

Since pressurized inhalers and aerosols are normally formulated with organic solvents as the propellant or the vehicle, leaching of extractables from the elastomeric and plastic components into the formulation is a potentially serious problem. Thus, the composition and the quality of materials used in the manufacture of the valve components (e.g., stem, gaskets, housing, etc.)

must be carefully selected and controlled. Their compatibility with formulation components should be well established so as to prevent distortion of the valve components and to minimize changes in the medication delivery, leak rate, and impurity profile of the drug product over time. The extractable profiles of a representative sample of each of the elastomeric and plastic components of the valve should be established under specified conditions and should be correlated to the extractable profile of the aged drug product or placebo, to ensure reproducible quality and purity of the drug product. Extractables, which may include polynuclear aromatics, nitrosamines, vulcanization accelerators, antioxidants, plasticizers, monomers, etc., should be identified and minimized wherever possible.

SOLID DOSAGE OF CAPSULES

Capsules are solid dosage forms in which the drug is enclosed within either a hard or soft soluble container or "shell." The shells are usually formed from gelatin; however, they also may be made from starch or other suitable substances. Hard-shell capsule sizes range from No. 5, the smallest, to No. 000, which is the largest, except for veterinary sizes. However, size No. 00 generally is the largest size acceptable to patients. Size 0 hard gelatin capsules having an elongated body (known as size OE) also are available, which provide greater fill capacity without an increase in diameter.

Hard gelatin capsules consist of two, telescoping cap and body pieces. Generally, there are unique grooves or indentations molded into the cap and body portions to provide a positive closure when fully engaged, which helps prevent the accidental separation of the filled capsules during shipping and handling. Positive closure also may be affected by spot fusion ("welding") of the cap and body pieces together through direct thermal means or by application of ultrasonic energy. Factory-filled hard gelatin capsules may be completely sealed by banding, a process in which one or more layers of gelatin are applied over the seam of the cap and body, or by a liquid fusion process wherein the filled capsules are wetted with a hydroalcoholic solution that penetrates into the space where the cap overlaps the body, and then dried. Hard-shell capsules made from starch consist of two, fitted cap and body pieces. Since the two pieces do not telescope or interlock positively, they are sealed together at the time of filling to prevent their separation. Starch capsules are sealed by the application of a hydroalcoholic solution to the recessed section of the cap immediately prior to its being placed onto the body.

The banding of hard-shell gelatin capsules or the liquid sealing of hard-shell starch capsules enhances consumer safety by making the capsules difficult to open without causing visible, obvious damage, and may improve the stability of contents by limiting O2 penetration. Industrially filled hard-shell capsules also are often of distinctive color and shape or are otherwise marked to identify them with the manufacturer. Additionally, such capsules may be printed axially or radially with strengths, product codes, etc. Pharmaceutical-grade printing inks are usually based on shellac and employ FDA-approved pigments and lake dyes.

In extemporaneous prescription practice, hard-shell capsules may be hand-filled; this permits the prescriber a latitude of choice in selecting either a single drug or a combination of drugs at the exact dosage level considered best for the individual patient. This flexibility gives hard-shell capsules an advantage over compressed tablets and soft-shell capsules as a dosage form. Hard-shell capsules are usually formed from gelatins having relatively high gel strength. Either type may be used, but blends of pork skin and bone gelatin are often used to optimize shell clarity and toughness. Hard-shell capsules also may be formed from starch or other suitable substances. Hard-shell capsules may also contain colorants, such as D&C and FD&C dyes or the various iron oxides, opaquing agents such as titanium dioxide, dispersing agents, hardening agents such as sucrose, and preservatives. They normally contain between 10% and 15% water.

Hard gelatin capsules are made by a process that involves dipping shaped pins into gelatin solutions, after which the gelatin films are dried, trimmed, and removed from the pins, and the body and cap pieces are joined. Starch capsules are made by injection molding a mixture of starch and water, after which the capsules are dried. A separate mold is used for caps and bodies, and the two parts are supplied separately. The empty capsules should be stored in tight containers until they are filled. Since gelatin is of animal origin and starch is of vegetable origin, capsules made with these materials should be protected from potential sources of microbial contamination.

Hard-shell capsules typically are filled with powder, beads, or granules. Inert sugar beads (nonpareils) may be coated with active ingredients and coating compositions that provide extended-release profiles or enteric properties. Alternatively, larger-dose active ingredients themselves may be suitably formed into pellets and then coated. Semisolids or liquids also may be filled into hard-shell capsules; however, when the latter are encapsulated, one of the sealing techniques must be employed to prevent leakage.

In hard gelatin capsule filling operations, the body and cap of the shell are separated prior to dosing. In hard starch shell filling operations, the bodies and caps are supplied separately and are fed into separate hoppers of the filling machine. Machines employing various dosing principles may be employed to fill powders into hard-shell capsules; however, most fully automatic machines form powder plugs by compression and eject them into empty capsule bodies. Accessories to these machines generally are available for the other types of fills. Powder formulations often require adding fillers, lubricants, and glidants to the active ingredients to facilitate encapsulation. The formulation, as well as the method of filling, particularly the degree of compaction, may influence the rate of drug release. The addition of wetting agents to the powder mass is common where the active ingredient is hydrophobic. Disintegrants also may be included in powder formulations to facilitate deaggregation and dispersal of capsule plugs in the gut. Powder formulations often may be produced by dry blending; however, bulky formulations may require densification by roll compaction or other suitable granulation techniques.

Powder mixtures that tend to liquefy may be dispensed in hard-shell capsules if an absorbent such as magnesium carbonate, colloidal silicon dioxide, or other suitable substance is used. Potent drugs are often mixed with an inert diluent before being filled into capsules. Where two mutually incompatible drugs are prescribed together, it is sometimes possible to place one in a small capsule and then enclose it with the second drug in a larger capsule. Incompatible drugs also can be separated by placing coated pellets or tablets, or soft-shell capsules of one drug into the capsule shell before adding the second drug.

Thixotropic semisolids may be formed by gelling liquid drugs or vehicles with colloidal silicas or powdered high molecular weight polyethylene glycols. Various waxy or fatty compounds may be used to prepare semisolid matrices by fusion.

Soft-shell capsules made from gelatin (sometimes called softgels) or other suitable material require large-scale production methods. The soft gelatin shell is somewhat thicker than that of hard-shell capsules and may be plasticized by the addition of a polyol such as sorbitol or glycerin. The ratio of dry plasticizer to dry gelatin determines the "hardness" of the shell and may be varied to accommodate environmental conditions as well as the nature of the contents.

Like hard shells, the shell composition may include approved dyes and pigments, opaquing agents such as titanium dioxide, and preservatives. Flavors may be added and up to 5% sucrose may be included for its sweetness and to produce a chewable shell. Soft gelatin shells normally contain 6% to 13% water. Soft-shell capsules also may be printed with a product code, strength, etc. In most cases, soft-shell capsules are filled with liquid contents. Typically, active ingredients are dissolved or suspended in a liquid vehicle. Classically, an oleaginous vehicle such as a vegetable oil was used; however, nonaqueous, water-miscible liquid vehicles such as the lower-molecular-weight polyethylene glycols are more common today due to fewer bioavailability problems.

Available in a wide variety of sizes and shapes, soft-shell capsules are both formed, filled, and sealed in the same machine; typically, this is a rotary die process, although a plate process or reciprocating die process also may be employed. Soft-shell capsules also may be manufactured in a bubble process that forms seamless spherical capsules. With suitable equipment, powders and other dry solids also may be filled into soft-shell capsules.

Liquid-filled capsules of either type involve similar formulation technology and offer similar advantages and limitations. For instance, both may offer advantages over dry-filled capsules and tablets in content uniformity and drug dissolution. Greater homogeneity is possible in liquid systems, and liquids can be metered more accurately. Drug dissolution may benefit because the drug may already be in solution or at least suspended in a hydrophilic vehicle. However, the contact between the hard or soft shell and its liquid content is more intimate than exists with dry-filled capsules, and this may enhance the chances for undesired interactions. The liquid nature of capsule contents presents different technological problems than dry-filled capsules in regard to disintegration and dissolution testing. From formulation, technological, and biopharmaceutical points of view, liquid-filled capsules of either type have more in common than liquid-filled and dry-filled capsules having the same shell composition. Thus, for compendial purposes, standards and methods should be established based on capsule contents rather than on whether the contents are filled into hard- or soft-shell capsules.

ADSORPTION IN GENERAL PHARMACEUTICAL PROCESSING

Because of the interactions existing between different materials as well as between like materials, the performance of excipients in a formulation could be different from the performance of the excipients themselves.

The most frequently used procedures in pharmaceutical processing for solid dosage formulations are mixing, granulation, and compaction, as well as storage of finished dosage forms. The effects of adsorption on these procedures have been studied, observed, and utilized widely in the pharmaceutical industry.

ADSORPTION IN MIXING

Adsorption of drugs onto the solid surfaces of excipients plays an important role in the preparation of pharmaceutical mixtures and can affect many aspects of the final formulations, especially the content uniformity of low-dose drugs. During mixing, there are many factors that can affect the content uniformity of mixtures, and adsorption generally plays a role when the average particle size is _400 *m*m. In drug manufacturing, it is a challenge to achieve content uniformity when mixing a small amount of one material with a much larger amount of another material. To achieve content uniformity for potent drugs in solid dosage forms, it is beneficial to produce mixes with good homogeneity if drugs can be dispersed very finely. If there are no interactions between mixed particles, ordered mixing and random mixing should achieve the same results.

However, due to interactions between particles, such as aggregation between small particles and adsorption of small particles onto large carrier particles, ordered mixing and random mixing may have significantly different outcomes. Adsorbing drugs onto carrier particles by ordered mixing can be very beneficial in several ways, such as facilitating the preparation of master batches, avoiding segregation in stable systems, promoting excellent homogeneity in mixtures, and improving content uniformity of tablets especially at low dosage.

For excipient particles, larger size and better flowability are beneficial for achieving content uniformity. Those excipients that can adsorb drugs more strongly exhibit a faster disintegration of agglomerates, and achieve higher homogeneity in both random and ordered mixing. Tablets of 120 mg containing 0.1 mg of digitoxin have been manufactured with very good content uniformity by adsorbing small particles onto larger carrier particles in ordered mixing. In the design, digitoxin, dissolved in a mixture of methylene chloride and methanol, is first deposited onto milled critical micellar concentration (CMC).

After mixing, the solvent is removed, and this dry blend is mixed with crystalline lactose. Other excipients are added later without affecting the content uniformity.

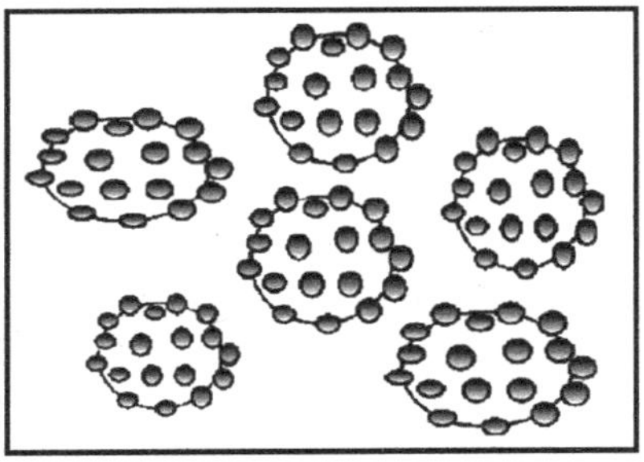

Fig. Adsorption of small Particles onto the Surfaces of Larger Carrier Particles.

The most important factors in the adsorption process are moisture content and particle size distribution. Moisture content and the state of adsorbed water affect the adsorption of fine antibiotic powders onto the surface of sorbitol by plasticizing sorbitol, forming an adherent monoparticle layer that affects the capillary forces and hydrogen bonding interactions between sorbitol and the drugs, as well as masking the interparticle forces. It was observed that high moisture content has caused higher adsorption onto sorbitol for those fine antibiotic powders with low interfacial energy, internal location of adsorbed water, and the ability to form hydrogen bonding with sorbitol.

When stored under high humidity, moisture adsorbed onto the surfaces of drugs and carriers could form a liquid bridge whose capillary interaction may significantly increase adhesion in the mixture. However, high moisture does not always increase the adsorption of drugs onto the solid surface of excipients. For example, as to antibiotics with high interfacial energy and external location of adsorbed water, low moisture content is beneficial for achieving higher adsorption onto sorbitol.

Overall, the mechanism by which moisture content affects the adsorption of drugs onto the solid surfaces of other excipients is very complex and can be affected by many factors. Other factors such as the surface structure of particles, reduction of interparticle distance (e.g., by means of intensive mixing processes), as well as particle size also affect surface adsorption. It was observed that irregular particles of sorbitol with a structured surface could adsorb more vitamins than regular particles.

By adjusting the ratio of three vitamins in the starting mixture, the proportions of vitamins B1, B2, and B6 in the adsorbed state can be changed.

Because of the high fraction of vitamins not adsorbed onto sorbitol surface, the ordered mixing was arranged, at first, by mixing each vitamin with sorbitol separately, then by combining the individual mixtures to make the final mixture.

Milling can affect the adsorption of actives on the carrier surface by altering the surface properties of excipients. It was noticed that milled actives frequently failed blend uniformity criteria, but unmilled active batches consistently met the blend uniformity criteria.

By adding lubricant magnesium stearate, the blending content uniformity of the milled batches can be significantly improved. A small amount of amorphous materials could affect the blending characteristics of a direct compression formulation.

Overall, factors that might affect the adsorption of actives on the carrier surface include surface properties, moisture content, the type and particle size/shape of carriers and actives, as well as the mixing ratio of actives and carriers.

Pharmaceutical processes (i.e., milling and granulation) could affect the adsorption process by altering carrier and active properties (i.e., surface properties, size, and shape), hence the characteristics of blending and the quality of the final dosage form.

Adsorption in Granulation

In solid dosage forms, granulation is frequently used to improve excipient properties such as flowability, compactibility, bulk density, granule strength, dissolution rates, and so on. The granulation process generally includes binder atomization, fluidization, adsorbing and spreading on powder surfaces, particle agglomeration, and so on.

In addition, binder adsorbed onto the particle surface can also provide solid bridges between particles. Of course, due to the complexity of the granulation process, many factors can affect the process and the final properties of the granules. However, the adsorption of the binder solution on solid surfaces, especially at the point of contact between particles or granules, is the key to the granulation process.

The adsorption process can be affected by the surface tension and the viscosity of a binder, and more energy may be consumed during granulation. As shown in Figure, the surface tension and the viscosity of a binder play important roles in granulation because these properties influence the liquid bridges between the particles, as well as the distribution of the binder during the wet massing stage.

For similar polymer binders, even at an equivalent viscosity, different molecular weights can affect binder surface tension and thus influence the granulation process and related granule properties.

It is worthwhile to note that source and batch variation of both drugs and excipients can have profound effects on the final product performance.

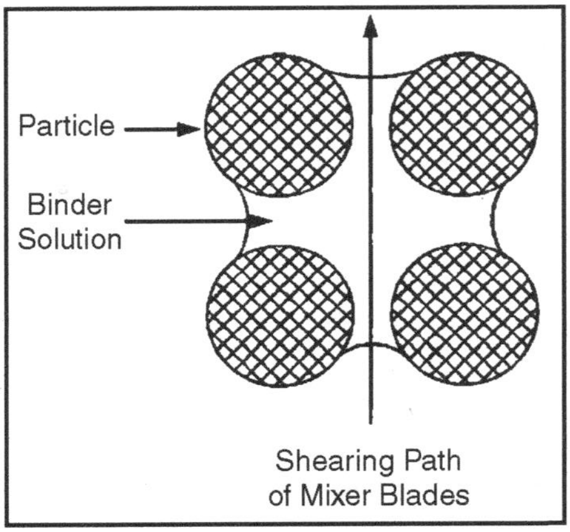

Fig. Shearing Path of Mixer Blades Through Wet Mass at Capillary State.

A suitable amount of binder adsorbed onto the granular surface at the points of contact, as well as their physical properties such as viscosity and flowability, are very important for granule growth. Four mechanisms of granule growth are shown in Figure, and the mechanism that plays the main role is determined by the degree of binder dispersion in the powder.

Many factors, such as binder atomization, addition rate, state of fluidization, and shear forces in the mixer, can affect the degree of binder dispersion.

In the "nucleation" mechanism, particles stick together with the help of the liquid binder adsorbed on the powder surface; in the "coalescence" mechanism, through deformation and bonding, two large agglomerates combine to form one granule with the help of a surface-adsorbed binder; and in the "layering" mechanism, fine particles stick to either a large granule or a binder droplet.

In a mixer, shear forces will cause particles to collide and bond together if some binder is present at the point of contact, as shown in Figure.

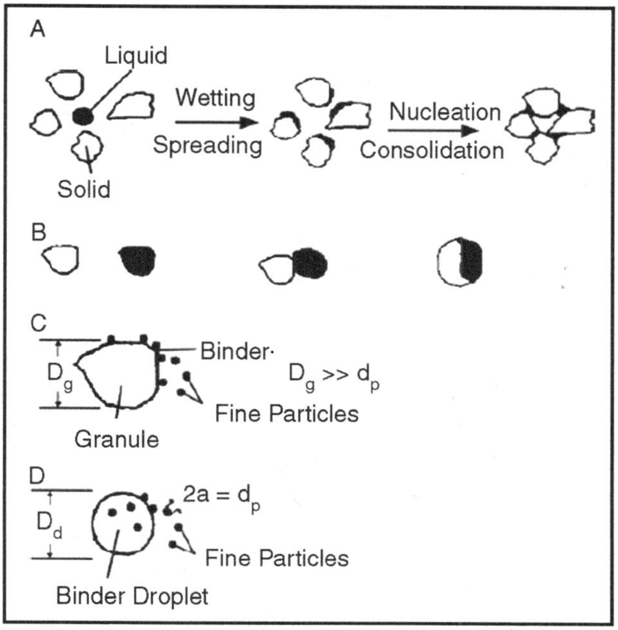

Fig. Granule Growth Mechanisms.

(A) Agglomerate formation by nucleation of particles; (B) Agglomerate growth by coalescence; (C) Layering of a binder-coated granule; and (D) Layering of a partially filled binder droplet.

Of course, two colliding particles may not always bond together, but may rebound and fail to achieve granule growth. Once granules have been formed and reach a certain size, they have to survive the shear forces in the mixer, which are determined by the relationship between shear forces of the outer shearing mass and the inner strength of the granules.

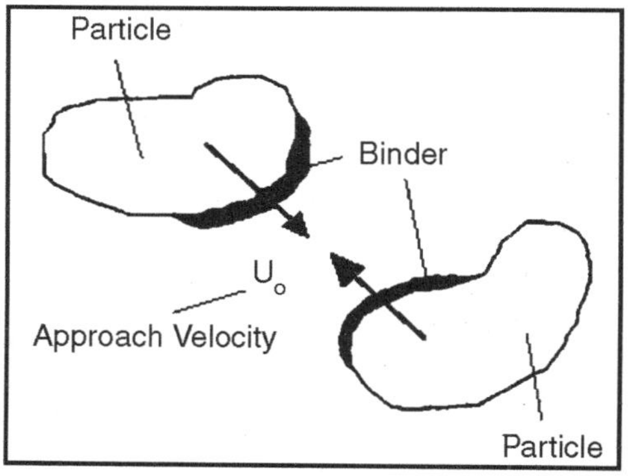

Fig. Coalescence of two Binder-covered Particles.

Besides the traditional wet granulation, surface adsorbents, especially porous materials, are very helpful for maintaining liquid in a granulation. For example, calcium silicate, a fine porous powder, has been successfully used to adsorb an oily medicine, tocopheryl nicotinate.

Due to the high capillarity of the pores inside calcium silicate, the adsorbent has an excellent liquid-holding ability. In the formulation preparation, after drug dissolved in ethanol was adsorbed on calcium silicate, hydroxypropylcellulose (HPC) was added to the mixture and granulated to improve flowability.

Adsorption in Compaction

In the manufacturing of tablets, compaction is a very important step, and many factors are involved in affecting the properties of excipients and final products, such as surface structure and wettability of excipients. For formulations, the larger is the surface area of compressible excipients, the greater is the compressibility.

When compressible and non-compressible excipients are mixed together, the adsorption of the noncompressible excipients onto the surface of the compressible excipients will reduce mixed powder compressibility.

Lerk et al. and teWierik et al. studied a new generation of starch products, which have a high surface area as excipients in pharmaceutical tablets, and confirmed that there is a positive relationship between the specific surface area and the binding capacity (i.e., compactibility).

The surface area that takes part in the interparticle attraction is small, and only a minor fraction of the geometrical surface area may be available for tablet bonding. The addition of some materials, such as binder and lubricant, may decrease particle fragmentation during compression and affect the "true" bonding surface area.

In tablet manufacturing, lubricants such as magnesium stearate act by adsorbing onto the surface of granules and forming a film, thus decreasing the crushing force and ejection force during compression. It was noticed that the distribution of adsorbed magnesium stearate was nonuniform on the surface of excipients, and much more lubricant existed on edges and defective parts of the sodium chloride crystals in their study, thus increasing the effective area for compaction. Besides the surface area of the excipients, moisture adsorbed to the excipients is one important factor in compaction. Nyqvist reported that by affecting the surface structure of excipients, a small amount of water in furosemide could significantly affect the compaction properties of the direct compressible formulation.

When water content was increased from 0.06% to 0.24%, the elastic recovery was increased and network was decreased at compaction. Besides, by adjusting the water content of furosemide, it was possible to avoid tablet capping as well. Because adsorption of water in furosemide did not change the internal structure of the crystals, water may be adsorbed onto the crystal surface with weak forces detectable with thermal analysis. The area of contact formed between particles was increased during compression due to the particle deformation facilitated by the higher moisture content.

Two types of interparticulate bonds, adsorption bonds and diffusional bonds, have been involved in the increased bond strength between particles. It has been observed that increased moisture content in amorphous lactose particles could increase tablet tensile strength and reduce tablet porosity by direct compression. Overall, moisture content can influence bonding forces by affecting both the contact area between particles and the relative fractions of adsorption and diffusion bonds in the tablet.

Adsorption in Storage

The physical and chemical stabilities of finished dosage forms stored at high humidity have always been concerns in the pharmaceutical industry and have been extensively studied. The adsorption of water molecules onto the solid surface can affect tablet strength. For tablets stored under humid conditions, the tensile strength of tablets increases initially when relative humidity (RH) increases; however, as RH further increases, the tensile strength of tablets starts to decrease. During storage, the sublimation of drugs such as bromhexine HCl (BHCl) can be prevented by selecting suitable excipients based on their surface adsorption capacity.

Hygroscopicity of Excipients

Based on hygroscopicity (i.e., the tendency of materials to absorb water under different conditions of humidity), solids have been classified into four categories, as follows:

- *Non-hygroscopic*: No increase in water content at RH below 90%.
- *Slightly hygroscopic*: No moisture increases at RH below 80% with less than 40% moisture increase at RH above 80% after 1 week.
- *Moderately hygroscopic*: Less than 5% water content increase at RH below 60%, with less than 50% water content increase at RH above 80% after 1 week.
- *Very hygroscopic*: Substantial water content increases at RH as low as 40–50%.

Different drugs and excipients have different hygroscopic properties, and the amount of moisture absorbed by drugs and excipients can affect not only their own properties but also the properties of the finished dosage forms, such as flowability, compression properties, and hardness of granules and tablets. The hygroscopicity of different excipients has been studied, and research on moisture adsorption may provide useful information in selecting excipients such as disintegrants, binders, and fillers in formulation design, as well as in choosing suitable manufacturing and storage conditions.

Effects of Moisture on Tablet Strength

Increasing RH during storage will increase tensile strength and decrease the surface area of tablets initially. Water can adsorb onto solid surfaces at three locations: monolayer-adsorbed moisture, normally condensed moisture, and adsorbed moisture, as shown in Figure.

Adsorbed water molecules affect tablet tensile strength by influencing interparticle separation and interparticle forces, especially van der Waals interactions, as well as the ratio of the binding to diffusion forces for water molecules on the solid surface. The initially adsorbed water molecules on the solid surface may form a monomolecular layer, increase the van der Waals interactions between particles, smooth out surface microirregularities, and reduce interparticle separation, thus increasing tensile strength. As more water molecules adsorb onto the particle surface, they are subjected to both surface binding and diffusion forces, and the diffusion forces induce water to penetrate into the particles.

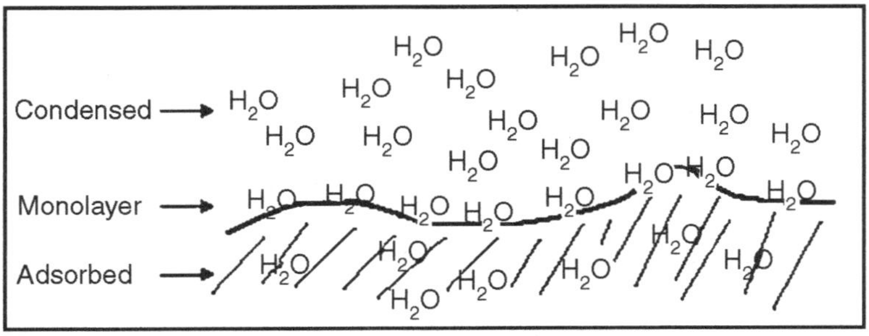

Fig. Three Locations of Adsorbed Water Molecules in Solid Materials: Monolayer Adsorbed Moisture, Condensed Moisture, and Moisture Adsorbed Inside Solid.

The penetrated water may soften the particle surface, increase contact area between particles under high pressure due to plastic deformation, and form more solid bonds. All those effects may explain the observed initial increase in tensile strength when RH increases.

However, as RH further increases, tablet strength decreases for most tested tablets, and it was suggested that condensed water on the solid surface at high RH weakens intermolecular attraction forces between particles in the tablets and the further softened particles and solid bonds cause the tensile strength to decrease. In addition, at high RH, water may form multilayers on the solid surface, which can act as a lubricant and reduce the frictional forces between particles, thus decreasing tensile strength as well. Ultimately, the effects of adsorbed moisture on particle surfaces are very complex, affected by many factors, especially the properties of the tablet excipients.

Preventing Drug Sublimation During Storage

During storage, BHCl can sublime away from solid dosage formulations and reduce the dosage potency. It was observed that BHCl prefers to adsorb to the surface of magnesium aluminum silicate (MAS) rather than to the package material polyethylene film. The adsorption to MAS was accelerated at high temperature and reduced pressure, and the BHCl adsorbed onto the surface of MAS was amorphous rather than crystalline. BHCl could also adsorb onto many other solid excipients, such as kaolin, Avicel PH101, and Avicel PH102, and many factors affect the adsorption. When pH increased, adsorption by microcrystalline cellulose (MCC) increased moderately.

As particle size increased, adsorption decreased. Within the range of 20–46_C, temperature did not affect the adsorption of BHCl on MCC. Thermodynamic analysis showed that both entropy and enthalpy changes favored the adsorption on MCC. Overall, selecting suitable excipients to keep BHCl from sublimating during storage proved a very useful application of adsorption at solid surfaces.

3

Antibiotics

In common usage, an antibiotic is a substance or compound that kills, or inhibits the growth of, bacteria. Antibiotics belong to the broader group of antimicrobial compounds, used to treat infections caused by micro-organisms, including fungi and protozoa. The term "antibiotic" was coined by Selman Waksman in 1942 to describe any substance produced by a micro-organism that is antagonistic to the growth of other micro-organisms in high dilution.

This original definition excluded naturally occurring substances that kill bacteria but are not produced by microorganisms (such as gastric juice and hydrogen peroxide) and also excluded synthetic antibacterial compounds such as the sulfonamides. Many antibiotics are relatively small molecules with a molecular weight less than 2000 Da. With advances in medicinal chemistry, most antibiotics are now semisynthetic—modified chemically from original compounds found in nature, as is the case with beta-lactams (which include the penicillins, produced by fungi in the genus *Penicillium*, the cephalosporins, and the carbapenems).

Some antibiotics are still produced and isolated from living organisms, such as the aminoglycosides, and others have been created through purely synthetic means: the sulfonamides, the quinolones, and the oxazolidinones.

In addition to this origin-based classification into natural, semisynthetic, and synthetic, antibiotics may be divided into two broad groups according to their effect on microorganisms: those that kill bacteria are bactericidal agents, while those that only impair bacterial growth are known as bacteriostatic agents.

HISTORY OF ANTIBIOTICS

Many cures for infectious diseases prior to the beginning of the twentieth century were based on medicinal folklore. Cures for infection in ancient Chinese medicine using plants with antibiotic-like properties began to be described over 2,500 years ago. Many other ancient cultures, including the ancient Egyptians, ancient Greeks and medieval Arabs already used molds and plants to treat infections. Cinchona bark was a widely effective treatment of malaria in the 17th century, the disease caused by protozoan parasites of the genus *Plasmodium*. Scientific endeavours to understand the science behind what caused these diseases, the development of synthetic antibiotic chemotherapy, the isolation of the natural antibiotics marked milestones in antibiotic development

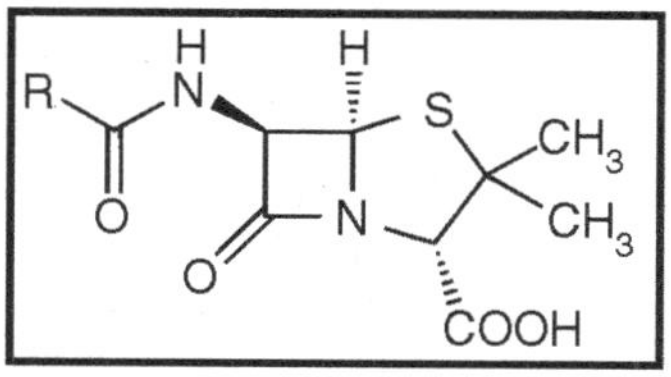

Fig. Penicillin, the First Natural Antibiotic Discovered by Alexander Fleming in 1928.

Originally known as antiobiosis, antibiotics were drugs that had actions against bacteria. The term antibiosis which means 'against life' was introduced by the French bacteriologist Vuillemin as a descriptive name of the phenomenon exhibited by these drugs. Antibiosis was first described in 1877 in bacteria when Louis Pasteur and Robert Koch observed that an airborne bacillus could inhibit the growth of *Bacillus anthracis*. These drugs were later renamed antibiotics by Selman Waksman, an American microbiologist in 1942.

Synthetic antibiotic chemotherapy as a science and the story of antibiotic development began in Germany with Paul Ehrlich, a German medical scientist in the late 1880s. Dr. Ehrlich noted that certain dyes would bind to and colour human, animal or bacterial cells, while others did not. He then extended the idea that it might be possible to make certain dyes or chemicals that would act as a magic bullet or selective drug that would bind to and kill bacteria while not harming the human host.

After much experimentation, screening hundreds of dyes against various organisms, he discovered a medicinally useful drug, the man-made antibiotic, Salvarsan. However, the adverse side-effect profile of salvarsan, coupled with the later discovery of the antibiotic penicillin, superseded its use as an antibiotic. The work of Ehrlich, which marked the birth of the antibiotic revolution, was followed by the discovery of Prontosil by Domagk in 1932.

Prontosil, the first commercially available antibacterial antibiotic was developed by a research team led by Gerhard Domagk (who received the 1939 Nobel Prize for Medicine for his efforts) at the Bayer Laboratories of the IG Farben conglomerate in Germany. Prontosil had a relatively broad effect against Gram-positive cocci but not against enterobacteria. The discovery and development of this first sulfonamide drug opened the era of antibiotics.

The discovery of natural antibiotics produced by microorganisms stemmed from earlier work on the observation of antibiosis between micro-organisms. Pasteur observed that "if we could intervene in the antagonism observed between some bacteria, it would offer 'perhaps the greatest hopes for therapeutics'". Bacterial antagonism of *Penicillium spp.* were first described in England by John Tyndall in 1875. However, his work went by without much notice from the scientific community until Alexander Fleming's discovery of Penicillin in 1928. Even then the therapeutic potential of penicillin was not pursued. More than ten years later, Ernst Chain and Howard Florey became interested in Fleming's work following the earlier discovery of another natural antibiotic like substance named gramicidin from B. brevis. In 1939, Rene Dubos isolated gramicidin, one of the first commercially manufactured antibiotics in use during World War II to prove highly effective in treating wounds and ulcers.Florey and Chain succeeded in purifying penicillin. The purified antibiotic displayed antibacterial activity against a wide range of bacteria. It also had low toxicity and could be taken without causing adverse effects.

Furthermore its activity was not inhibited by biological constituents such as pus, unlike the synthetic antibiotic class available at the time the sulfonamides. The discovery of such a powerful antibiotic was unprecedented. The development of penicillin led to renewed interest in the search for antibiotic compounds with similar capabilities. Because of their discovery of penicillin Ernst Chain, Howard Florey and Alexander Fleming shared the 1945 Nobel Prize in Medicine. Florey credited Dubos with pioneering the approach of deliberately, systematically searching for antibacterial compounds. Such a methodology had led to the discovery of gramicidin, which revived Florey's research in penicillin.

SPECIFIC TYPE OF DRUG RESISTANCE

ANTIBIOTIC RESISTANCE

Antibiotic resistance is a specific type of drug resistance when a microorganism has the ability of withstanding the effects of antibiotics. Antibiotic resistance evolves via natural selection acting upon random mutation, but it can also be engineered by applying an evolutionary stress on a population. Once such a gene is generated, bacteria can then transfer the genetic information in a horizontal fashion (between individuals) by plasmid exchange. If a bacterium carries several resistance genes, it is called multiresistant or, informally, a superbug. The term antimicrobial resistance is sometimes used to explicitly encompass organisms other than bacteria.

Antibiotic resistance can also be introduced artificially into a microorganism through transformation protocols. This can aid in implanting artificial genes into the microorganism. If the resistance gene is linked with the gene to be implanted, the antibiotic can be used to kill off organisms that lack the new gene.

CAUSES

The widespread use of antibiotics both inside and outside of medicine is playing a significant role in the emergence of resistant bacteria. They are often used in animals but also in other industries which at least in the case of agricultural use lead to the spread of resistant strains to human populations. In some countries antibiotics are sold over the counter without a prescription which compounds the problem. In human medicine the major problem of the emergence of resistant bacteria is due to misuse and overuse of antibiotics by doctors as well as patients. Other practices contributing towards resistance include the addition of antibiotics to the feed of livestock.

Household use of antibacterials in soaps and other products, although not clearly contributing to resistance, is also discouraged (as not being effective at infection control). Also unsound practices in the pharmaceutical manufacturing industry can contribute towards the likelihood of creating antibiotic resistant strains. Certain antibiotic classes are highly associated with colonisation with superbugs compared to other antibiotic classes.

The risk for colonisation increases if there is a lack of sensitivity (resistance) of the superbugs to the antibiotic used and high tissue penetration as well as broad spectrum activity against "good bacteria". In the case of MRSA, increased rates of MRSA infections are seen with glycopeptides, cephalosporins and especially quinolones. In the case of colonisation with C difficile the high risk antibiotics include cephalosporins and in particular quinolones and clindamycin.

In Medicine

The volume of antibiotic prescribed is the major factor in increasing rates or bacterial resistance rather than compliance with antibiotics. Inappropriate prescribing of antibiotics has been attributed to a number of causes including: people who insist on antibiotics, physicians simply prescribe them as they feel they do not have time to explain why they are not necessary, physicians who do not know when to prescribe antibiotics or else are overly cautious for medical legal reasons. A third of people for example believe that antibiotics are effective for the common cold and 22% of people do not finish a course of antibiotics primarily due to that fact that they feel better (varying from 10% to 44% depending on the country). Compliance with once daily antibiotics is better than with twice daily antibiotics. Sub optimum antibiotic concentrations in critically ill people increase the frequency of antibiotic resistance organisms. While taking antibiotics doses less than those recommended may increase rates of resistance, shortening the course of antibiotics may actually decrease rates of resistance. Poor hand hygiene by hospital staff has been associated with the spread of resistant organisms and an increase in hand washing compliance results in decreased rates of these organisms.

Role of Other Animals

Drugs are used in animals that are used as human food, such as cows, pigs, chickens, fish, etc, and these drugs can affect the safety of the meat, milk, and eggs produced from those animals and can be the source of superbugs. For example, farm animals, particularly pigs, are believed to be able to infect people with MRSA. The resistant bacteria in animals due to antibiotic exposure can be transmitted to humans via three pathways, those being through the consumption of meat, from close or direct contact with animals, or through the environment.

The World Health Organization concluded that antibiotics as growth promoters in animal feeds should be prohibited (in the absence of risk assessments). In 1998, European Union health ministers voted to ban four antibiotics widely used to promote animal growth (despite their scientific panel's recommendations). Regulation banning the use of antibiotics in European feed, with the exception of two antibiotics in poultry feeds, became effective in 2006. In Scandinavia, there's evidence that the ban has led to a lower prevalence of antimicrobial resistance in (non-hazardous) animal bacterial populations.

In the USA federal agencies do not collect data on antibiotic use in animals but animal to human spread of drug resistant organisms has been demonstrated in research studies. Antibiotics are still used in U.S. animal feed-along with other ingredients which have safety concerns. Growing U.S. consumer concern about using antibiotics in animal feed has led to a niche market of "antibiotic-free" animal products, but this small market is unlikely to change entrenched industry-wide practices.

In 2001, the Union of Concerned Scientists estimated that greater than 70% of the antibiotics used in the US are given to food animals (e.g. chickens, pigs and cattle) in the absence of disease. In 2000 the US Food and Drug Administration (FDA) announced their intention to revoke approval of fluoroquinolone use in poultry production because of substantial evidence linking it to the emergence of fluoroquinolone resistant campylobacter infections in humans. The final decision to ban fluoroquinolones from use in poultry production was not made until five years later because of challenges from the food animal and pharmaceutical industries. Today, there are two federal bills (S. 549 and H.R. 962) aimed at phasing out "non-therapeutic" antibiotics in US food animal production.

Mechanisms

The top section represents a population of bacteria before exposure to an antibiotic. The middle section shows the population directly after exposure, the phase in which selection took place. The last section shows the distribution of resistance in a new generation of bacteria. The legend indicates the resistance levels of individuals. Antibiotic resistance can be a result of horizontal gene transfer, and also of unlinked point mutations in the pathogen genome and a rate of about 1 in 10^8 per chromosomal replication.

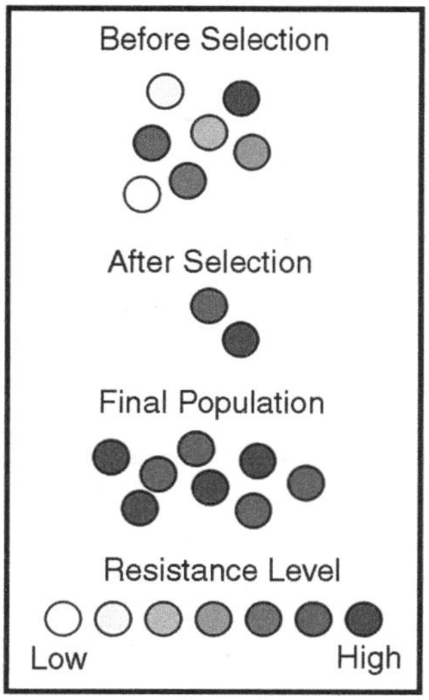

Fig. Schematic Representation of How Antibiotic Resistance Evolves Via Natural Selection.

The antibiotic action against the pathogen can be seen as an environmental pressure; those bacteria which have a mutation allowing them to survive will live on to reproduce. They will then pass this trait to their offspring, which will result in a fully resistant colony.

The four main mechanisms by which microorganisms exhibit resistance to antimicrobials are:

- *Drug inactivation or modification*: e.g. enzymatic deactivation of *Penicillin* G in some penicillin-resistant bacteria through the production of β-lactamases.
- *Alteration of target site*: e.g. alteration of PBP—the binding target site of penicillins—in MRSA and other penicillin-resistant bacteria.
- *Alteration of metabolic pathway*: e.g. some sulfonamide-resistant bacteria do not require para-aminobenzoic acid (PABA), an important precursor for the synthesis of folic acid and nucleic acids in bacteria inhibited by sulfonamides. Instead, like mammalian cells, they turn to utilizing preformed folic acid.
- *Reduced drug accumulation*: By decreasing drug permeability and/or increasing active efflux (pumping out) of the drugs across the cell surface.

There are three known mechanisms of fluoroquinolone resistance. Some types of efflux pumps can act to decrease intracellular quinolone concentration. In gram-negative bacteria, plasmid-mediated resistance genes produce proteins that can bind to DNA gyrase, protecting it from the action of quinolones. Finally, mutations at key sites in DNA gyrase or Topoisomerase IV can decrease their binding affinity to quinolones, decreasing the drug's effectiveness. Research has shown that the bacterial protein LexA may play a key role in the acquisition of bacterial mutations giving resistance to quinolones and rifampicin.

RESISTANT PATHOGENS

Staphylococcus Aureus

Staphylococcus aureus (colloquially known as "Staph aureus" or a *Staph infection*) is one of the major resistant pathogens. Found on the mucous membranes and the human skin of around a third of the population, it is extremely adaptable to antibiotic pressure. It was the first bacterium in which penicillin resistance was found—in 1947, just four years after the drug started being mass-produced. Methicillin was then the antibiotic of choice, but has since been replaced by oxacillin due to significant kidney toxicity. MRSA (methicillin-resistant *Staphylococcus aureus*) was first detected in Britain in 1961 and is now "quite common" in hospitals. MRSA was responsible for 37% of fatal cases of blood poisoning in the UK in 1999, up from 4% in 1991. Half of all *S. aureus* infections in the US are resistant to penicillin, methicillin, tetracycline and erythromycin.

This left vancomycin as the only effective agent available at the time. However, strains with intermediate (4-8 ug/ml) levels of resistance, termed GISA (glycopeptide intermediate *Staphylococcus aureus*) or VISA (vancomycin intermediate *Staphylococcus aureus*), began appearing in the late 1990s. The first identified case was in Japan in 1996, and strains have since been found in hospitals in England, France and the US. The first documented strain with complete (>16 ug/ml) resistance to vancomycin, termed VRSA (Vancomycin-resistant *Staphylococcus aureus*) appeared in the United States in 2002. A *new* class of antibiotics, oxazolidinones, became available in the 1990s, and the first commercially available oxazolidinone, linezolid, is comparable to vancomycin in effectiveness against MRSA. Linezolid-resistance in *Staphylococcus aureus* was reported in 2003.

CA-MRSA (Community-acquired MRSA) has now emerged as an epidemic that is responsible for rapidly progressive, fatal diseases including necrotizing pneumonia, severe sepsis and necrotizing fasciitis. Methicillin-resistant *Staphylococcus aureus* (MRSA) is the most frequently identified antimicrobial drug-resistant pathogen in US hospitals. The epidemiology of infections caused by MRSA is rapidly changing. In the past 10 years, infections caused by this organism have emerged in the community. The 2 MRSA clones in the United States most closely associated with community outbreaks, USA400 (MW2 strain, ST1 lineage) and USA300, often contain Panton-Valentine leukocidin (PVL) genes and, more frequently, have been associated with skin and soft tissue infections. Outbreaks of community-associated (CA)-MRSA infections have been reported in correctional facilities, among athletic teams, among military recruits, in newborn nurseries, and among men who engage in frequent homosexual activities. CA-MRSA infections now appear to be endemic in many urban regions and cause most CA-S. aureus infections.

Streptococcus *and* Enterococcus

Streptococcus pyogenes (Group A Streptococcus: GAS) infections can usually be treated with many different antibiotics. Early treatment may reduce the risk of death from invasive group A streptococcal disease. However, even the best medical care does not prevent death in every case. For those with very severe illness, supportive care in an intensive care unit may be needed. For persons with necrotizing fasciitis, surgery often is needed to remove damaged tissue. Strains of *S. pyogenes* resistant to macrolide antibiotics have emerged, however all strains remain uniformly sensitive to penicillin.

Resistance of *Streptococcus pneumoniae* to penicillin and other beta-lactams is increasing worldwide. The major mechanism of resistance involves the introduction of mutations in genes encoding penicillin-binding proteins. Selective pressure is thought to play an important role, and use of beta-lactam antibiotics has been implicated as a risk factor for infection and colonization. Streptococcus pneumoniae is responsible for pneumonia, bacteremia, otitis media, meningitis, sinusitis, peritonitis and arthritis.

Penicillin-resistant pneumonia caused by *Streptococcus pneumoniae* (commonly known as *pneumococcus*), was first detected in 1967, as was penicillin-resistant gonorrhea. Resistance to penicillin substitutes is also known as beyond *S. aureus*. By 1993 *Escherichia coli* was resistant to five fluoroquinolone variants.

Mycobacterium tuberculosis is commonly resistant to isoniazid and rifampin and sometimes universally resistant to the common treatments. Other pathogens showing some resistance include *Salmonella, Campylobacter,* and *Streptococci.*

Enterococcus faecium is another superbug found in hospitals. Penicillin-Resistant Enterococcus was seen in 1983, vancomycin-resistant enterococcus (VRE) in 1987, and Linezolid-Resistant Enterococcus (LRE) in the late 1990s.

Pseudomonas Aeruginosa

Pseudomonas aeruginosa is a highly prevalent opportunistic pathogen. One of the most worrisome characteristics of *P. aeruginosa* consists in its low antibiotic susceptibility. This low susceptibility is attributable to a concerted action of multidrug efflux pumps with chromosomally-encoded antibiotic resistance genes (e.g. *mexAB-oprM, mexXY* etc) and the low permeability of the bacterial cellular envelopes. Besides intrinsic resistance, *P. aeruginosa* easily develop acquired resistance either by mutation in chromosomally-encoded genes, or by the horizontal gene transfer of antibiotic resistance determinants. Development of multidrug resistance by *P. aeruginosa* isolates requires several different genetic events that include acquisition of different mutations and/or horizontal transfer of antibiotic resistance genes.

Hypermutation favours the selection of mutation-driven antibiotic resistance in *P. aeruginosa* strains producing chronic infections, whereas the clustering of several different antibiotic resistance genes in integrons favours the concerted acquisition of antibiotic resistance determinants. Some recent studies have shown that phenotypic resistance associated to biofilm formation or to the emergence of small-colony-variants may be important in the response of *P. aeruginosa* populations to antibiotics treatment.

Clostridium Difficile

Clostridium difficile is a nosocomial pathogen that causes diarrheal disease in hospitals world wide. Clindamycin-resistant *C. difficile* was reported as the causative agent of large outbreaks of diarrheal disease in hospitals in New York, Arizona, Florida and Massachusetts between 1989 and 1992. Geographically dispersed outbreaks of *C. difficile* strains resistant to fluoroquinolone antibiotics, such as Cipro (ciprofloxacin) and Levaquin (levofloxacin), were also reported in North America in 2005.

Salmonella *and* E. coli

E. coli and Salmonella come directly from contaminated food. Of the meat that is contaminated with E. coli, eighty per cent of the bacteria are resistant to one or more drugs made; it causes bladder infections that are resistant to antibiotics ("HSUS Fact Sheet"). Salmonella was first found in humans in the 1970s and in some cases is resistant to as many as nine different antibiotics ("HSUS Fact Sheet"). When both bacterium are spread, serious health conditions arise. Many people are hospitalized each year after becoming infected, and some die as a result.

Acinetobacter Baumannii

On November 5, 2004, the Centers for Disease Control and Prevention (CDC) reported an increasing number of *Acinetobacter baumannii* bloodstream infections in patients at military medical facilities in which service members injured in the Iraq/Kuwait region during Operation Iraqi Freedom and in Afghanistan during Operation Enduring Freedom were treated. Most of these showed multidrug resistance (MRAB), with a few isolates resistant to all drugs tested.

Research

New Medications

Until recently, research and development (R&D) efforts have provided new drugs in time to treat bacteria that became resistant to older antibiotics.

That is no longer the case. The potential crisis at hand is the result of a marked decrease in industry R&D, and the increasing prevalence of resistant bacteria. Infectious disease physicians are alarmed by the prospect that effective antibiotics may not be available to treat seriously ill patients in the near future. The pipeline of new antibiotics is drying up. Major pharmaceutical companies are losing interest in the antibiotics market because these drugs may not be as profitable as drugs that treat chronic (long-term) conditions and lifestyle issues.

The resistance problem demands that a renewed effort be made to seek antibacterial agents effective against pathogenic bacteria resistant to current antibiotics. One of the possible strategies towards this objective is the rational localization of bioactive phytochemicals. Plants have an almost limitless ability to synthesize aromatic substances, most of which are phenols or their oxygen-substituted derivatives such as tannins. Most are secondary metabolites, of which at least 12,000 have been isolated, a number estimated to be less than 10% of the total. In many cases, these substances serve as plant defence mechanisms against predation by microorganisms, insects, and herbivores.

Many of the herbs and spices used by humans to season food yield useful medicinal compounds including those having antibacterial activity. Traditional healers have long used plants to prevent or cure infectious conditions. Many of these plants have been investigated scientifically for antimicrobial activity and a large number of plant products have been shown to inhibit growth of pathogenic bacteria.

A number of these agents appear to have structures and modes of action that are distinct from those of the antibiotics in current use, suggesting that cross-resistance with agents already in use may be minimal. For example the combination of 5'-methoxyhydnocarpine and berberine in herbs like Hydrastis canadensis and Berberis vulgaris can block the MDR-pumps that cause multidrug resistance. This has been shown for Staphylococcus aureus. Archaeocins is the name given to a new class of potentially useful antibiotics that are derived from the Archaea group of organisms. Eight archaeocins have been partially or fully characterized, but hundreds of archaeocins are believed to exist, especially within the haloarchaea. The prevalence of archaeocins is unknown simply because no one has looked for them.

The discovery of new archaeocins hinges on recovery and cultivation of archaeal organisms from the environment. For example, samples from a novel hypersaline field site, Wilson Hot Springs, recovered 350 halophilic organisms; preliminary analysis of 75 isolates showed that 48 were archaeal and 27 were bacterial. In research published on October 17, 2008 in *Cell*, a team of scientists pinpointed the place on bacteria where the antibiotic myxopyronin launches its attack, and why that attack is successful. The myxopyronin binds to and inhibits the crucial bacterial enzyme, RNA polymerase. The myxopyronin changes the structure of the switch-2 segment of the enzyme, inhibiting its function of reading and transmitting DNA code. This prevents RNA polymerase from delivering genetic information to the ribosomes, causing the bacteria to die. One of the major causes of antibiotic resistance is the decrease of effective drug concentrations at their target place, due to the increased action of ABC transporters. Since ABC transporter blockers can be used in combination with current drugs to increase their effective intracellular concentration, the possible impact of ABC transporter inhibitors is of great clinical interest. ABC transporter blockers that may be useful to increase the efficacy of current drugs have entered clinical trials and are available to be used in therapeutic regimes.

Applications

Antibiotic resistance is an important tool for genetic engineering. By constructing a plasmid which contains an antibiotic resistance gene as well as the gene being engineered or expressed, a researcher can ensure that when bacteria replicate, only the copies which carry along the plasmid survive. This ensures that the gene being manipulated passes along when the bacteria replicates.

The most commonly used antibiotics in genetic engineering are generally "older" antibiotics which have largely fallen out of use in clinical practice.

These include:

- Ampicillin
- Kanamycin
- Tetracycline
- Chloramphenicol

Industrially the use of antibiotic resistance is disfavored since maintaining bacterial cultures would require feeding them large quantities of antibiotics. Instead, the use of auxotrophic bacterial strains (and function-replacement plasmids) is preferred.

Alternatives

Prevention

Rational use of antibiotics may reduce the chances of development of opportunistic infection by antibiotic-resistant bacteria due to dysbacteriosis. In one study the use of fluoroquinolones are clearly associated with *Clostridium difficile* infection, which is a leading cause of nosocomial diarrhea in the United States, and a major cause of death, worldwide. There is clinical evidence that topical dermatological preparations containing tea tree oil and thyme oil may be effective in preventing transmittal of CA-MRSA. Vaccines do not suffer the problem of resistance because a vaccine enhances the body's natural defenses, while an antibiotic operates separately from the body's normal defenses. Nevertheless, new strains may evolve that escape immunity induced by vaccines. While theoretically promising, anti-staphylococcal vaccines have shown limited efficacy, because of immunological variation between *Staphylococcus* species, and the limited duration of effectiveness of the antibodies produced. Development and testing of more effective vaccines is under way. The Australian Commonwealth Scientific and Industrial Research Organization (CSIRO), realizing the need for the reduction of antibiotic use, has been working on two alternatives. One alternative is to prevent diseases by adding cytokines instead of antibiotics to animal feed.

These proteins are made in the animal body "naturally" after a disease and are not antibiotics so they do not contribute to the antibiotic resistance problem. Furthermore, studies on using cytokines have shown that they also enhance the growth of animals like the antibiotics now used, but without the drawbacks of non-therapeutic antibiotic use. Cytokines have the potential to achieve the animal growth rates traditionally sought by the use of antibiotics without the contribution of antibiotic resistance associated with the widespread non-therapeutic uses of antibiotics currently utilized in the food animal production industries. Additionally, CSIRO is working on vaccines for diseases.

Phage Therapy

Phage therapy, an approach that has been extensively researched and utilized as a therapeutic agent for over 60 years, especially in the Soviet Union, is an alternative that might help with the problem of resistance. Phage therapy was widely used in the United States until the discovery of antibiotics, in the early 1940s. Bacteriophages or "phages" are viruses that invade bacterial cells and, in the case of lytic phages, disrupt bacterial metabolism and cause the bacterium to lyse. Phage therapy is the therapeutic use of lytic bacteriophages to treat pathogenic bacterial infections.

Bacteriophage therapy is an important alternative to antibiotics in the current era of multidrug resistant pathogens. A review of studies that dealt with the therapeutic use of phages from 1966–1996 and few latest ongoing phage therapy projects via internet showed: phages were used topically, orally or systemically in Polish and Soviet studies. The success rate found in these studies was 80–95% with few gastrointestinal or allergic side effects. British studies also demonstrated significant efficacy of phages against *Escherichia coli, Acinetobacter* spp., *Pseudomonas* spp and *Staphylococcus aureus*. US studies dealt with improving the bioavailability of phage. Phage therapy may prove as an important alternative to antibiotics for treating multidrug resistant pathogens.

CHEMICAL SUBSTANCES OF ANTIBIOTIC

Antibiotics are chemical substances that can inhibit the growth of, and even destroy, harmful microorganisms. They are derived from special microorganisms or other living systems, and are produced on an industrial scale using a fermentation process. Although the principles of antibiotic action were not discovered until the twentieth century, the first known use of antibiotics was by the Chinese over 2,500 years ago. Today, over 10,000 antibiotic substances have been reported. Currently, antibiotics represent a multibillion dollar industry that continues to grow each year.

Antibiotics are used in many forms-each of which imposes somewhat different manufacturing requirements. For bacterial infections on the skin surface, eye, or ear, an antibiotic may be applied as an ointment or cream. If the infection is internal, the antibiotic can be swallowed or injected directly into the body. In these cases, the antibiotic is delivered throughout the body by absorption into the bloodstream. Antibiotics differ chemically so it is understandable that they also differ in the types of infections they cure and the ways in which they cure them. Certain antibiotics destroy bacteria by affecting the structure of their cells. This can occur in one of two ways.

First, the antibiotic can weaken the cell walls of the infectious bacteria, which causes them to burst. Second, antibiotics can cause the contents of the bacterial cells to leak out by damaging the cell membranes. Another way in which antibiotics function is by interfering with the bacteria's metabolism. Some antibiotics such as tetracycline and erythromycin interfere with protein synthesis. Antibiotics like rifampin inhibit nucleic acid biosynthesis. Still other antibiotics, such as sulfonamide or trimethoprim have a general blocking effect on cell metabolism.

The commercial development of an antibiotic is a long and costly proposal. It begins with basic research designed to identify organisms, which produce antibiotic compounds. During this phase, thousands of species are screened for any sign of antibacterial action. When one is found, the species is tested against a variety of known infectious bacteria. If the results are promising, the organism is grown on a large scale so the compound responsible for the antibiotic effect can be isolated.

This is a complex procedure because thousands of antibiotic materials have already been discovered. Often, scientists find that their new antibiotics are not unique. If the material passes this phase, further testing can be done. This typically involves clinical testing to prove that the antibiotic works in animals and humans and is not harmful. If these tests are passed, the Food and Drug Administration (FDA) must then approve the antibiotic as a new drug. This whole process can take many years.

The large-scale production of an antibiotic depends on a fermentation process. During fermentation, large amounts of the antibiotic-producing organism are grown. During fermentation, the organisms produce the antibiotic material, which can then be isolated for use as a drug. For a new antibiotic to be economically feasible, manufacturers must be able to get a high yield of drug from the fermentation process, and be able to easily isolate it. Extensive research is usually required before a new antibiotic can be commercially scaled up.

History

While our scientific knowledge of antibiotics has only recently been developed, the practical application of antibiotics has existed for centuries. The first known use was by the Chinese about 2,500 years ago. During this time, they discovered that applying the moldy curd of soybeans to infections had certain therapeutic benefits. It was so effective that it became a standard treatment. Evidence suggests that other cultures used antibiotic-type substances as therapeutic agents. The Sudanese-Nubian civilization used a type of tetracycline antibiotic as early as 350 A.D.

In Europe during the Middle Ages, crude plant extracts and cheese curds were also used to fight infection. Although these cultures used antibiotics, the general principles of antibiotic action were not understood until the twentieth century. The development of modern antibiotics depended on a few key individuals who demonstrated to the world that materials derived from microorganisms could be used to cure infectious diseases. One of the first pioneers in this field was Louis Pasteur. In 1877, he and an associate discovered that the growth of disease-causing anthrax bacteria could be inhibited by a saprophytic bacteria.

They showed that large amounts of anthrax bacilli could be given to animals with no adverse affects as long as the saprophytic bacilli were also given. Over the next few years, other observations supported the fact that some bacterially derived materials could prevent the growth of disease-causing bacteria. In 1928, Alexander Fleming made one of the most important contributions to the field of antibiotics. In an experiment, he found that a strain of green *Penicillium* mold inhibited the growth of bacteria on an agar plate. This led to the development of the first modern era antibiotic, penicillin. A few years later in 1932, a paper was published which suggested a method for treating infected wounds using a penicillin preparation. Although these early samples of penicillin were functional, they were not reliable and further refinements were needed. These improvements came in the early 1940s when Howard Florey and associates discovered a new strain of *Penicillium,* which produced high yields of penicillin. This allowed large-scale production of penicillin, which helped launch the modern antibiotics industry.

After the discovery of penicillin, other antibiotics were sought. In 1939, work began on the isolation of potential antibiotic products from the soil bacteria streptomyces. It was around this time that the term antibiotic was introduced. Selman Waxman and associates discovered streptomycin in 1944. Subsequent studies resulted in the discovery of a host of new, different antibiotics including actinomycin, streptothricin, and neomycin all produced by *Streptomyces.* Other antibiotics that have been discovered since include bacitracin, polymyxin, viomycin, chloramphenicol and tetracyclines. Since the

1970s, most new antibiotics have been synthetic modifications of naturally occurring antibiotics.

Raw Materials

The compounds that make the fermentation broth are the primary raw materials required for antibiotic production. This broth is an aqueous solution made up of all of the ingredients necessary for the proliferation of the microorganisms. Typically, it contains a carbon source like molasses, or soy meal, both of which are made up of lactose and glucose sugars. These materials are needed as a food source for the organisms. Nitrogen is another necessary compound in the metabolic cycles of the organisms. For this reason, an ammonia salt is typically used. Additionally, trace elements needed for the proper growth of the antibiotic-producing organisms are included.

These are components such as phosphorus, sulfur, magnesium, zinc, iron, and copper introduced through water soluble salts. To prevent foaming during fermentation, anti-foaming agents such as lard oil, octadecanol, and silicones are used.

THE MANUFACTURING PROCESS

Although most antibiotics occur in nature, they are not normally available in the quantities necessary for large-scale production. For this reason, a fermentation process was developed.

It involves isolating a desired microorganism, fueling growth of the culture and refining and isolating the final antibiotic product. It is important that sterile conditions be maintained throughout the manufacturing process, because contamination by foreign microbes will ruin the fermentation.

Fermentation

- The fermentation tank is essentially a larger version of the steel, seed tank, which is able to hold about 30,000 gallons. It is filled with the same growth media found in the seed tank and also provides an environment inducive to growth. Here the microorganisms are allowed to grow and multiply. During this process, they excrete large quantities of the desired antibiotic. The tanks are cooled to keep the temperature between 73-81° F (23-27.2 ° C). It is constantly agitated, and a continuous stream of sterilized air is pumped into it. For this reason, anti-foaming agents are periodically added. Since pH control is vital for optimal growth, acids or bases are added to the tank as necessary.

Isolation and purification

- After three to five days, the maximum amount of antibiotic will have been produced and the isolation process can begin. Depending on the specific antibiotic produced, the fermentation broth is processed by various purification methods. For example, for antibiotic compounds that are water soluble, an ion-exchange method may be used for purification. In this method, the compound is first separated from the waste organic materials in the broth and then sent through equipment, which separates the other water-soluble compounds from the desired one. To isolate an oil-soluble antibiotic such as penicillin, a solvent extraction method is used. In this method, the broth is treated with organic solvents such as butyl acetate or methyl isobutyl ketone, which can specifically dissolve the antibiotic. The dissolved antibiotic is then recovered using various organic chemical means. At the end of this step, the manufacturer is typically left with a purified powdered form of the antibiotic, which can be further refined into different product types.

Starting the Culture

- Before fermentation can begin, the desired antibiotic-producing organism must be isolated and its numbers must be increased by many times. To do this, a starter culture from a sample of previously isolated, cold-stored organisms is created in the lab. In order to grow the initial culture, a sample of the organism is transferred to an agar-containing plate. The initial culture is then put into shake flasks along with food and other nutrients necessary for growth. This creates a suspension, which can be transferred to seed tanks for further growth.
- The seed tanks are steel tanks designed to provide an ideal environment for growing microorganisms. They are filled with the all the things the specific microorganism would need to survive and thrive, including warm water and carbohydrate foods like lactose or glucose sugars. Additionally, they contain other necessary carbon sources, such as acetic acid, alcohols, or hydrocarbons, and nitrogen sources like ammonia salts. Growth factors like vitamins, amino acids, and minor nutrients round out the composition of the seed tank contents. The seed tanks are equipped with mixers, which keep the growth medium moving, and a pump to deliver sterilized, filtered air. After about 24-28 hours, the material in the seed tanks is transferred to the primary fermentation tanks.

Refining

- Antibiotic products can take on many different forms. They can be sold in solutions for intravenous bags or syringes, in pill or gel capsule form, or they may be sold as powders, which are incorporated into topical ointments. Depending on the final form of the antibiotic, various refining steps may be taken after the initial isolation. For intravenous bags, the crystalline antibiotic can be dissolved in a solution, put in the bag, which is then hermetically sealed. For gel capsules, the powdered antibiotic is physically filled into the bottom half of a capsule then the top half is mechanically put in place. When used in topical ointments, the antibiotic is mixed into the ointment.

- From this point, the antibiotic product is transported to the final packaging stations. Here, the products are stacked and put in boxes. They are loaded up on trucks and transported to various distributors, hospitals, and pharmacies. The entire process of fermentation, recovery, and processing can take anywhere from five to eight days.

QUALITY CONTROL

Quality control is of utmost importance in the production of antibiotics. Since it involves a fermentation process, steps must be taken to ensure that absolutely no contamination is introduced at any point during production. To this end, the medium and all of the processing equipment are thoroughly steam sterilized. During manufacturing, the quality of all the compounds is checked on a regular basis. Of particular importance are frequent checks of the condition of the microorganism culture during fermentation.

These are accomplished using various chromatography techniques. Also, various physical and chemical properties of the finished product are checked such as pH, melting point, and moisture content. In the United States, antibiotic production is highly regulated by the Food and Drug Administration (FDA). Depending on the application and type of antibiotic, more or less testing must be completed. For example, the FDA requires that for certain antibiotics each batch must be checked by them for effectiveness and purity. Only after they have certified the batch can it be sold for general consumption.

The Future

Since the development of a new drug is a costly proposition, pharmaceutical companies have done very little research in the last decade. However, an alarming development has spurred a revived interest in the development of new antibiotics. It turns out that some of the disease-causing bacteria have mutated and developed a resistance to many of the standard antibiotics. This could have grave consequences on the world's public health unless new antibiotics are discovered or improvements are made on the ones that are available. This challenging problem will be the focus of research for many years to come.

Antibiotic Resistance

The emergence of antibiotic resistance is an evolutionary process that is based on selection for organisms that have enhanced ability to survive doses of antibiotics that would have previously been lethal.

Antibiotics like Penicillin and Erythromycin which used to be one-time miracle cures are now less effective because bacteria have become more resistant. Antibiotics themselves act as a selective pressure which allows the growth of resistant bacteria within a population and inhibits susceptible bacteria. Antibiotic selection of pre-existing antibiotic resistant mutants within bacterial populations was demonstrated in 1943 by the Luria-Delbrück experiment. Survival of bacteria often results from an inheritable resistance.

Any antibiotic resistance may impose a biological cost and the spread of antibiotic resistant bacteria may be hampered by the reduced fitness associated with the resistance which proves disadvantageous for survival of the bacteria when antibiotic is not present. Additional mutations, however, may compensate for this fitness cost and aids the survival of these bacteria.

The underlying molecular mechanisms leading to antibiotic resistance can vary. Intrinsic resistance may naturally occur as a result of the bacteria's genetic makeup. The bacterial chromosome may fail to encode a protein which the antibiotic targets.

Acquired resistance results from a mutation in the bacterial chromosome or the acquisition of extra-chromosomal DNA. Antibiotic-producing bacteria have evolved resistance mechanisms which have been shown to be similar to and may have been transferred to antibiotic resistant strains.

The spread of antibiotic resistance mechanisms occurs through vertical transmission of inherited mutations from previous generations and genetic recombination of DNA by horizontal genetic exchange. Antibiotic resistance is exchanged between different bacteria by plasmids that carry genes which encode antibiotic resistance which may result in co-resistance to multiple antibiotics. These plasmids can carry different genes with diverse resistance mechanisms to unrelated antibiotics but because they are located on the same plasmid multiple antibiotic resistance to more than one antibiotic is transferred. Alternatively, cross-resistance to other antibiotics within the bacteria results when the same resistance mechanism is responsible for resistance to more than one antibiotic is selected for.

Antibiotic Misuse

Inappropriate antibiotic treatment and overuse of antibiotics have been a contributing factor to the emergence of resistant bacteria. The problem is further exacerbated by self-prescribing of antibiotics by individuals without the guidelines of a qualified clinician and the non-therapeutic use of antibiotics as growth promoters in agriculture. Antibiotics are frequently prescribed for indications in which their use is not warranted, an incorrect or sub-optimal antibiotic is prescribed or in some cases for infections likely to resolve without treatment.The overuse of antibiotics like penicillin and erythromycin which used to be one-time miracle cures were associated with emerging resistance since the 1950s.

Therapeutic usage of antibiotics in hospitals has been seen to be associated with increases in multi-antibiotic resistant bacteria. Common forms of antibiotic misuse include excessive use of prophylactic antibiotics in travelers, failure to take into account the patient's weight and history of prior antibiotic use when prescribing, since both can strongly affect the efficacy of an antibiotic prescription, failure to take the entire prescribed course of the antibiotic, failure to prescribe or take the course of treatment at fairly precise correct daily intervals (e.g. "every 8 hours" rather than merely "3x per day"), or failure to rest for sufficient recovery to allow clearance of the infecting organism. These practices may facilitate the development of bacterial populations with antibiotic resistance. Inappropriate antibiotic treatment is another common form of antibiotic misuse. A common example is the prescription and use of antibiotics to treat viral infections such as the common cold that have no effect.

One study on respiratory tract infections found "physicians were more likely to prescribe antibiotics to patients who they believed expected them, although they correctly identified only about 1 in 4 of those patients". Multifactorial interventions aimed at both physicians and patients can reduce inappropriate prescribing of antibiotics. Delaying antibiotics for 48 hours while observing for spontaneous resolution of respiratory tract infections may reduce antibiotic usage; however, this strategy may reduce patient satisfaction.

Several organizations concerned with antimicrobial resistance are lobbying to improve the regulatory climate. Approaches to tackling the issues of misuse and overuse of antibiotics by the establishment of the U.S. Interagency Task Force on Antimicrobial Resistance which aims actively address the problem antimicrobial resistance are being organised and coordinated by the US Centers for Disease Control and Prevention, the Food and Drug Administration (FDA), and the National Institutes of Health (NIH) and also includes several other federal agencies.

An NGO campaign group is Keep Antibiotics Working. In France, an "Antibiotics are not automatic" government campaign starting in 2002 led to a marked reduction of unnecessary antibiotic prescriptions, especially in children. In the United Kingdom, there are NHS posters in many doctors surgeries indicating that 'unfortunately, no amount of antibiotics will get rid of your cold', following on from many patients specifically requesting antibiotics from their doctor inappropriately, believing they will help treat viral infections.

In agriculture, associated antibiotic resistance with the non-therapeutic use of antibiotics as growth promoters in animals resulted in their restricted use in the UK in the 1970. Currently there is a EU wide ban on the non-therapeutic use of antibiotics as growth promoters. It is estimated that greater than 70% of the antibiotics used in U.S. are given to feed animals (e.g. chickens, pigs and cattle) in the absence of disease. Antibiotic use in food animal production has been associated with the emergence of antibiotic-resistant strains of bacteria including Salmonella spp., Campylobacter spp., Escherichia coli, and Enterococcus spp. Evidence from some US and European studies suggest that these resistant bacteria cause infections in humans that do not respond to commonly prescribed antibiotics. In response to these practices and attendant problems, several organizations (e.g. The American Society for Microbiology (ASM), American Public Health Association (APHA) and the American Medical Association (AMA)) have called for restrictions on antibiotic use in food animal production and an end to all non-therapeutic uses.

However, delays in regulatory and legislative actions to limit the use of antibiotics are common, and may include resistance to these changes by industries using or selling antibiotics, as well as time spent on research to establish causal links between antibiotic use and emergence of untreatable bacterial diseases. Two federal bills (S.742 and H.R. 2562) aimed at phasing out non-therapeutic antibiotics in US food animal production were proposed but not passed. These bills were endorsed by public health and medical organizations including the American Holistic Nurses' Association, the American Medical Association, and the American Public Health Association (APHA). The EU has banned the use of antibiotics as growth promotional agents since 2003.

Resistance Modifying Agents

One solution to combat resistance currently being researched is the development of pharmaceutical compounds that would revert multiple antibiotic resistance. These so called resistance modifying agents may target and inhibit MDR mechanisms, rendering the bacteria susceptible to antibiotics to which they were previously resistant.

These compounds targets include among others:

- Efflux inhibition(Phe-Arg-β-naphthylamide)
- Beta Lactamase inhibitors - Including Clavulanic acid and Sulbactam

Beyond Antibiotics: Treating Non-bacterial Infections

The comparative ease of identifying compounds which safely cured bacterial infections was more difficult to duplicate in treatments of fungal and viral infections. Antibiotic research led to great strides in the knowledge of biochemistry, establishing large differences between the cellular and molecular physiology of the bacterial cell and that of the mammalian cell. This explained the observation that many compounds that are toxic to bacteria are non-toxic to human cells. In contrast, the basic biochemistries of the fungal cell and the mammalian cell are much more similar. This restricts the development and use of therapeutic compounds that attack a fungal cell, while not harming mammalian cells. Similar problems exist in antibiotic treatments of viral diseases. Human viral metabolic biochemistry is very closely similar to human biochemistry, and the possible targets of antiviral compounds are restricted to very few components unique to a mammalian virus.

Beyond Antibiotics: Treating Multi-drug Resistant Bacteria

Multi-drug resistant organisms (MDRO) generally refer to bacteria that are not affected by the clinical doses of classical antibiotics, particularly the antibiotics which were being used to treat them until recently. The rise of these organisms has created a need for alternative antibacterial therapies.

Phage therapy, the use of particular viruses to attack bacteria, was in use during the 1920s and 1930s on humans in the US, Western and Eastern Europe. Phage are commonly a part of the ecology surrounding bacteria and provide substantial population control of bacteria in the intestine, the ocean, the soil and other environments. The success of these therapies are largely anecdotal or otherwise poorly controlled. The original publications are also generally inaccessible, even to persons with Russian language fluency. With the discovery of penicillin in the 1940s, Europe and the US changed therapeutic strategies to the use of antibiotics. However, in the former Soviet Union phage therapies continued to be studied.

In the Republic of Georgia, the Eliava Institute of Bacteriophage, Microbiology & Virology continues to research the use of phage therapy. Various companies (Intralytix, among others), universities, and foundations in North America and Europe are currently researching phage therapies. However, concerns about genetic engineering in freely released viruses currently limit certain aspects of phage therapy.

One result is attempts to use phage in ways other than to directly infect the bacteria. While bacteriophage and related therapies provide a possible solution to aspects of antibiotic resistance, their place in clinical therapy is still in question. Different classes of bacteriocins have different potential as therapeutic agents. Small molecule bacteriocins (microcins, for example, and lantibiotics) may be similar to the classic antibiotics; colicin-like bacteriocins are more likely to be narrow-spectrum, demanding new molecular diagnostics prior to therapy but also not raising the spectre of resistance to the same degree.

One drawback to the large molecule antibiotics is that they will have relative difficulty crossing membranes and travelling systemically throughout the body. For this reason, they are most often proposed for application topically or gastrointestinally. Because bacteriocins are peptides, they are more readily engineered than small molecules. This may permit the generation of cocktails and dynamically improved antibiotics that are modified to overcome resistance.

Nutrient withdrawal is a potential strategy for replacing or supplementing antibiotics. The restriction of iron availability is one way the human body limits bacterial proliferation. Mechanisms for freeing iron from the body (such as toxins and siderophores) are common among pathogens. Building on this dynamic, various research groups are attempting to produce novel chelators which would withdraw iron otherwise available to pathogens (bacterial, fungal and parasitic). This is distinct from chelation therapy for conditions other than bacterial infections - including successful treatment for iron overload.

Vaccines are a commonly suggested method for combating MDRO infections. They actually fit within a larger class of therapies that rely on immune modulation or augmentation. These therapies either excite or reinforce the natural immune competency of the infected or susceptible host, leading to the activity of macrophages, the production of antibodies, inflammation, or other classic immune reactions. Just as the macrophage engulfs and consumes bacteria, various forms of biotherapy have been suggested which employ organisms to consume the pathogens. This includes the employment of protozoa and maggot therapy. Probiotics are another alternative that goes beyond traditional antibiotics by employing a live culture which may in theory establish itself as a symbiont, competing, inhibiting, or simply interfering with colonization by pathogens.

ONCOLOGY ANTIBIOTICS DRUGS

Antibiotics are drugs that are used to treat infections caused by bacteria and other organisms, including protozoa, parasites, and fungi.

PURPOSE

Many treatments for cancer destroy disease-fighting white blood cells, thereby reducing the body's ability to fight infection. For example, bladder, pulmonary, and urinary tract infections may occur with chemotherapy. Single-celled organisms called protozoa are rarely a problem for healthy individuals. However, they can cause serious infections in individuals with low white blood cell counts. Because of the dangers that infections present for cancer patients, antibiotic treatment often is initiated before the exact nature of the infection has been determined; instead, the choice of antibiotic may depend on the site of the infection and the organism that is likely to be the cause. Often, an antibiotic that kills a broad spectrum of bacteria is chosen and several antibiotics may be used together.

The common antibiotics that are used during cancer treatment include:

- *Atovaquone (Mapren)*: antiprotozoal drug used to prevent and treat a very serious type of pneumonia called *Pneumocystis carinii* pneumonia (PCP), in individuals who experience serious side effects with SMZ-TMP (Sulfamethoxazole/Trimethoprim, brand name Bactrim).
- *Aztreonam (Azactam)*: monobactam antibiotic used to treat gram-

negative bacterial infections of the urinary and lower respiratory tracts and the female organs, and infections that are present throughout the body (systemic infections or septicemia).

- Cefepime (Maxipime), ceftazidime (Ceptaz, Fortaz, Tazicef, Tazidime), and ceftriaxone sodium (Rocephin): members of a group of antibiotics called cephalosporins used to treat bacterial infections of the urinary and lower respiratory tracts, and infections of the skin, bones, joints, pelvis, and abdomen.
- *Ciprofloxacin (Cipro)*: fluoroquinolone antibiotic used to treat certain gram-negative and gram-positive bacteria and some mycobacteria.
- *Clindamycin phosphate (Cleocin)*: used to treat gram-positive and gram-negative bacterial infections and, in individuals who are allergic to sulfadiazine, toxoplasmosis caused by a parasitic protozoa.
- Gentamicin (gentamycin) sulfate (generic name product, Garamycin, G-Mycin, Jenamicin): amino-glycoside antibiotic used to treat serious infections by many gram-negative bacteria that cannot be treated with other medicines.
- Metronidazole hydrochloride (Flagyl, Metric 21, Metro I.V., Protostat): used for anaerobic bacteria and protozoa.
- Pentamidine (generic name product, Pentam 300): used to treat PCP if serious side effects develop with SMZ-TMP.
- *Pyrimethamine (Daraprim)*: antiprotozoal medicine used together with sulfadiazine to treat toxo-plasmosis; or in combination with other medicines for treating mild to moderate PCP, in individuals who cannot tolerate the standard treatment.
- *Sulfadiazine (generic name product)*: sulfonamide antibiotic used with pyrimethamine to treat toxoplasmosis.
- *Sulfamethoxazole*: Trimethoprim (SMZ-TMP) (generic name product, Bactrim, Cofatrim Forte, Cotrim, Septra, Sulfatrim): the sulfonamide antibiotic, sulfamethoxazole, used in combination with trimethoprim, to prevent and treat PCP and bacterial infections, such as bronchitis and middle ear and urinary tract infections.
- Trimethoprim (generic name product, Proloprim, Trimpex): primarily used to prevent or treat urinary tract infections.
- Vancomycin hydrochloride (generic name product, Vancocin): glycopeptide antibiotic used to treat a variety of serious gram-positive bacterial infections for which other medicines are ineffective, including strains of *Staphylococcus* that are resistant to most oral antibiotics.

Most of these antibiotics kill bacteria by preventing them from making protein for their cell walls. Ciprofloxacin and metronidazole prevent bacteria from reproducing by interfering with their ability to make new DNA. All of these drugs are approved for prescription by the U.S. Food and Drug Administration.

Recommended Dosage

Dosages of antibiotics depend on the individual, the infection that is being treated, and the presence of other medical conditions. For children, the dosage usually is based on body weight and is lower than the adult dosage. To be effective, an entire treatment with antibiotics must be completed, even if the symptoms of infection have disappeared.

Furthermore, it is important to keep the level of antibiotic in the body at a constant level during treatment. Therefore, the drug should be taken on a regular schedule. If a dose is missed, it should be taken as soon as possible. If it is almost time for the next dose, the missed dose should be skipped. Doubling up doses is generally not recommended.

Average adult dosages of common antibiotics for cancer patients are as follows:

- *Atovaquone*: For PCP treatment, 750 mg oral suspension twice a day, or tablets three times per day, for 21 days; for PCP prevention, 1,500 mg oral suspension, once a day; must be taken with balanced meals.
- *Aztreonam*: 1–2 gm every 6–12 hours, injected into a vein, over a 20–60 minute-period.
- *Cefepime*: 500 mg to 2 gm, injected into a vein or muscle, every 8–12 hours for 7–10 days.
- *Ceftazidime*: 250 mg to 2 gm, injected into a vein or muscle, every 8–12 hours.
- *Ceftriaxone*: 1–2 gm, injected into a vein or muscle, every 24 hours.
- *Ciprofloxacin*: 500–750 mg of the tablet or suspension, every 12 hours, for 3–28 days, taken two hours after meals with 8 oz of water; bone and joint infections usually are treated for at least 4–6 weeks; 200–400 mg injected every 8–12 hours.
- *Clindamycin*: 150–300 mg of capsule or solution, every six hours; 300–600 mg every six to eight hours or 900 mg every eight hours, injected into a vein or muscle.
- *Gentamicin*: Dosage determined by body weight, every 8–24 hours for at least 7–10 days, injected into a vein or muscle.
- *Metronidazole*: For bacterial infections, 7.5 mg per kg (3.4 mg per lb) of body weight up to a maximum of 1 gm, every six hours for at least seven days (capsules or tablets); 15 mg per kg (6.8 mg per lb) for the first dose, followed by half that dosage every six hours for at least seven days (injected into a vein); for protozoal infections caused by amebas, 500–750 mg of oral medicine, three times per day for 5–10 days; for trichomoniasis, 2 gm for one day or 250 mg three times per day for seven days (oral medicine); extended-release tablets for vaginal bacterial infections, 750 mg once a day for seven days.
- *Pentamidine*: For treating PCP, 4 mg per kg (1.8 mg per lb) of body

weight, once per day for 14–21 days, injected into a vein over one to two hours, while lying down.

- *Pyrimethamine*: For toxoplasmosis, 25–200 mg tablets, taken with other medicine, for several weeks.
- *Sulfadiazine*: For bacterial and protozoal infections, 2–4 gm for the first dose, followed by 1 gm every four to six hours (tablets).
- *SMZ-TMP*: 800 mg of sulfamethoxazole and 160 mg of trimethoprim, (tablet or oral suspension), every 12 hours for bacterial infections and every 24 hours for prevention of PCP; dosage based on body weight for PCP treatment; injections based on body weight, every six, eight or 12 hours for bacterial infections and every six hours for PCP treatment.
- *Trimethoprim*: 100 mg tablet every 12 hours for 10 days; for prevention of urinary tract infections, once a day for a long period.
- *Vancomycin*: 7.5 mg per kg (3.4 mg per lb) of body weight, or 500 mg–1 gram, injected or taken orally, every 6–12 hours.

Precautions

Stomach or intestinal problems or colitis (inflammation of the colon) may affect the use of:

- Atovaquone
- Cephalosporins
- Clindamycin

Central nervous system or seizure disorders may affect the use of:

- Ciprofloxacin
- Metronidazole
- Pyrimethamine

Anemia (low red blood cell count) or other blood disorders may affect the use of:

- Metronidazole
- Pentamidine
- Pyrimethamine
- Sulfadiazine
- SMZ-TMP
- Trimethoprim

Kidney or liver disease may affect the use of:

- Aztreonam
- Cefepime
- Ceftazidime
- Ciprofloxacin
- Clindamycin
- Gentamicin
- Metronidazole
- Pentamidine

- Pyrimethamine
- Sulfadiazine
- SMZ-TMP
- Trimethoprim
- Vancomycin

Ciprofloxacin may not be suitable for individuals with tendinitis or with skin sensitivities to sunlight. Gentamicin may not be suitable for people with hearing problems, myasthenia gravis, or Parkinson's disease. Metronidazole may not be suitable for individuals with heart disease, oral or vaginal yeast infections, or a history of alcoholism.

Pentamidine may not be suitable for individuals with heart disease, bleeding disorders, or low blood pressure. Pentamidine may affect blood sugar levels, making control of diabetes mellitus or hypoglycemia (low blood sugar) difficult. Vancomycin may not be appropriate for individuals with hearing problems. Many antibiotics should not be taken during pregnancy or while breast-feeding. Older individuals may be more susceptible to the side effects of sulfadiazine, SMZ-TMP, or trimethoprim.

Side Effects

Some individuals may have allergic reactions to antibiotics. If symptoms of an allergic reaction (such as rash, shortness of breath, swelling of the face and neck), severe diarrhea, or abdominal cramping occur, the antibiotic should be stopped and the individual should seek medical advice. Because antibiotics can affect bacteria that are beneficial, as well as those that are harmful, women may become susceptible to infections by fungi when taking antibiotics. Vaginal itching or discharge may be symptoms of such infections. All patients may develop oral fungal infections of the mouth, indicated by white plaques in the mouth.

Injected antibiotics may result in irritation, pain, tenderness, or swelling in the vein used for injection. Antibiotics used in cancer patients may have numerous side effects, both minor and severe; however, most side effects are uncommon or rare. The more common side effects of atovaquone, aztreonam, cephalosporins, ciprofloxacin, clindamycin, gentamicin, metronidazole, and SMZ-TMP include:

- Nausea and vomiting
- Diarrhea
- Loss of appetite Eating active cultured yogurt may help counteract diarrhea, but if a patient has low white blood cells, this remedy is not recommended. For mild diarrhea with cephalosporins, only diarrhea medicines containing kaolin or attapulgite should be taken. With clindamycin, diarrhea medicines containing attapulgite should be taken several hours before or after the oral antibiotic. Diarrhea following antibiotics like clindamycin may indicate a bacterial

infection that needs additional therapy, and a physicians hould be consulted.

Other side effects of atovaquone may include:

- fever
- skin rash
- cough
- headache
- insomnia

Other side effects of ciprofloxacin may include:

- Abdominal pain
- Increase in blood tests for kidney function
- Dizziness or light-headedness
- Inflammation or tearing of a tendon
- Drowsiness
- Insomnia

Other common side effects of clindamycin include abdominal pain and fever. Side effects may occur up to several weeks after treatment with this medicine. Gentamicin and vancomycin may cause serious side effects, particularly in elderly individuals and newborn infants. These include kidney damage and damage to the auditory nerve that controls hearing.

When gentamicin is injected into a muscle, vein, or the spinal fluid, the following side effects may occur:

- Leg cramps
- Skin rash
- Fever
- Seizures Side effects from gentamicin may develop up to several weeks after the medicine is stopped.

Other, more common side effects of gentamicin may include:

- Changes in urination
- Increased thirst
- Muscle twitching or seizures
- Headache
- Lethargy

More common side effects of metronidazole include:

- Mouth dryness
- Unpleasant or metallic taste
- Dizziness or light-headedness
- Headache
- Stomach pain Sugarless candy or gum, bits of ice, or a saliva substitute may relieve symptoms of dry mouth.

Pentamidine, pyrimethamine, sulfonamides, SMZTMP, and trimethoprim can lower the number of white blood cells, resulting in an increased risk of infection. These drugs also can lower the number of blood platelets that are important for blood clotting. Thus, there is an increased risk of bleeding or bruising while taking these drugs.

Serious side effects of pentamidine may include:

- Heart problems
- Low blood pressure
- High or low blood sugar
- Other blood problems
- Decrease in urination
- Sore throat and fever
- Sharp pain in upper abdomen Some of these symptoms may not occur until several months after treatment with pentamidine.

Pyrimethamine and trimethoprim may lower the red blood cell count, causing anemia. Leucovorin or the vitamin folic acid may be prescribed for anemia. Some individuals become more sensitive to sunlight when taking sulfonamides, SMZ-TMP, or trimethoprim. Other common side effects of sulfonamides and SMZTMP include:

- Dizziness
- Itching
- Skin rash
- Headache
- Mouth sores or swelling of the tongue
- Fatigue

If vancomycin is injected into a vein too quickly, it can cause flushing and a rash over the neck, face, and chest, wheezing or difficulty breathing, and a dangerous decrease in blood pressure.

Interactions

Many prescription and non-prescription medicines can interact with these antibiotics. Therefore, it is important to consult a complete list of known drug interactions. Among the more common or dangerous interactions:

- Antibiotics that lower the number of blood platelets, with blood thinners (anticoagulants), such as warfarin
- Aztreonam and metronidazole with alcohol; it is important not to consume alcohol until at least three days after treatment with these antibiotics
- Ciprofloxacin with antacids, iron supplements, or caffeine
- Pentamidine or pyrimethamine with previous treatments with x rays or cancer medicines (increased risk of blood cell damage)
- Trimethoprim with diuretics to remove excess fluid in the elderly

Many medicines can increase the risk of hearing or kidney damage from gentamicin.

These include:

- Cisplatin

- Combination pain medicine with acetaminophen and aspirin or other salicylates (taken regularly in large amounts)
- Cyclosporine
- Inflammation or pain medicine, except narcotics
- Lithium
- Methotrexate
- Other medicines for infection

CLASSIFICATIONS OF ANTIBIOTIC

Chemical substance that in dilute solutions can inhibit the growth of microorganisms or destroy them with little or no harm to the infected host. Early antibiotics were natural microbial products, but chemists have modified the structures of many to produce semisynthetic and even wholly synthetic ones. Since the discovery of penicillin (1928), antibiotics have revolutionized the treatment of bacterial, fungal, and some other diseases. They are produced by many actinomycetes (e.g., streptomycin, tetracycline) and other bacteria (e.g., polypeptides such as bacitracin) and by fungi (e.g., penicillin). Antibiotics may be broad-spectrum (active against a wide range of pathogens) or specific (active against one, or one class). Drawbacks include activity against beneficial microorganisms, often causing diarrhea; allergies; and development of drug-resistant strains of the targeted microorganisms.

Antibiotics may be informally defined as the sub-group of anti-infectives that are derived from bacterial sources and are used to treat bacterial infections. Other classes of drugs, most notably the sulfonamides, may be effective antibacterials. Similarly, some antibiotics may have secondary uses, such as the use of demeclocycline (Declomycin, a tetracycline derivative) to treat the syndrome of inappropriate antidiuretic hormone (SIADH) secretion. Other antibiotics may be useful in treating protozoal infections.

Although there are several classification schemes for antibiotics, based on bacterial spectrum (broad versus narrow) or route of administration (injectable versus oral versus topical), or type of activity (bactericidal vs. bacteriostatic), the most useful is based on chemical structure. Antibiotics within a structural class will generally show similar patterns of effectiveness, toxicity, and allergic potential.

- *Penicillins*: The penicillins are the oldest class of antibiotics, and have a common chemical structure which they share with the cephalopsorins. The two groups are classed as the beta-lactam antibiotics, and are generally bacteriocidal—that is, they kill bacteria rather than inhibiting growth. The penicillins can be further subdivided. The natural pencillins are based on the original penicillin G structure; penicillinase-resistant penicillins, notably methicillin and oxacillin, are active even in the presence of the bacterial enzyme that inactivates most natural penicillins. Aminopenicillins such as ampicillin and amoxicillin have an extended spectrum of action compared with the natural penicillins; extended spectrum penicillins are effective against a wider range of bacteria. These generally include coverage for *Pseudomonas aeruginaosa* and may provide the penicillin in combination with a penicillinase inhibitor.

- *Cephalosporins*: Cephalosporins and the closely related cephamycins and carbapenems, like the pencillins, contain a beta-lactam chemical structure. Consequently, there are patterns of cross-resistance and cross-allergenicity among the drugs in these classes. The "cepha" drugs are among the most diverse classes of antibiotics, and are themselves subgrouped into 1st, 2nd and 3rd generations. Each generation has a broader spectrum of activity than the one before. In addition, cefoxitin, a cephamycin, is highly active against anaerobic bacteria, which offers utility in treatment of abdominal infections. The 3rd generation drugs, cefotaxime, ceftizoxime, ceftriaxone and others, cross the blood-brain barrier and may be used to treat meningitis and encephalitis. Cephalopsorins are the usually preferred agents for surgical prophylaxis.
- *Fluroquinolones*: The fluroquinolones are synthetic antibacterial agents, and not derived from bacteria. They are included here because they can be readily interchanged with traditional antibiotics. An earlier, related class of antibacterial agents, the quinolones, was not well absorbed, and could be used only to treat urinary tract infections.

  The fluroquinolones, which are based on the older group, are broad-spectrum bacteriocidal drugs that are chemically unrelated to the penicillins or the cephaloprosins. They are well distributed into bone tissue, and so well absorbed that in general they are as effective by the oral route as by intravenous infusion.
- *Tetracyclines*: Tetracyclines got their name because they share a chemical structure that has four rings. They are derived from a species of *Streptomyces* bacteria. Broad-spectrum bacteriostatic agents, the tetracyclines may be effective against a wide variety of microorganisms, including rickettsia and amebic parasites.
- *Macrolides*: The macrolide antibiotics are derived from *Streptomyces* bacteria, and got their name because they all have a macrocyclic lactone chemical structure. Erythromycin, the prototype of this class, has a spectrum and use similar to penicillin. Newer members of the group, azithromycin and clarithyr-omycin, are particularly useful for their high level of lung penetration. Clarithromycin has been widely used to treat *Helicobacter pylori* infections, the cause of stomach ulcers.
- *Others*: Other classes of antibiotics include the aminoglycosides,

which are particularly useful for their effectiveness in treating *Pseudomonas aeruginosa* infections; the lincosamindes, clindamycin and lincomycin, which are highly active against anaerobic pathogens. There are other, individual drugs which may have utility in specific infections.

RECOMMENDED DOSAGE

Dosage varies with drug, route of administration, pathogen, site of infection, and severity. Additional considerations include renal (kidney) function, age of patient, and other factors. Patients should consult manufac turers' recommendations or ask their doctors.

Side Effects

All antibiotics cause risk of overgrowth by non-susceptible bacteria. Manufacturers list other major hazards by class; however, the health care provider should review each drug individually to assess the degree of risk.

Generally, breastfeeding is not recommended while taking antibiotics because of risk of alteration to infant's intestinal flora, and risk of masking infection in the infant. Excessive or inappropriate use may promote growth of resistant pathogens.

- *Penicillins*: Hypersensitivity may be common, and cross allergenicity with cephalosporins has been reported. Penicillins are classed as category B during pregnancy.
- *Cephalosporins*: Several cephalosporins and related compounds have been associated with seizures. Cefmetazole, cefoperazone, cefotetan and ceftriaxone may be associated with a fall in prothrombin activity and coagulation abnormalities. Pseudomembranous colitis (inflammation of the colon) has been reported with cephalosporins and other broad spectrum antibiotics. Some drugs in this class may cause renal toxicity. Pregnancy category B.
- *Fluoroquinolones*: Lomefloxacin has been associated with increased photosensitivity. All drugs in this class have been associated with convulsions. Pregnancy category C.
- *Tetracyclines*: Demeclocycline may cause increased photosensitivity. Minocycline may cause dizziness. hildren under the age of eight should not use tetracyclines, and specifically during periods of tooth development. Oral tetracyclines bind to anions such as calcium and iron. Although doxycycline and minocycline may be taken with meals, patients are advised to take other tetracycline antibiotics on an empty stomach, and not to take the drugs with milk or other calcium-rich foods. Expired tetracycline should never be administered. Pregnancy category D; use during pregnancy may cause alterations in bone development.

- *Macrolides*: Erythromycin may aggravate the weakness of patients with myasthenia gravis. Azithromycin has, rarely, been associated with allergic reactions, including angioedema, anaphylaxis, and dermatologic reactions, including Stevens-Johnson syndrome and toxic epidermal necrolysis. Oral erythromycin may be highly irritating to the stomach and may cause severe phlebitis (inflammation of the vein) when given by injection. These drugs should be used with caution in patients with liver dysfunction. Pregnancy category B: Azithromycin, erythromycin. Pregnancy category C: Clarithromycin, dirithromycin, troleandomycin.
- *Aminoglycosides*: This class of drugs causes kidney and hearing problems. These problems can occur even with normal doses. Dosing should be based on renal function, with periodic testing of both kidney function and hearing. Pregnancy category D.

Interactions

Use of all antibiotics may temporarily reduce the effectiveness of birth control pills; alternative birth control methods should be used while taking these medications. Antacids should be avoided while on tetracyclines as the calcium can impair absorption of this antibiotic class.

For this reason, tetracyclines should not be taken just before or after consuming foods rich in calcium or iron. Consult specialized references for additional interactions to specific antibiotics.

Recommended Usage

To minimize risk of adverse reactions and development of resistant strains of bacteria, antibiotics should be restricted to use in cases where there is either known or a reasonable presumption of bacterial infection. The use of antibiotics in viral infections is to be avoided. Avoid use of fluroquinolones for trivial infections. In severe infections, presumptive therapy with a broad-spectrum antibiotic such as a third generation cephalosporin may be appropriate. Treatment should be changed to a narrow spectrum agent as soon as the pathogen has been identified. After 48 hours of treatment, if there is clinical improvement, an oral antibiotic should be considered.

When the pathogen is known or suspected to be *Pseudomonas,* a suitable beta-lactam drug is often prescribed in combination with an aminoglycoside. A single agent cannot be relied upon for treatment of *Pseudomonas*. When the patient has renal insufficiency, azactam should be considered in place of the aminoglycoside. In treatment of children with antibiotic suspensions, caregivers should be instructed in use of oral syringes or measuring teaspoons. Household teaspoons are not standardized and will give unreliable doses.

4

Pharmacogenomics and DNA Microarrays

Pharmacogenomics is the study of genes and the gene products (proteins) essential for pharmacological or toxicological responses to pharmaceutical agents. Oligonucleotide-based DNA chip technology or cDNA microarray can be used to analyse gene expression profiles that are induced or repressed by xenobiotics. An attempt has been made to develop microarray genotyping system for multiplex analysis of a panel of single nucleotide polymorphisms (SNPs) in genes encoding proteins involved in blood pressure regulation, and to apply this system in a pilot study demonstrating its feasibility in the pharmacogenetics of anti-hypertensive drug response.

DNA microarray techniques might prove to be a reliable basis for predicting response (or lack of response) of individuals to herbal drugs. Technologies designed to characterize genes and their products on a discovery scale are now having an impact on many areas of biology, including toxicology. Toxicogenomics is the sub-discipline that merges genomics with toxicology. In toxicology research, gene expression profiling facilitates mechanism-based research on toxicant action by comparing results for an experimental compound with a database.

Recently several studies have demonstrated the utility of microarray analysis for studying genome-wide effects of xenobiotics and the rapid identification of toxic hazards for novel drug candidates. An example of such a platform is ToxBlot II, a custom microarray containing cDNAs representing 12564 human genes chosen on the basis of their potential relevance to a broad range of toxicities. ToxBlot II allows the simultaneous expression profiling of genes representing entire cellular pathways, facilitating a very detailed investigation of potential mechanisms of toxicity.

cDNA microarray analysis was used to study the expression level of genes in oral fibroblast cell lines in response to exposure to ripe areca nut extract. The results showed up-regulation of IL-6 expression and downregulation of PDGFR, APP-1 and KGF-1 expressions in multiple cell lines assayed. The down-regulation of KGF-1 expression in oral fibroblast cell lines potentially impairs the proliferation of overlying keratinocytes, which could partially explain the frequent epithelial atrophy observed in chronic areca chewers in vivo. This study established a novel toxicogenomic database for areca nut extract. cDNA microarray analysis was also used to analyse the mRNA expression patterns of 1177 genes in ten oral cancer patients with betel quid chewing history. This study provides pilot data for understanding the pathogenesis of oral cancer in countries like Taiwan where betel quid chewing is prevalent. Recent studies have highlighted that concurrent use of herbs may mimic, magnify, or oppose the effect of drugs. DNA microarray can be used for studying herb–drug interactions, and the mechanisms underlying these interactions.

Several botanical constituents in PC-SPES, a botanical preparation, inhibit tumor growth through cell cycle arrest and apoptosis. LNCaP prostate carcinoma cells were treated with PC-SPES, and changes in gene expression were determined by complementary DNA (cDNA) microarray hybridization and northern blot analyses. mRNA levels of a-tubulin decreased sevenfold. The results show that PC-SPES may interfere with microtubule polymerization. This activity has implications for the clinical management of patients with advanced prostate cancer who may be taking PC-SPES concurrently with microtubule-modulating chemotherapeutic agents, such as paclitaxel. cDNA microarray analysis was used to identify gene expression changes in LNCaP prostate carcinoma cells exposed to PC-SPES and estrogenic agents including diethylstilbestrol.

Transcripts encoding cell cycleregulatory proteins, α- and β-tubulins, and the androgen receptor were down-regulated by PC-SPES. A comparison of gene expression profiles resulting from these treatments indicates that PC-SPES exhibits activities distinct from those attributable to diethylstilbestrol and suggests that alterations in specific genes involved in modulating the cell cycle, cell structure, and androgen response may be responsible for PC-SPES-mediated cytotoxicity.

THE DNA MICROARRAY

DNA microarray is an orderly arrangement of thousands of oligonucleotides or identified sequenced genes printed on an impermeable solid support, usually glass, silicon chips or nylon membrane. A number of terms such as DNA arrays, gene chips and biochips are often used to describe these devices.

The DNA microarray field is a combination of several technologies, including automated DNA sequencing, DNA amplification by PCR, oligonucleotide synthesis, nucleic acid labeling chemistries and bioinformatics. Two most common types of DNA microarrays include one in which the DNA (in the form of a single stranded short oligonucleotide) is synthesized in situ using photolithographic or other techniques and another where the DNA (usually in the form of a cDNA or full-length ORF) is post-synthetically attached to a solid support.

Microarray type	Brief description	Applications
Comparative genomic hybridization (CGH)	Detect chromosomal aberrations	Tumor classification Disease diagnosis Risk assessment Expression analysis
Analysis of gene	Drug development expression levels	 Drug response Tracking disease
progression		Therapy
development		Drug mechanism of
action		
Mutation/ polymorphism analysis	Detect mutations/ polymorphisms in gene sequence	Drug development Tracking disease progression Disease risk
assessment		Genotyping Species
identification		Population genetics

Depending on the kind of probe (immobilized DNA) used to generate the array and, ultimately, the information that is derived from it, DNA microarrays are used for different applications. As a definition in the context of microarrays, a 'probe' is the (partial) genomic sequence of a gene deposited and fixed on the Microarray, whereas the 'target' is the biological sample material.

Although there are many protocols and types of microarray experiments, the basic steps involved in any microarray experiment are isolating RNA or mRNA from appropriate biological samples; applying a fluorescent tag to the RNA or cDNA copy of it; hybridizing the labeled RNA or cDNA (target) to a microarray (probe) for a period of time after which the excess is washed off; scanning the microarray under laser light and data analysis using appropriate software.

A brief description of three basic types of DNA microarray experiments and their applications is provided in Table.

Applications of DNA Microarrays in Herbal Drug Research and Development

Natural product research is often based on ethnobotanical information and many of the drugs used today were employed in indigenous societies. One of the aims of ethnopharmaceutical research is better understanding of the pharmacological effects of different medicinal plants traditionally used in healthcare. Plants are regarded as a promising source of novel therapeutic agents due to their higher structural diversity as compared to standard synthetic chemistry.

Plants have applications in the development of therapeutic agents: as a source of bioactive compounds for possible use as drugs. There are three approaches to natural product-based drug discovery: screening of crude extracts; screening of prefractionated extracts; screening of pure compounds.

In the past decade there has been a paradigm shift from single-target drugs to multi-target drugs. Multi-target approaches are directed toward activation of defence, protective and repair mechanisms of the body rather than destruction of the damage-causing agent.

This may be achieved by the use of a combination of drugs as in AIDS, cancer, atherosclerosis or use of single compounds that affect many targets simultaneously, e.g. GleevecTM, salicylate, metformin, Pemetrexed (Alimta) and a multi-targeted antifolate.

The concept of multi-targeted therapy exists in traditional medical treatments that employ multi-component extracts of natural products. With the exception of pure substances, botanical derivatives obtained from medicinal plants usually contain several classes of compounds, which simultaneously act on multiple targets. However, very few herbal medicinal products have documentation comparable to synthetic drugs.

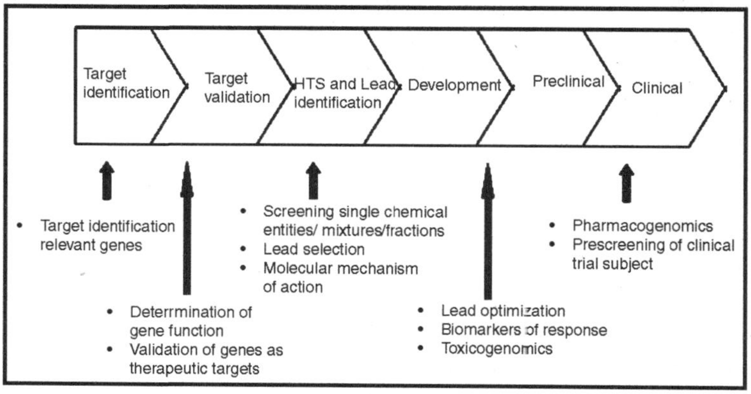

Fig. DNA Microarray: Applications in Natural Product Drug Discovery and Development.

Preparation of highly standardized herbal products with respect to chemical composition and biological activity is complicated due to the myriad range of chemicals.

This also makes it difficult to precisely predict the molecular targets, mechanism of action and side effects of such products. Use of validated analytical methods to quantify specific chemical markers remains the key point to obtain chemically standardized extracts. However many minor components with potential biological effects remain neglected from quantitative analysis of chemical markers. In the past years the development of assays on the level of molecules (receptor binding and enzyme inhibition) opened complete new perspectives for plant-based products as source of new drugs.

Advances in highthroughput experimentation have resulted in massive databases of genomic, proteomic and chemical data which in combination with efficient separation methods and powerful spectrometric methods for identification and structure elucidation can be used for identification of active compounds. In addition, the past two decades has witnessed a number of citations that describe a diverse range of molecular mechanisms that govern cellular behaviour. A systems biology approach that integrates such large and diverse sources of information together will serve to make useful biological predictions about the pharmacological effects of natural products.

DNA microarrays may provide a suitable high-throughput platform for research and development of drugs from natural products. In natural products a broad repertoire of chemical entities act together on multiple targets that makes it necessary to study the changes in expression of multiple genes simultaneously.

Novel technologies such as SAGE and DNA microarrays allow rapid and detailed analysis of thousands of transcripts, providing a revolutionary approach to the investigation of gene expression. In this review, we discuss the applications of DNA microarray technology in herbal drug research and development with suitable examples.

There are three main applications of DNA microarrays: first, in pharmacodynamics for discovery of new diagnostic and prognostic indicators and biomarkers of therapeutic response; elucidation of molecular mechanism of action of a herb, its formulations or its phytochemical components and identification and validation of new molecular targets for herbal drug development. Second, in pharmacogenomics for prediction of potential side-effects of the herbal drug during preclinical activity and safety studies; identification of genes involved in conferring drug sensitivity or resistance and prediction of patients most likely to benefit from the drug and use in general pharmacogenomic studies.

Third, in pharmacognosy for correct botanical identification and authentication of crude plant materials as part of standardization and quality control. These applications are described here with some examples. Figure shows applications of DNA microarrays at various stages of a typical drug discovery pipeline.

PHARMACOGNOSY AND DNA MICROARRAY

Use of authentic herbal materials is the first step to ensure quality, safety and efficacy of herbal medicines. DNA polymorphism-based assays have been developed for the identification of herbal medicines. In this approach, small amounts of DNA are amplified by the polymerase chain reaction and the reaction products are analysed by gel electrophoresis, sequencing, or hybridization with species-specific probes. Recently, microarrays have been applied for the DNA sequence-based identification of medicinal plants. To utilize DNA microarrays for identification and authentication of herbal material, it is necessary to identify a distinct DNA sequence that is unique to each species of medicinal plant.

The DNA sequence information is then used to synthesize a corresponding probe on a silicon-based gene chip. These probes are capable of detecting complementary target DNA sequences if present in the test sample being analysed. Oligonucleotide probes specific for polymorphisms in the D2 and D3 regions of 26S rDNA gene of several Fritillaria species were designed and printed on the poly-lysine coated slides to prepare a DNA chip. Differentiation of the various Fritillaria species was accomplished based on hybridization of fluorescent labeled PCR products with the DNA chip.

The results demonstrated the reliability of using DNA chips to identify different species of Fritillaria, and that the DNA chip technology can provide a rapid, high throughput tool for genotyping and plant species authentication. Similarly, using fluorescence-labeled ITS2 sequences as probes, distinctive signals were obtained for the five medicinal Dendrobium species listed in the Chinese Pharmacopoeia. The established microarray was able to detect the presence of D. nobile in a Chinese medicinal formulation containing nine herbal components. A silicon-based DNA microarray using species-specific oligonucleotide probes is designed and fabricated to identify multiple toxic traditional Chinese medicinal plant species by parallel genotyping.

Chip-based authentication of medicinal plants may be useful as economical, precise tool for quality control and safety monitoring of herbal pharmaceuticals and neutraceuticals. Identification of herbal materials, which commonly consist of dried or processed parts, is difficult. This is particularly true for similar looking herbal materials that can often vary greatly in their medicinal properties and market value.

DNA microarray based technology can provide an efficient, accurate and cheaper means of testing the authenticity of hundreds of samples simultaneously while conventional chemical methodologies usually take several days for verification. Chip-based authentication of medicinal plants can be useful as a tool for quality control and safety monitoring of herbal pharmaceuticals and neutraceuticals and will significantly add to the medical potential and commercial profitability of herbal products.

This application of DNA microarrays will not only benefit the herbal drug industry but can also facilitate the identification of herbal products by regulatory authorities. An international initiative, the Consortium for the Barcode of Life (CBOL) is devoted to developing DNA barcoding as a tool for correct taxonomic identification of species, including medicinal plants. DNA microarrays can provide a suitable technology platform for such an initiative.

THE PHARMACEUTICAL INDUSTRY

PERSPECTIVE

Traditionally, the pharmaceutical industry has focused on modifying one gene or protein at a time as the basis for identifying new drugs. The quest for a molecule to modify a single key factor in a disease process may not provide a solution for many complex diseases where multiple cell types and/or multiple pathways contribute to disease. Herbal extracts or mixtures represent combinatorial chemistry of nature with vast repertoire of chemical entities that have a complex effect on numerous cellular components and functions. They have great potential in the multi-target approach to diseases. However, potential of herbal drugs as defined therapeutic agents is undermined by the difficulty in standardization, pharmacodynamics and pharmacokinetics of these multicomponent mixtures. Microarray analysis of gene expression can be useful for elucidating the molecular mechanisms and networks underlying the complex pharmacological function of herbal extracts and mixtures.

Studying the patterns of gene expression at many stages during the treatment process can reveal mechanism from a modern genetic perspective and will help to identify biomarkers of adverse or favorable response. A positive correlation between the transcriptional response induced by an herbal drug and a database profile of an existing therapeutic agent can provide insight into target specificity, mechanism of action, as well as facilitate analysis of pathways downstream of the target. It will also help in identification of novel therapeutic applications of a herbal drug. DNA microarrays can also be used for activity-guided fractionation of herbal extracts thereby helping in narrowing down upon the active principle producing the desired effect. Microarrays have applications in the entire drug discovery pipeline to improve selection of biological targets and lead compounds. Correlation of gene expression data of herbal drug candidates with clinical outcome or biomarkers of response in biological system will facilitate selection of best candidate for drug development.

In the area of pharmacogenomics, DNA microarrays will facilitate development of individually optimized drugs based on differential gene expression patterns. Genetic polymorphism studies can be done to classify individuals according to their drug metabolizing capacities or response to disease. SNP screening of patients undergoing treatment with a particular herbal drug and correlating these data with recovery profile would help to understand which individuals or groups are likely to be benefited by the drug. Microarray analysis of gene expression has become a suitable approach for the biological effects of drugs and other chemicals. Gene expression profile screens focused on genes that are relevant to toxicity can be useful for prediction of potential herbal drug side effects during pre-clinical development and toxicology studies. They can also be applied for studying herb–drug interactions and their underlying mechanisms. Medicinal plants are an integral part of TM/CAM such as Ayurvedic, Chinese, Native American and many other traditional systems of medicine. Clinical experiences, observations or documented data becomes a starting point for traditional knowledge based drug development.

Although the clinical existence of traditional medicine spans several generations, it may be hazardous to extrapolate from such uses to current practice, since herbs and or their formulations used may be very different from the commercial herbal products available today. Many questions remain concerning quality, efficacy, safety, bioavailability and mechanisms of action of even the most widely used herbs. These questions can be answered by rigorous application of scientific principles in high quality basic and clinical research. There is a need for scientific validation of safety, efficacy, therapeutic claims, bioavailability and mechanism of action of these medicines that are being used in the population since centuries.

Microarray based approach can facilitate high throughput, fast track evidence-based herbal drug development following Reverse Pharmacology path. Phytomics, a technology platform for characterization of herbal compositions wherein, Herbal Bio Response Arrays (HBR Arrays) are used to determine bioactive constituents and biological activities of an herbal composition has recently been developed and patented. Further, CAM disciplines owing to their holistic nature may be impacting gene expression and signaling pathways. Application of such novel expression analysis tools may help to understand and characterize the genes and signaling pathways involved in CAM modalities. DNA microarrays have the potential for application in different phases of herbal drug discovery and development.

This includes quality control and standardization of the herbal drugs, identification and validation of new targets, the profiling of on-target and off-target effects during the optimization of new therapeutic agents, understanding molecular mechanisms of action, structure–activity relationships and the prediction of side-effects, and the discovery of diagnostic, prognostic, and pharmacodynamic biomarkers. At present the major practical applicability of DNA microarrays remains in SNP mapping, genotyping and pharmacogenetics. In recent years, array-based sequencing (minisequencing/resequencing arrays) that combines target hybridization with enzymatic primer extension reactions has emerged as a powerful means to scan for all possible DNA sequence variations.

Highly multiplexed microarray systems offering potential for mutation analysis and genome-wide SNP mapping have been developed. DNA microarrays are a effective new strategy for global gene expression profiling, but reliable detection of expression changes and understanding the functional relevance of these expression profiles is a challenge.

Other problems in microarray experiments include non-reproducibility, difficulty in statistical analysis and comparability of data from different experiments. Incons-istencies in data may arise due to low specificity of cDNA probes, discrepancy in fold-change calculations, variations in laboratories, technicians, probe labeling, biochemical reactions, scanners, lasers etc. Also the sensitivity and specificity of DNA microarrays depends on the nucleotide composition of the target and probe. Repetitive sequence elements and highly structured nucleic acids can greatly decrease the sensitivity of hybridization based analysis.

In its current stage, a microarray experiment needs to be validated by independent confirmation of results by laboratory-based analysis and in silico comparison with information available in literature and expression databases. Although DNA microarrays have huge potential for pharmacodynamics and toxigenomics applications, these are still in exploratory stage and need validation by other biological experiments.

Bioinformatics and statistical tools have a major role to play in the microarray field, whereby data from multiple experiments can be integrated to answer complex biological questions. Another factor limiting microarray application is the cost of the technology. The challenge today is to provide standardized, sensitive, reproducible microarray platforms, databases and visualization methods for expression profiles that are also affordable to scientists. With the development of new, uniform and more sophisticated experimental designs, data management systems statistical tools and algorithms for data analysis DNA microarrays can be optimally used in herbal drug research. In spite of the huge potential offered by microarray technology, the importance of in vitro biological assays, cell-line studies and in vivo animal studies cannot be ignored. A comprehensive strategy integrating information from diverse scientific experiments and technologies will lead to molecular evidence-based herbal medicine.

DNA MICROARRAYS IN PHARMACODYNAMICS

Herbal products are usually whole herbs, their formulations or extracts consisting of several bioactive compounds. With the increased demand for scientifically validated and standardized herbal product there is a need for better understanding of the molecular mechanisms underlying their biological activity. Although, the physiological actions of many herbal drugs are being studied at the molecular level, it remains unclear what are the targets of individual phytochemical components of herbs and how these molecules contribute to biological activities. Examples of the applications of microarray-based gene expression studies in elucidating molecular mechanism of action of pure compounds, different phytochemical groups and herbal extracts are discussed here. Figure shows microarray based gene expression analysis of multi-component mixtures and the information that can be obtained from such studies.

Purified compounds/specific phytochemical groups The molecular pathways underlying diverse biological activity of the triterpenoid compounds isolated from the tropical medicinal plant Centella asiatica when studied with gene microarrays showed that Centella triterpenes evoke a geneexpression response consistent with their prevailing medical uses in the treatment of connective tissue disorders such as wound healing and microangiopathy. The identification of genes modulated by these compounds provides the basis for a molecular understanding of Centella's bioactivity, and opportunities for the quantitative correlation of this activity with clinical effectiveness at a molecular level.

Similarly, the antiproliferative activity of Coptidis rhizoma, a medicinal herb, and its major component berberine was investigated in human pancreatic cancer cell lines. Gene expression patterns associated with sensitivities to each agent were analysed with oligonucleotide arrays that comprised approximately 11,000 genes.

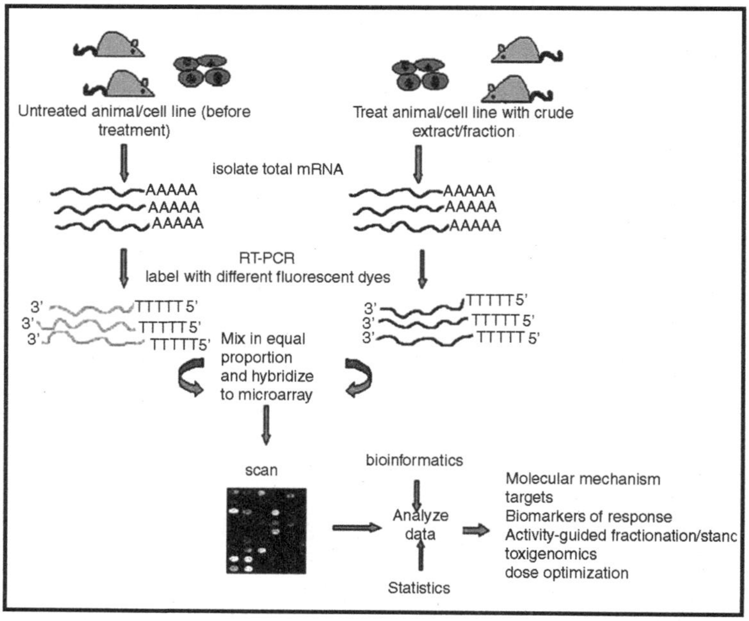

Fig. Microarray-based Gene Expression Analysis of Multi-component Mixtures.

It was possible to identify common and distinct genes related to anti-proliferative activities of purified berberine and C. rhizoma. Involvement of genes in apoptosis inducing activity of alkaloids from Chinese medicinal herb

Tripterygium hypoglaucum (levl.) Hutch (Celastraceae) root was examined using cDNA microarrays containing 3000 human genes derived from a leukocyte cDNA library. Induction of apoptosis by Tripterygium hypoglaucum alkaloids was found to be through c-myc and NF-kappa B signaling pathways. Gene expression pattern of inferior colliculus from DBA/2J mice with audiogenic seizure and those treated with Qingyangshenylycosides, a traditional Chinese medicine, were examined. Gene expression analysis using Agilent oligo microarray showed that total of 134 genes were either up- or down-regulated during audiogenic seizure. Qingyangshenylycosides prevented many of the audiogenic seizure induced gene expression changes. Nevertheless, some of the audiogenic seizure induced genes were further enhanced or reversed by Qingyangshenylycosides treatment.

The data provided important information regarding the molecular mechanisms of audiogenic seizure and the mechanism of action of Qingyangshenylycosides. Oligonucleotide microarrays consisting of 87 probes representing 87 human cardiovascular disease-related genes were used to investigate the molecular mechanisms of saponins from the rhizome of Anemarrhena asphodeloides Bunge. Gene expression patterns in saponin treated and untreated human umbilical vein endothelial cells were analysed. The results suggest that saponins may have beneficial effect on cardiovascular diseases by modulating the function of vein endothelial cells.

A gene chip, (Rat Genome U34A) was used to elucidate the gene regulatory pattern of Epimedium flavonoids in immune homeostasis remodeling in the aged rats. Expression pattern characterized by up-regulation of apoptosis promoting gene expression and down-regulation of apoptosis inhibiting genes expression, is the important gene background of immunohomeostasis imbalance in aged. Epimedium flavonoids reverse these abnormal changes to reconstruct a beneficial equilibrium of gene expression and thus to further remodel the immunohomeostasis in the aged. Mechanism of herbal glycoside recipes retrieving deficient ability of spatial learning memory in mice suffering from cerebral ischemia/repurfusion was studied using DNA micoarray system.

Gene expression pattern was analysed in the groups that showed increased ability of spatial learning. A 1.8-fold increase in expression was observed for many genes including genes in cell cycle regulation, signal transduction, nerve system transcription factors, DNA-binding protein, etc. Nine genes related to retrieving deficient ability of spatial learning memory when treated with glycoside recipes were found in this study. Chronic cocaine use is known to elicit changes in the pattern of gene expression within the brain. The hippocampus plays a critical role in learning and memory and may also play a role in mediating behaviors associated with cocaine abuse. To profile the gene expression response of the hippocampus to chronic cocaine treatment, cDNA hybridization arrays were used to illuminate cocaine-regulated genes in rats treated noncontingently with a binge model of cocaine.

Similarly, using DNA microarray analysis gene expression changes with chronic morphine and antagonist-precipitated withdrawal in two brain regions involved in behavioral effects of morphine in both mice and rats, were characterized. Si-Jun-Zi decoction (SJZD), a traditional Chinese herbal prescription, has been used clinically for treating patients with disorders of the digestive system. Previous studies indicated that the polysaccharides of SJZD are the active components contributing towards its pharmacological effects in improving gastrointestinal function and immunity.

SJZD polysaccharide was found to have protective effect and enhanced re-epithelialization on wounded IEC-6 cells. To elucidate the modulatory effect of polysaccharides of SJZD on wounded IEC-6 cells at the molecular level, an oligonucleotide microarray was employed to study differential gene expression of treated IEC-6 cells and the candidate genes were validated by RT–PCR.

There was increased expression of genes coding for ion channels and transporters, which are critical to cell migration and restoration of wounded intestinal cells, suggesting a possible mechanism for re-epithelialization.

Clinical trials on antimicrotubule agents, docetaxel and estramustine combination treatment have suggested improved antitumor activity in hormone-refractory prostate cancer. However, the molecular mechanisms involved in the combination treatment with docetaxel and estramustine have not been fully elucidated. In order to establish such molecular mechanisms in both hormone insensitive and sensitive prostate cancer cells, gene expression profiles of docetaxel- and estramustinetreated prostate cancer cells were obtained by using Affymetrix Human Genome U133A Array.

Clustering analysis based on biological function showed that docetaxel and estramustine combination treatment down-regulated some genes that are known to regulate cell proliferation, transcription, translation, and oncogenesis. In contrast, docetaxel and estramustine combination treatment up-regulated some genes related to induction of apoptosis, cell cycle arrest, and tumor suppression. Docetaxel and estramustine also showed differential effects on gene expression between mono- and combination treatment.

Combination treatment with docetaxel and estramustine caused alternations of a large number of genes, many of which may contribute to the molecular mechanisms by which docetaxel and estramustine inhibit the growth of prostate cancer cells. These results provide novel molecular targets of docetaxel and estramustine combination treatment in prostate cancer cells.

EXTRACTS

High-density oligonucleotide microarrays have been used for pioneer studies on the multiple gene expression effects exhibited by Ginkgo biloba leaf extract EGb 761, changing traditional pharmacology and medicine concepts. EGb 761 has been known to have neuroprotective effects ranging from molecular and cellular, to animal and human studies, however, the mechanisms remain unclear. DNA microarray based analyses has largely helped in identifying its targets and mechanism of action.

High-density oligonucleotide microarrays were used to define the transcriptional effects in the cortex and hippocampus of mice whose diets were supplemented with the herbal extract.

This study reveals that diets supplemented with Ginkgo biloba extract have notable neuromodulatory effects in vivo and illustrates the utility of genome-wide expression monitoring to investigate the biological actions of complex extracts. Further, DNA microarray analyses revealed that transcription of multiple apoptosisrelated genes is either up- or down-regulated in cells treated with EGb 761. These results suggest that inhibition of apoptotic machinery may, at least in part, mediate multiple neuroprotective effects of EGb 761. Mechanism of cytostatic action of EGb 761 was elucidated with the help of DNA microarray and the genes important for tumor growth were identified. Ginkgo extract and one of its terpenoid constituents, ginkgolide B, inhibited the proliferation of a highly aggressive human breast cancer cell line and xenografts of this cell line in nude mice. cDNA microarray analyses have shown that exposure of human breast cancer cells to a Ginkgo extract altered the expression of genes that are involved in the regulation of cell proliferation, cell differentiation or apoptosis, and that exposure of human bladder cancer cells to a Ginkgo extract produced an adaptive transcriptional response that augments antioxidant status and inhibits DNA damage.

A human genome screen for TNFa-inducible genes was used to identify the genetic basis of the anti-inflammatory effects of standardized Boswellia serrata extract, 5-Loxin in microvascular endothelial cells. Acutely, TNFa-induced 522 genes; of which 113 genes were clearly sensitive to 5-Loxin treatment. Such genes directly related to inflammation, cell adhesion, and proteolysis. The robust 5-Loxin -sensitive candidate genes were then subjected to further processing which led to the recognition of the primary 5-Loxin -sensitive TNFa-inducible pathways. One mechanism by which TNFa, causes inflammation is by potently inducing the expression of adhesion molecule VCAM-1. 5-Loxin was found to completely inhibit VCAM-1 expression. Water-extract of Paeoniae Radix has inhibitory effect on the growth of both HepG2 and Hep3B cell lines. Using cDNA microarray technology and RT–PCR analysis the drug targets of Paeoniae Radix in inhibition of tumor cells growth were elucidated.

Several studies have indicated that extracts of S. barbata have growth inhibitory effects on a number of human cancers. However, the mechanism underlying the antitumor activity was unclear. cDNA microarray analysis showed that 16 genes, involved in DNA damage, cell cycle control, nucleic acid binding and protein phosphorylation, underwent more than 5-fold change. These data indicated that these processes are involved in S. barbata-mediated killing of cancer cells. The effect of Syzygium aromaticum (L.) Merrill and Perry (clove) extract on tissues that regulate glucose metabolism was studied by analysis of gene expression using DNA microarray. The data showed that the extract acts like insulin in hepatocytes and hepatoma cells by reducing phosphoenolpyruvate carboxykinase and glucose 6-phosphatase gene expression. Much like insulin, clove-mediated repression was reversed by PI3K inhibitors and N-acetylcysteine.

Moreover, clove and insulin were found to regulate the expression of many of the same genes in a similar manner. These results indicate a potential role for compounds derived from clove as insulin-mimetic agents. A cDNA microarray study demonstrated significant inhibition of inducible endothelial CD36 expression, a novel cardioregulatory gene, by IH636 grape seed proanthocyanidin extract.

Both the prototypic tricyclic antidepressant imipramine (IMI) and the extract of St John's wort (SJW) can be effective in the treatment of major depressive disorder. Affymetrix chips were used to study hypothalamic gene expression in rats treated with SJW or IMI to test the hypothesis that chronic antidepressant treatment by various classes of drugs results in shared patterns of gene expression that may underlie their therapeutic effects. SJW treatment differentially regulated 66 genes and expression sequence tags (ESTs) and IMI treatment differentially regulated 74 genes and ESTs. Six common transcripts in response to both treatments were found. Both treatments also affected different genes that are part of the same cell function processes, such as glycolytic pathways and synaptic function. The data support the hypothesis that chronic antidepressant treatment by drugs of various classes may result in a common, final pathway of changes in gene expression in a discrete brain region.

Recently, a rapid throughput screening system involving gene expression analysis was developed in order to investigate the potential of bioactive chemicals contained in natural health products as effective drug therapy, in particular the ability of these chemicals to alleviate the inflammatory response in human airway epithelial cells. The pharmacogenomic activities of Anoectochilus formosanus extract as a crude phytocompound mixture were compared to those conferred by the single-compound drug, plumbagin in MCF-7 cancer cells. This study offers evidence to support the search for fractionated medicinal herb extracts or phytocompound mixtures, in addition to single-compound drugs, as defined therapeutic agents. DNA microarray can be used for activity-guided fractionation of herbal extracts in order to determine active principles.

5

Absorption Mechanisms of Drug

OBJECTIVES OF DRUG ABSORPTION

Absorption may be defined as the process by which a compound penetrates one or more biological membranes to gain entry into the body. Absorption is not to be confused with bioavailability, which describes entry of administered compounds into the systemic circulation.

For some drugs and dosage routes, absorption and bioavailability may be identical, *i.e.*, after intravenous (IV) dosing. However, in many cases they are not. For a drug that does not undergo any metabolic transformation between an immediate postabsorption site and entry into the systemic circulation, absorption and bioavailability are likely to be the same.

All of the absorbed drug enters the systemic circulation. This is regardless of any drug that may be degraded or changed in some other way, *i.e.*, preabsorption. On the other hand, for any drug that is degraded at a point between the postabsorption site and entry into the systemic circulation, the systemic availability— bioavailability—will be less than the absorption.

An orally administered drug that undergoes extensive firstpass hepatic clearance may give rise to poor oral bioavailability despite being efficiently absorbed from the gastrointestinal (GI) tract into the splanchnic circulation.

The pharmacologic activity profile of a systemically active drug is a function of its intrinsic activity and of the concentration profile that is achieved in the circulation. The speed of onset of action and the intensity and duration of activity are functions of the drug concentration profile.

The speed of onset of drug action is determined by the rate of drug absorption. Extreme cases are the use of bolus IV injection, which yields immediate and usually maximal pharmacologic effect, and slow controlled release, not necessarily by the oral route, where the onset of action is deliberately prolonged to achieve a desired therapeutic profile.

The intensity of pharmacologic effect is generally a function of the concentration of drug achieved in the circulation. Actual pharmacokinetic/ pharmacodynamic relationships are often complex, but it is reasonable to generalize that higher circulating drug concentrations yield greater effect.

The levels of circulating drug that are achieved are a function of dose, absorption efficiency, overall bioavailability, distribution, and also clearance. The major determinant of drug distribution volume is its lipophilicity. As lipophilicity increases, so does the ability of the drug to cross biological membranes and move into extravascular environments, particularly into fatty tissue and the central nervous system (CNS).

Many drugs bind to plasma proteins, in particular to plasma albumin. Although binding of drugs to plasma proteins is dynamic and reversible, any drug that is bound at a particular time is necessarily confined to the plasma volume and thus cannot participate in extravascular distribution.

The last factor affecting circulating drug levels is clearance. The faster a drug is cleared from the circulation as a result of metabolism or any other process (*i.e.*, the shorter its elimination half-life) the lower are its circulating levels. High circulating levels are less likely to be achieved with a high clearance drug than with a low clearance drug, and accumulation of a high clearance drug in the circulation with repeated dosing is unlikely.

Thus, the phenomenon of drug absorption is only one, albeit an important one, of several factors that determine a drug profile in the circulation. It is important to understand all of these factors before drug profiles, and in particular pharmacokinetic/pharmacodynamic relationships, can be fully characterized, particularly in a predictive sense.

All of the above factors are functions of the physical and chemical properties of a drug. While distribution and clearance are affected only by drug properties and cannot generally be altered except by introducing some kind of interaction, drug absorption and bioavailability are often markedly influenced by route of administration, dosage form, and coadministration of other substances. Some of the major thrusts of pharmaceutical research during the last two decades have been devoted to these latter issues.

DESICCANT FOR STABILIZING MOISTURE-SENSITIVE DRUG

Colloidal silica has been successfully used by Gore and Banker to improve the chemical stability of aspirin, whose hydrolytic tendency is well-known. Colloidal silicas have a large surface area and a highly polar silanol surface, which imparts a high moisture adsorption capacity. Water adsorbs onto the surface of colloidal silicas by forming hydrogen bonds with silanol. Colloidal silicas are commonly used as desiccants in the pharmaceutical industry, especially for hygroscopic or hydrolabile drugs. Even though the desiccant effect of colloidal silica is not the only factor that helps to stabilize aspirin in the studied formulation, its strong moisture adsorption capacity has played a very important role.

Dry Powder Inhalation

The surface adsorption approach has been used to make a DPI device. First, a fatty acid or fatty alcohol derivative or a poloxamer is dissolved or dispersed in a solvent in which drugs and carriers are insoluble. The preferred solvent is n-hexane or cyclohexane. Second, the fatty acid solution or suspension is adsorbed onto the surface of soluble micronized drugs and/or acceptable carriers such as lactose monohydrate, anhydrous glucose, polylactides, or PVA. The surface modification of drugs via adsorption has several advantages in making DPI formulations, such as the reduction of electrostatic charge during processing and handling, reduction of adhesion on contact surfaces, improvement of powder flowability in pneumatic transport, and improvement of drug content uniformity. This approach has made the scaling up of DPI formulations easier, and has significantly improved the inhalation properties of powders.

Sublingual Nitroglycerin Tablets

Waaler *et al.* have successfully applied the adsorption approach to make sublingual nitroglycerin tablets by direct compression with better content uniformity, lower weight variation, and much higher stability than tablets made from a molding approach.

In the molding approach, nitroglycerin will migrate to the surface of a tablet during the drying process. During manufacturing as well as storage, there may be potency loss due to the volatility of nitroglycerin. Tablets can also be made by wet granulation, where drying may also induce unblending. In the direct compression approach, nitroglycerin dissolved in alcohol was first added to MCC, and triturated slowly for about 5 min to assure content uniformity. The powder is left at room temperature for 12 hr with occasional stirring. Then alcohol is evaporated and nitroglycerin becomes adsorbed onto MCC. Afterwards, MCC/nitroglycerin is mixed with other excipients to achieve suitable properties for the finished tablets, including solubility and disintegration rate.

Surface Area Measurement

Besides liquid adsorption on solid surfaces, gas adsorption on solid surfaces is also very useful in measuring surface area in the pharmaceutical industry. Based on the Brunauer–Emmet–Teller (BET) equation, Vertommen, Rombaut, and Kinget used krypton adsorption to measure the specific surface area of pellets made from wet granulation, extrusion, and spheronization. Westermarck *et al.* noticed that nitrogen adsorption was more sensitive to changes in the surface area of mannitol tablets caused by compression. In measuring pore size, nitrogen adsorption can measure smaller pores (diameter range 3–200 nm) than high-pressure mercury porosimetry, with a range of 7 nm–14 *m*m.

It is the large number of small pores with diameters <200 nm that contribute to the increased compactibility of granules and, ultimately, greater tablet strength. The surface morphology of five commonly used excipients has been studied using both particle size distribution and nitrogen adsorption methods. The five tableting excipients were unmilled dicalcium phosphate dihydrate (Di-Tab), MCC (Avicel PH102), corn starch, croscarmellose sodium (Ac-di-sol), and sodium starch glycolate (Primojel).

A surface irregularity index (SII) was established to indicate surface roughness due to porosity, and its value was consistent with the direct microscopic observation of a powder sample. The nitrogen adsorption–desorption isotherms showed certain hystereses, especially for Avicel PH102, Di-Tab, and Ac-di-sol, and the remaining nitrogen adsorbed on the surface during desorption process was believed to be due to pores on the powder surface.

PARENTERAL ROUTES OF DRUG

Parenteral delivery routes are those that do not give rise to drug absorption into the splanchnic circulation. Thus, they avoid the possibility of hepatic first-pass metabolism. It should be noted that some parenteral routes do not avoid other first-pass metabolism effects (*e.g.*, pleural metabolism for some inhaled drugs). Some major parenteral drug delivery routes are intraarterial, intrathecal, intravenous, intramuscular, transdermal, intranasal, buccal, inhalation, intraperitoneal, vaginal, and rectal.

INTRA-ARTERIAL INJECTION

Intra-arterial injection is used to deliver drugs directly to organs, for example, in cancer chemotherapy, and in the use of vasopressin for GI bleeding. Intra-arterial carmustine is effective to treat brain tumors and pelvic intra-arterial actinomycin D is used for malignant trophoblastic disease. Intra-arterial drug administration has potential safety implications. Embolization, arterial occlusion, and localized drug toxicity have been reported.

Intrathecal

Injection directly into the cerebrospinal fluid (CFS) ensures complete CNS bioavailability for drugs that cannot cross the blood-brain barrier. This dosage route is used to treat serious CNS infections such as meningitis and ventriculitis, and with such agents as mepivacaine and prilocaine for spinal anesthesia. Drugs injected intrathecally initially distribute into approximately 140 ml of CSF, thus producing high concentrations in the CNS with low risk of systemic toxicity.

Intravenous (IV)

IV administration introduces drug directly into the venous circulation. The shape of the resulting circulating drug profile is determined by the size, rate, and duration of injection. IV bolus is used for immediate therapeutic effect, typically for general anesthesia and for treatment of cardiac arrhythmia. IV dosing is popular for preclinical testing of compounds during drug development and also as a standard to determine absolute bioavailability from other dosage routes.

Intramuscular (IM)

Following intramuscular (IM) administration, drugs must cross one or more biological membranes in order to enter the systemic circulation. Intramuscular injection is used mainly for drugs and vaccines that are not absorbed orally, for example, aminoglycosides, insulin, and hepatitis vaccine. The IM route is often used for sustained medication and specialized vehicles, such as aqueous suspensions, oily vehicles, complexes and microencapsulation, which has been developed for slow delivery of drugs by this route.

Transdermal

Since the introduction of transdermal scopolamine, many transdermal delivery systems have been developed for systemic activity. Major advantages claimed for this drug delivery route include continuous release of drug over a specified period, low presystemic clearance, facile drug withdrawal by simply removing the device, and good patient convenience and compliance. Some disadvantages relate to barrier properties of the skin, skin reactions, and the relatively large dose size. Transdermal delivery is a realistic option only for drugs generally given in small doses (<10mg) and which have good membrane penetration. Drugs currently approved for transdermal delivery include clonidine, estradiol, nicotine, nitroglycerin, and scopolamine.

Intranasal

Intranasal administration may be used for local or systemic effects. Local effects include treatment of nasal allergies, rhinitis, and nasal congestion. Nasal delivery for systemic effects is established for a small number of drugs and is being examined for many others. The sophisticated structure and specialized function of airways and membranes in the nasal cavity, and also the small surface area of this region, may limit its capacity for drug delivery. The effect of chronic drug exposure on the integrity of nasal membranes must also be considered. This problem may be compounded by the evident need for surfactants to achieve good systemic penetration with this dosage route.

Notwithstanding these factors, the physical characteristics of compounds for optimal intranasal absorption are the same as for other absorption routes. The drug must dissolve in the fluids of the nasal mucosa and must be sufficiently lipophilic to cross the membranes of the nasal epithelium. Nasal absorption is facilitated by the high permeability of small venules and capillaries associated with the nasal mucusa. A variety of delivery systems have been described for nasal drug delivery, including drops, aerosols, nebulizers, and soluble matrices. Small peptide molecules seem to be ideally suited for intranasal drug delivery. Vasopressin analogues and oxytocin are commercially available for intranasal dosage. Thyrotropin-releasing hormone agonists and antagonists, other vasopressin analogs, and peptides are being examined. Intranasal delivery of sex hormones has produced interesting results in animals. Intranasal delivery of insulin has been examined, but only with moderate success.

Buccal

Early recognition of buccal and sublingual absorption was manifested in the use of nitroglycerin by these dosage routes to treat severe headache and to relieve angina pectoris. Drugs can be absorbed from the oral cavity itself or sublingually. Absorption from either route is rapid, sublingual more so apparently because of greater permeability of sublingual membranes and rich blood supply. The mean pH of saliva is approximately 6 so that drug absorption, predominantly passive in nature, is favoured for unchanged molecules, acids with pKa values >3, and bases with pKa values <9. Compounds that are currently marketed or are being considered for buccal or sublingual routes include organic nitrates, barbiturates, papaverine, trypsin, prochlorperazine, benzodiazepines, buprenorphine, captopril, isoprenalin, oxytocin, and nifedipine. Oxytocin is currently the only peptide marketed in sublingual form. Sublingual steroids have been examined with moderate success.

Inhalation

When a substance is inhaled, it is exposed to membranes of the mouth or nose, pharynx, trachea, bronchi, bronchioles, alveolar sacs, and alveoli. The lung has a potential absorption surface of some 70m2, a much larger surface than the small intestine.

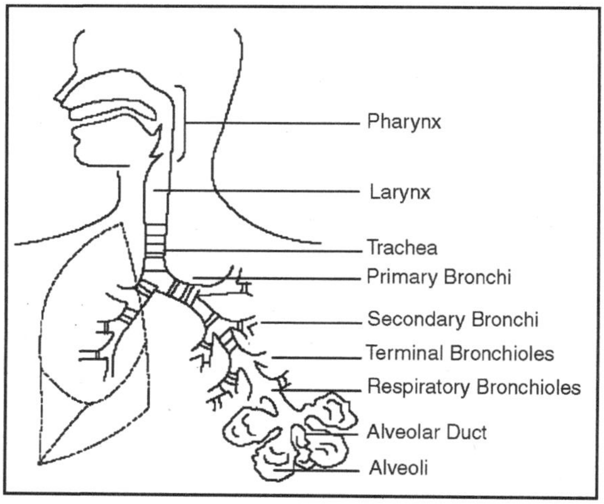

Fig. The Human Respiratory Tract.

However, the lungs and their associated airways are designed to deny access of administered compounds to the highly absorptive peripheral lung surfaces. The system is designed to deny access to particulate matter. However, if compounds can reach the peripheral region of the lung, absorption can be very efficient. Particle (droplet) size and velocity of application control the extent to which inhaled substances penetrate into airway spaces. Optimum size for deep airway penetration is 3–5 *m*M. Large particles tend to deposit in upper airways.

Most inhalation devices deliver approximately 10 per cent of the administered dose to the lower respiratory tract. A number of devices have been developed to increase lung delivery, and delivery of up to 21 per cent has been reported with a pressurized metered-dose inhaler. Despite these advances, drug delivery via the lung is still inefficient. Systemic availability of inhaled drugs may be inhibited by first-pass pulmonary metabolism. The lungs contain many drug metabolizing enzymes, including mixed function oxidases, monoamine oxidase, and esterases. Several animal models principally the rat, rabbit, and dog, are used to study drug inhalation.

Intraperitoneal

Intraperitoneal drug administration is not common. It is used pedominantly to administer compounds during preclinical discovery and development. Its clinical use is generally limited to chemotherapy for tumors with peritoneal involvement. Peritonitis occurs frequently in renally impaired patients who are receiving continuous ambulatory peritoneal dialysis (CAPD). Peritonitis is often accompanied by systemic infection so that therapeutic levels of antibiotic are needed both in the peritoneal cavity and in the systemic circulation. Drugs may be given orally or by injection in order to achieve adequate systemic levels in the hope of also achieving therapeutic levels in the peritoneal cavity. Alternatively, drugs may be administered directly into the peritoneal cavity with the objective of also achieving systemic levels by intraperitoneal absorption. To the author's knowledge, there is little definitive information on the relative merits of these alternatives.

Vaginal

The human vagina, a fibromuscular tube 10–15 cm long, extends upwards and backwards from the vulva to the lower uterine cervix. Blood is supplied to the vagina via the uterine and pudendal arteries, and is drained from the vagina by a rich plexus, which flows into the internal iliac veins. The surface of the vaginal epithelium is kept moist by cervical secretions. The pH of vaginal fluid is 4–5. Vaginal drug delivery is used mostly for local effects, but vaginal absorption can give rise to rapid and efficient systemic delivery. Good systemic absorption, and also the ability of the vagina to retain delivery devices, has given rise to many vaginal dosage forms, in particular for steroid contraceptives.Alarge number of vaginal controlled release dosage forms are available, including vaginal rings and biodegradable microspheres. Rectal

The human rectum is 15–20 cm long. It is normally empty and contains 2–3 ml of mucous fluid with pH 7–8. There are no villi and only a limited surface area of 200–400 cm2 is available for absorption. Blood and lymph vessels are abundant in the rectal submucosa. Veins from the upper rectum drain into the portal circulation, while veins from the middle and lower rectum drain directly into the inferior vena cava.

However, there are extensive anastomoses among these veins so that precise anatomical differentiation is difficult. It appears that compounds absorbed from the lower rectum, in contrast to those absorbed from the upper rectum, avoid hepatic first-pass metabolism. Rectal absorption is generally slower than oral absorption, but for some drugs, rectal absorption exceeds oral absorption presumably due to avoidance of first-pass metabolism after rectal delivery. This has been reported for morphine, metoclopramide, ergotamine, lidocaine, and propranolol. Human rectal systemic availability of the extensively metabolized drug lidocaine is 65 per cent as compared to 30 per cent after oral administration.

Rectal absorption of drugs from aqueous or alcoholic solutions is generally much faster than from suppositories. Non-surfactant adjuvants, such as salicylates, increase rectal absorption of water-soluble drugs and also of high molecular weight compounds, such as insulin, heparin, and gastrin.

ENTERAL ROUTES

Enteral routes of drug absorption are from the stomach and the small and large intestine. Substances absorbed from these areas enter the splanchnic circulation and pass through the liver before entering the systemic circulation.

The GI tract is the site of absorption for most nutrients. Thus, the GI tract has evolved to facilitate absorption of substances. The peristaltic action of the stomach, secretion of enzymes and hydrochloric acid, the villi and microvilli of the intestine, as well as the rich blood supply and lymphatics in this region, all facilitate absorption.

Enteral absorption is generally by far the most effective drug delivery route and, whenever possible, drugs are administered in this way. Any orally absorbed compound is exposed to an absorption environment that is both friendly and hostile, depending on the compound and the patient. A brief review of the structure and physiology of the GI tract relative to drug absorption follows. Acidity is a consequence of hydrochloric acid secretion by the parietal cells of the stomach. This plays an important role in food digestion by facilitating conversion of pepsinogens and zymogens to active proteolytic enzymes. The acidic environment in the stomach tends to favour gastric absorption of acidic drugs provided the drugs are in solution. On the other hand, basic drugs tend to dissolve readily in the stomach but absorption may be prevented because the drug will be ionized and therefore not sufficiently fat soluble for efficient membrane penetration. The acidic environment in the stomach may give rise to reduced drug absorption due to acid-catalyzed degradation.

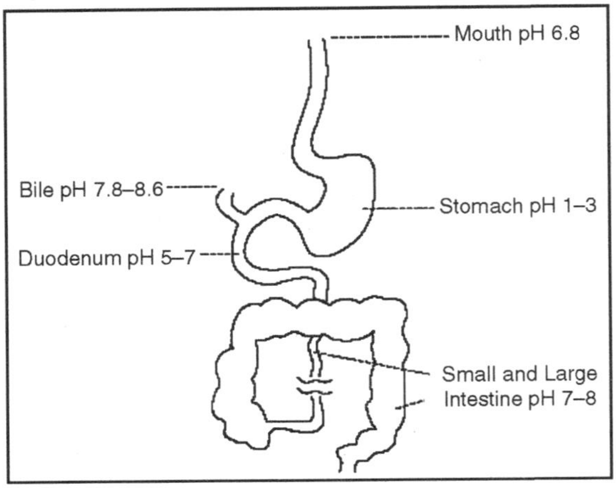

Fig. Approximate pH Values in the Human Gastrointestinal Tract.

If a drug dissolves in the stomach or is a liquid, and if it is fat-soluble and acid stable, then it is likely to be absorbed efficiently from the stomach. Ethyl alcohol is a liquid that is completely miscible with water and sufficiently lipophilic to cross biological membranes. It is efficiently absorbed from the stomach. After passing through the pyloric sphincter, a compound reaches the duodenum, jejunum, and ileum. These regions of the small intestine differ from the stomach with respect to pH, the presence of digestive enzymes, and the absorptive surface area. Excretion of alkaline bile into the duodenum raises the pH of the duodenal and more distal intestinal contents to 5–7.

The change in pH from acidic to essentially neutral causes many changes that affect drug absorption. Enteric coatings that were impermeable in the stomach will dissolve. Acidic drugs will dissolve more rapidly and yet the pH will not be sufficiently high to prevent dissolution or cause precipitation of weakly basic drugs.

GI Structure and Motility

The stomach is a pouch-like organ lined with a relatively smooth epithelial surface. Although compounds can be absorbed from the stomach, the contribution of this organ to overall enteral drug absorption is modest. The absorptive properties of the proximal small intestine are superior to those of the stomach or any other region of the GI tract. The rate at which compounds pass from the stomach into the small intestine is a rate-limiting step controlling drug absorption. Stomach motility is complex and is influenced by nervous and hormonal stimuli. Stomach emptying rate is a function of rhythmic contractions that have a frequency of approximately three per minute in a fasted person, and less when food enters the stomach.

Food passes from the stomach into the duodenum as a result of these rhythmic contractions. The heavier the meal and the higher the fat content, the longer it will take for a meal, and any drug that may be ingested with it, to pass into the small intestine. This process acts as a defence mechanism by which substances are prevented from entering the proximal small intestine and injuring the delicate absorptive surface of this region until they are reduced to a suitable consistency in the stomach. Whereas solid food delays stomach emptying, liquids tend to accelerate the process. Acceleration results from activation of stretch receptors in the stomach wall. When the fluid is water, activation of the inhibitory receptors is stopped. This results in rapid emptying of stomach contents into the duodenum.

By far the most important difference between the proximal small intestine and the stomach is the nature ofthe mucosal surface of the epithelium. The mucosal surface of the small intestine is increased by fingerlikeprojections, or villi, that arise from the folds of Kerckring, and in turn by microvilli that arise from the villi. These invaginations increase the surface area of the intestinal mucosa some 600-fold to approximately 200m2, and 1.0–1.5 L of blood passes through intestinal capillaries each minute. Corresponding values for the stomach are only 100m2 of surface area and a blood flow rate of 150 ml/min.

Thus, the small intestine has a surface areaapproximately double that of the stomach and a blood perfusion rate 6–10 times faster. Both factors strongly favour more efficient absorption from the small intestine. The villi and microvilli of the small intestine are lined by a sulphated mucoprotein, glycocalyx. Fluid trapped within the glycocalyx is stationary, and a series of thin layers, each progressively more stirred, extends to the bulk phase of the intestinal lumen. This series of unstirred layers has an effective thickness of 0.01–1.0 mm. Molecules move within the unstirred layers by diffusion at a rate inversely proportional to the square root of molecular weight below 450, and inversely proportional to the cube root of molecular weight above 450. The glycocalyx is negatively charged, with counterions in the unstirred layer. If a substantial proportion of these cations is composed of hydrogen ions, as is often the case, then the microclimate within the brush border of the epithelium is likely to be acidic relative to the bulk phase. This may influence drug ionization at the membrane surface, and provides a basis for the "acid microclimate" frequently associated with the GI mucosa.

The length of time during which material stays in the small intestine is approximately 5 min in the duodenum, 2 h in the jejunum, and 3–6 h in the ileum. Material then enters the large intestine. The large intestine does not have villi or microvilli at its mucosal surface. Its contents are neutral or alkaline. Therefore, absorption of drugs from the large intestine is less efficient than from the small intestine. The large intestine and colon contain an active bacterial microflora that can degrade foreign molecules that also tends to reduce absorption of drugs from this region of the GI tract.

GI Secretions

The rate of acid secretion into the stomach is controlled by parietal cells. Acetylcholine, histamine, and gastrin are important for regulation of hydrochloric acid secretion, and they act directly on parietal cells to enhance acid secretion rate.

Phases of Gastric Acid Secretion

The basal rate of hydrochloric acid secretion varies diurnally, being highest in the evening and lowest in the morning. After ingestion of a meal, the rate of acid secretion in the stomach increases. The three phases of increased acid secretion in response to food are the cephalic phase (before food reaches the stomach), the gastric phase (elicited by the presence of food in the stomach), and the intestinal phase (elicited by input from the duodenum and upper jejunum).

Cephalic Phase

The sight, smell, and taste of food elicit this phase. Acid secretion during this phase can be as much as 40 per cent of the maximum rate. Other stimuli sensed in the brain, in addition to those related to the presence of food, may evoke acid secretion through vagal impulses.

Gastric Phase

The presence of food in the stomach evokes gastric secretion. The principal stimuli include distension of the stomach and the presence of amino acids and peptides. Distension of the stomach stimulates mechanoreceptors that bring about secretion of acetylcholine, hydrochloric acid, and gastrin.

Intestinal Phase

The presence of chyme in the duodenum stimulates neuronal and endocrine responses that stimulate and later inhibit secretion of acid into the stomach. The stimulatory influences dominate when the pH of gastric chyme is above 3. However, when the buffer capacity is exhausted and the pH falls below 2, inhibitory influences dominate. Gastric acid secretion may be regulated by several brain peptides, some of which may enhance secretion, while others may act centrally to inhibit secretion.

Gastric Juice

Gastric juice contains a mixture of secretions from surface epithelial cells and gastric glands. Salts, water, pepsins, intrinsic factor, and mucus are main components of gastric juice. Gastric secretions increase after a meal. Ionic composition of gastric juice is related to the rate of secretion.

The higher the secretory rate, the higher the hydrogen ion concentration.. The rate of gastric acid secretion varies among individuals. In humans, the basal rate is 1–5mEq/h. With histamine or pentagastrin stimulation, acid output rises to 5–40mEq/h.

Other Secretions

Bile, pH 7.8–8.6, is produced continuously in humans. Hepatic bile is concentrated and stored in the gall bladder between meals. It is ejected from the gall bladder and flows into the duodenum when food enters the intestine. The main constituents of bile are bile salts, bilirubin, end products of hemoglobin breakdown, the electrolytes sodium, chloride, and bicarbonate, cholesterol, phospholipids, and lecithin. The gall bladder contracts within 30 min after eating due to liberation of cholecystokinin. The most effective stimulus to this is food high in fat. Bile salts, which are surface active, promote dissolution of lipophilic drugs and lipophilic drug formulations, enteric coatings, and waxy drug matrices. Bile salts may also promote membrane permeability of lipophilic molecules through micelle formation and solubilization. Pancreatic juice contains an alkaline fluid and enzymes, both of which empty into the duodenum. The alkaline pH contributes to neutralization of the acid that empties from the stomach. The enzymes amylase, lipase, trypsin, and chymotrypsin play major roles in the digestion of carbohydrates, fats, and proteins. Trypsin and chymotrypsin are secreted as inactive precursors and are converted to the active forms enzymatically.

As a result of proteolytic enzyme secretion into the duodenum, protein or peptide drugs, such as corticotropin, vasopressin, and insulin, are rapidly degraded and generally cannot be given orally. Secretory activity of the pancreas is under hormonal and neuronal control. Intestinal secretions do not exist in the same sense as gastric, pancreatic, or biliary secretions. Nonetheless, large fluid fluxes take place throughout the intestine. Any secretions from the intestinal mucosa appear to have a lubricant and protective effect.

GI Blood Flow in Relation to Drug Absorption

Drugs may be transported away from the serosal side of the GI tract by one or both of two mechanisms. The GI tract is supplied by a blood capillary network from the splanchnic circulation. Drugs may also be taken up by the lymph vessels in the GI epithelium and carried by the lymphatic system that drains the abdominal area into the thoracic duct.

Any drug that is absorbed via this system enters the systemic circulation directly and is not susceptible to first-pass hepatic metabolism. Despite the presence of both the blood capillary network and the lymphatic system, absorption of the great majority of drugs appears to occur predominantly via the capillary system associated with the splanchnic circulation. The reason for this appears to lie in the relative flow rates of blood and lymph. The rate of blood flow in the splanchnic circulation is 1.0–1.5 L/min, or 30 per cent of cardiac output. This rate may increase to 2 L/min after a meal. Lymph flow through the same region is only 1–2 ml/min, but may increase to 5–20 ml/min after a meal. Lymph flow in this region is thus 500– 700 times slower than blood flow. Relatively fast splanchnic blood flow establishes virtual sink conditions on the serosal side of the GI epithelium and ensures a steep concentration gradient. These conditions promote efficient absorption into the bloodstream rather than into lymph. Only a small number of drugs are absorbed via the lymph system. These include drugs with high molecular weights that cannot enter the capillaries and specific molecules such as steroids.

Hepatic First-Pass Metabolism

The majority of compounds absorbed from the stomach and intestines enter the splanchnic circulation. This leads to the portal vein, the liver, and then to the general circulation. Compounds absorbed via this route must therefore pass through the liver and will do so initially at a higher concentration relative to when they eventually distribute into the general circulation and elsewhere. As hepatic metabolism is generally first order in nature, a large proportion of any orally administered drug that is highly and efficiently metabolized in the liver will be cleared during the initial first pass. A drug could be efficiently absorbed from the GI tract and yet poorly available to the general circulation as a consequence of first-pass hepatic clearance. Such high extraction drugs include acebutolol, alprenolol, desipramine, isoproterenol, and lidocaine.

ABSORPTION MECHANISMS

An orally administered drug must pass through a number of membranes in order to be absorbed into the systemic circulation. Many physiological membranes differ in structure and function. Despite this, there is general consensus regarding the basic structure of the cell membrane.

DOSAGE ROUTES AND DRUG ABSORPTION

A required drug absorption profile is achieved by a variety of dosage routes. These routes may be divided into parenteral and enteral. Due to of the importance of the various routes of administration in drug delivery, and of recent advances in optimizing routedependent drug delivery, they are briefly reviewed here.

FACTORS MODIFYING DIGESTIVE ABSORPTION AND BIOAVAILABILITY

Form or Presentation of the Drug

The digestive absorption of the same molecule, taken at the same dose, by the same patient, but under a different presentation, drops or tablets for example, will not have necessarily the same kinetics of absorption, nor the same bioavailability. Drugs supplied in the form of drops are immediately available for absorption whereas tablets must be first disintegrated so as to release the powder which is emulsified. In general, drops give a more rapid and higher Cmax than tablets, the bioavailability being identical or not. This difference between drops and tablets explains the greater severity of poisonings due to the intake of drops than the intake of tablets which can stagnate a long time in the digestive tract before being absorbed.

There are preparations known as sustained-release which gradually release the active molecule into the digestive tract, which delays and extends its absorption. This artifice makes it possible to reduce the frequency of administration compared to that which one could calculate according to the real elimination half-life. The persistence of digestive absorption avoids the fast decrease of the plasma concentration.

Foods

When should a drug be taken: under fasting conditions, before, during, after meals? There is no simple response to this question. It is necessary to consider pharmacokinetic and pharmacodynamic parameters.

The bioavailability of certain drugs when they are taken during meals can be reduced, unchanged or increased:

- reduced: it is the case of tetracyclines, isoniazid, penicillamine, captopril.
- unchanged or little modified: it is the case of amoxicillin.
- increased: it is the case of propranolol; the increase in its bioavailability comes from a decrease of first-pass intestinal and hepatic metabolism. It is the case of griseofulvine when it is taken with lipids, but the responsible mechanism is not well understood.

The desired effect or the tolerance of a drug can depend on the moment of its administration compared to meals:

- An hypoglycemic drug is generally given before meals to compensate for the hyperglycemia caused by food.
- A gastric protectant is taken apart from meals and in the evening at bedtime for a better adhesion to the gastric mucous membrane.
- Nonsteroidal antiinflammatory drugs are taken during meals to reduce the gastric irritation which they elicit.

The responses to these questions are generally found in the user information sheet accompanying each drug and in the "summary of product characteristics" of each drug.

Digestive Transit

Any modification of the transit of the bolus, either of pathological origin (vomiting, diarrhea, etc), or drug-induced (acceleration or slowing of transit), is likely to modify the kinetics of absorption and the bioavailability. Atropine slows down the transit and the absorption; metoclopramide and cisapride accelerate it.

Drug interaction in the digestive tract

In addition to the modifications of transit elicited by certain drugs, there are also direct chemical interactions between drugs:

- Cholestyramine is an anion exchanging resin, able to bind drugs, such as diuretics, and reduce their absorption.
- Metals such as iron and aluminum can reduce the bioavailability of certain antibiotics, like tetracyclines, with which they form organometallic complexes.
- Activated charcoal has been known for a long time for its capacity to adsorb a great number of molecules, in particular drugs. It is used, by oral or gastric administration, to reduce or prevent the absorption of toxic products or drugs taken in excess by oral route. Preparations of activated charcoal have a large area of adsorption which 2000 m2 by gram. The activated charcoal must be given as soon as possible after ingestion of the supposed poison.

Preparations available for oral route

Knowledge of the various available preparations of the drugs is necessary to physicians. Indeed, if their prescription contains an error, for example "capsule" instead of "tablet", the pharmacist who dispenses the drug and the patient who takes it can wonder about the error, if it concerns the presentation or the drug itself.

The principal available preparations containing powder (active substance+ excipient) are as follows:

- Packages and sachets: containing generally about 10 to 20g of product
- Capsules: made of two often coloured encasable cylindrical parts of a mixture of gelatin and carboxymethyl cellulose, containing the active product. They can be made resistant to gastric acidity.
- Tablets: obtained by compression of the powder. There have various forms, and can be effervescent.
- Sugar-coated tablets: covered by a layer of sugar often aromatized and coloured.

These "dry" forms (, capsules, tablets...) must be taken with a glass of water in a sitting or upright position, but not lying, to facilitate their esophageal transit and to prevent their binding to the esophageal wall which they could damage with possible severe ulcerations.

The available preparations containing fluid are as follows:

- Drinkable vials in coloured glass to distinguish them from the injectable vials in transparent glass except if the active product must be protected from light.
- Aqueous or alcoholic solutions, in bottle with a graduated dropper or mesurette.
- Syrups: concentrated aqueous solution of sucrose containing the active products. The syrups are administered by spoonfuls and preferably by a spoon/dose provided with each bottle.
- Suspensions: the granules contained in a bottle are dispersed in a given quantity of water before use.
- Hydrosols are pseudo-solutions containing water-soluble and liposoluble molecules.

Preparations for rectal route

They are suppositories. One of disadvantages of suppositories is the variability of the resorption of the active product.

Pulmonary route

It is a route of fast absorption avoiding the liver: drugs absorbed by the lungs, reach the left atrium and the left ventricle, and then the general circulation. The pulmonary route is used:

- for local treatment: bronchial indications but with possibility of partial absorption and general effects. The forms used are aerosols conveying drugs such as antibiotics, mucolytic, beta-adrenergic mimetic, muscarinic receptor antagonists. A lot of devices like pressurized atomizers, aerosol-batchers or sprays, inhalers of dry powder are used to introduce drugs into bronchial airways.
- for general treatment: medical gases and general anesthetics by inhalation. Oxygen, nitric oxide and nitrous oxide which are in gas form are administered naturally by pulmonary route. The anesthetics such as fluothane, liquids easily volatile, are administered by inhalation

The interest of the pulmonary route is to avoid intestinal and hepatic first-pass metabolism. Its efficacy makes it a route used for cannabis, cocaine and, of course, of nicotine (tobacco). It could be used for drugs such as heparin and insulin.

Nasal route

It is used:

- for local treatment with vasoconstrictive and antiallergic drugs but with possibility of absorption and general effects.
- for general treatment: the nasal route can be used for the administration of polypeptide hormones such as desmopressin.

The nasal route also avoids the first-pass metabolism.

Cutaneous or transdermal route

The permeability of the skin to drugs depends on the drug itself, in particular its liposolubility and on the vehicle in which the drug is introduced. Resorption varies with several parameters:

- localisation: it is low at the level of the plant of the feet, of the palm of the hands, important at the level of armpits, of the angle of the jaw and the scrotum.
- temperature and cutaneous circulation: if the temperature is high, there is a vasodilation which facilitates absorption.
- state of the skin: the existence of lesions (burns, for example) increases the absorption.
- age: it decreases with the age, i.e. it is more important in the new-born baby. In addition the body area compared to the weight is more important in the new-born baby and the infant that in adults.

The cutaneous route is used for local treatments by disinfectants, antimycotics, antibiotics or glucocorticoids which can sometimes be absorbed, diffuse in the whole body and be at the origin of general effects. The severe poisonings which occurred in France in 1972 in infants following applications of talc contaminated by hexachlorophene, neurotoxic product, shows that the skin, especially in new-born babies and infants, can absorb various molecules.

The cutaneous route called transdermal route is used for introducing drugs into the body, in particular for avoiding hepatic catabolism. There are various devices for cutaneous administration, from simple application to the use of complex transdermal devices. It is used for various drugs: hormones like androstanolone, estradiol and progesterone, others such as nitroglycerin, scopolamine, fentanyl.

The cutaneous route like the pulmonary and the nasal routes avoid first-pass metabolism.

Other routes: ocular, ear and vaginal

These routes are used for local treatment but a diffusion in the body is possible, with adverse effects, for example, after prescription of ophthalmic solutions, containing beta-blockers. The drugs administered by these routes are called ophthalmic solutions for the eyes, drops for the ears, and ovules for the vagina.

PARTICULAR PREPARATIONS

Liposomes

Liposomes are microscopic vesicles, generally made up of phospholipids, in which an active drug is introduced to protect it and allow its absorption or its specific tissue distribution. There are uni-lamellate and plurilamellate preparations. They have several advantages:

- a progressive release of the active product,
- the absorption of a drug which, if not, would be degraded or would not cross the membranes,
- a specific distribution in a particular organ or a type of cells. In this last case, the lipid membranes could carry antibodies able to bind specifically to such or such tissue.

Preparations containing microparticles or nanoparticles in which the drug is inserted are under development.

Virus vectors

The use of viruses as means of introduction of a drug in the body and of its penetration into cells is a relatively recent technique, developed for gene therapy. The drug used can be DNA itself which, to be active, must penetrate in the nucleus of the cell and use the cellular machinery for the synthesis of a desired protein. The virus vectors are retroviruses or adenoviruses. The virus vector, like the other drugs, can be given by various routes, parenteral, oral, etc.

PARENTERAL ROUTES

Parenteral delivery routes are those that do not give rise to drug absorption into the splanchnic circulation. Thus, they avoid the possibility of hepatic first-pass metabolism. It should be noted that some parenteral routes do not avoid other first-pass metabolism effects (e.g., pleural metabolism for some inhaled drugs).

Some major parenteral drug delivery routes are intraarterial, intrathecal, intravenous, intramuscular, transdermal, intranasal, buccal, inhalation, intraperitoneal, vaginal, and rectal.

Intravenous (IV)

IV administration introduces drug directly into the venous circulation. The shape of the resulting circulating drug profile is determined by the size, rate, and duration of injection. IV bolus is used for immediate therapeutic effect, typically for general anesthesia and for treatment of cardiac arrhythmia.

IV dosing is popular for preclinical testing of compounds during drug development and also as a standard to determine absolute bioavailability from other dosage routes.

Intramuscular (IM)

Following intramuscular (IM) administration, drugs must cross one or more biological membranes in order to enter the systemic circulation. Intramuscular injection is used mainly for drugs and vaccines that are not absorbed orally, for example, aminoglycosides, insulin, and hepatitis vaccine.

The IM route is often used for sustained medication and specialized vehicles, such as aqueous suspensions, oily vehicles, complexes and microencapsulation, which has been developed for slow delivery of drugs by this route.

Intra-arterial

Intra-arterial injection is used to deliver drugs directly to organs, for example, in cancer chemotherapy, and in the use of vasopressin for GI bleeding. Intra-arterial carmustine is effective to treat brain tumors and pelvic intra-arterial actinomycin D is used for malignant trophoblastic disease.

Intra-arterial drug administration has potential safety implications. Embolization, arterial occlusion, and localized drug toxicity have been reported.

Intrathecal

Injection directly into the cerebrospinal fluid (CFS) ensures complete CNS bioavailability for drugs that cannot cross the blood-brain barrier. This dosage route is used to treat serious CNS infections such as meningitis and ventriculitis, and with such agents as mepivacaine and prilocaine for spinal anesthesia.

Drugs injected intrathecally initially distribute into approximately 140 ml of CSF, thus producing high concentrations in the CNS with low risk of systemic toxicity.

Transdermal

Since the introduction of transdermal scopolamine, many transdermal delivery systems have been developed for systemic activity. Major advantages claimed for this drug delivery route include continuous release of drug over a specified period, low presystemic clearance, facile drug withdrawal by simply removing the device, and good patient convenience and compliance.

Some disadvantages relate to barrier properties of the skin, skin reactions, and the relatively large dose size. Transdermal delivery is a realistic option only for drugs generally given in small doses (<10mg) and which have good membrane penetration.

Drugs currently approved for transdermal delivery include clonidine, estradiol, nicotine, nitroglycerin, and scopolamine.

Intranasal

Intranasal administration may be used for local or systemic effects. Local effects include treatment of nasal allergies, rhinitis, and nasal congestion. Nasal delivery for systemic effects is established for a small number of drugs and is being examined for many others. The sophisticated structure and specialized function of airways and membranes in the nasal cavity, and also the small surface area of this region, may limit its capacity for drug delivery. The effect of chronic drug exposure on the integrity of nasal membranes must also be considered. This problem may be compounded by the evident need for surfactants to achieve good systemic penetration with this dosage route.

Notwithstanding these factors, the physical characteristics of compounds for optimal intranasal absorption are the same as for other absorption routes. The drug must dissolve in the fluids of the nasal mucosa and must be sufficiently lipophilic to cross the membranes of the nasal epithelium.

Nasal absorption is facilitated by the high permeability of small venules and capillaries associated with the nasal mucusa. A variety of delivery systems have been described for nasal drug delivery, including drops, aerosols, nebulizers, and soluble matrices.

Small peptide molecules seem to be ideally suited for intranasal drug delivery. Vasopressin analogues and oxytocin are commercially available for intranasal dosage. Thyrotropin-releasing hormone agonists and antagonists, other vasopressin analogs, and peptides are being examined. Intranasal delivery of sex hormones has produced interesting results in animals. Intranasal delivery of insulin has been examined, but only with moderate success.

Buccal

Early recognition of buccal and sublingual absorption was manifested in the use of nitroglycerin by these dosage routes to treat severe headache and to relieve angina pectoris.

Drugs can be absorbed from the oral cavity itself or sublingually. Absorption from either route is rapid, sublingual more so apparently because of greater permeability of sublingual membranes and rich blood supply. The mean pH of saliva is approximately 6 so that drug absorption, predominantly passive in nature, is favored for unchanged molecules, acids with pKa values >3, and bases with pKa values <9.

Compounds that are currently marketed or are being considered for buccal or sublingual routes include organic nitrates, barbiturates, papaverine, trypsin, prochlorperazine, benzodiazepines, buprenorphine, captopril, isoprenalin, oxytocin, and nifedipine. Oxytocin is currently the only peptide marketed in sublingual form. Sublingual steroids have been examined with moderate success.

Inhalation

When a substance is inhaled, it is exposed to membranes of the mouth or nose, pharynx, trachea, bronchi, bronchioles, alveolar sacs, and alveoli. The lung has a potential absorption surface of some 70m2, a much larger surface than the small intestine.

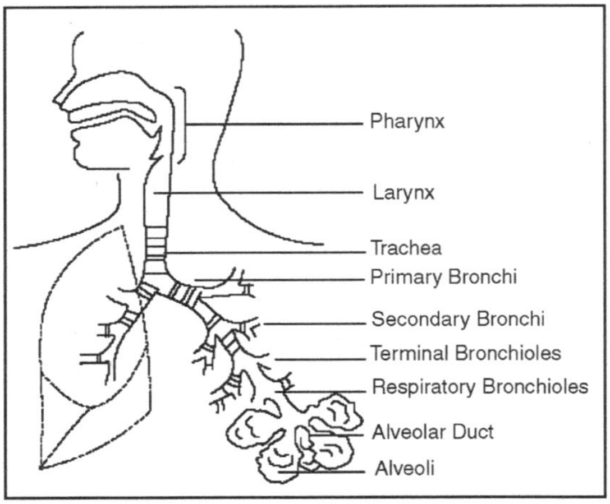

Fig. The Human Respiratory Tract.

However, the lungs and their associated airways are designed to deny access of administered compounds to the highly absorptive peripheral lung surfaces. The system is designed to deny access to particulate matter. However, if compounds can reach the peripheral region of the lung, absorption can be very efficient.

Particle (droplet) size and velocity of application control the extent to which inhaled substances penetrate into airway spaces. Optimum size for deep airway penetration is 3–5 *m*M. Large particles tend to deposit in upper airways.

Most inhalation devices deliver approximately 10% of the administered dose to the lower respiratory tract. A number of devices have been developed to increase lung delivery, and delivery of up to 21% has been reported with a pressurized metered-dose inhaler. Despite these advances, drug delivery via the lung is still inefficient.

Systemic availability of inhaled drugs may be inhibited by first-pass pulmonary metabolism. The lungs contain many drug metabolizing enzymes, including mixed function oxidases, monoamine oxidase, and esterases. Several animal models principally the rat, rabbit, and dog, are used to study drug inhalation.

Intraperitoneal

Intraperitoneal drug administration is not common. It is used pedominantly to administer compounds during preclinical discovery and development. Its clinical use is generally limited to chemotherapy for tumors with peritoneal involvement. Peritonitis occurs frequently in renally impaired patients who are receiving continuous ambulatory peritoneal dialysis (CAPD). Peritonitis is often accompanied by systemic infection so that therapeutic levels of antibiotic are needed both in the peritoneal cavity and in the systemic circulation.

Drugs may be given orally or by injection in order to achieve adequate systemic levels in the hope of also achieving therapeutic levels in the peritoneal cavity. Alternatively, drugs may be administered directly into the peritoneal cavity with the objective of also achieving systemic levels by intraperitoneal absorption. To the author's knowledge, there is little definitive information on the relative merits of these alternatives.

Vaginal

The human vagina, a fibromuscular tube 10–15 cm long, extends upwards and backwards from the vulva to the lower uterine cervix. Blood is supplied to the vagina via the uterine and pudendal arteries, and is drained from the vagina by a rich plexus, which flows into the internal iliac veins. The surface of the vaginal epithelium is kept moist by cervical secretions. The pH of vaginal fluid is 4–5.

Vaginal drug delivery is used mostly for local effects, but vaginal absorption can give rise to rapid and efficient systemic delivery. Good systemic absorption, and also the ability of the vagina to retain delivery devices, has given rise to many vaginal dosage forms, in particular for steroid contraceptives.Alarge number of vaginal controlled release dosage forms are available, including vaginal rings and biodegradable microspheres. Rectal

The human rectum is 15–20 cm long. It is normally empty and contains 2–3 ml of mucous fluid with pH 7–8. There are no villi and only a limited surface area of 200–400 cm2 is available for absorption. Blood and lymph vessels are abundant in the rectal submucosa.

Veins from the upper rectum drain into the portal circulation, while veins from the middle and lower rectum drain directly into the inferior vena cava. However, there are extensive anastomoses among these veins so that precise anatomical differentiation is difficult. It appears that compounds absorbed from the lower rectum, in contrast to those absorbed from the upper rectum, avoid hepatic first-pass metabolism.

Rectal absorption is generally slower than oral absorption, but for some drugs, rectal absorption exceeds oral absorption presumably due to avoidance of first-pass metabolism after rectal delivery. This has been reported for morphine, metoclopramide, ergotamine, lidocaine, and propranolol. Human rectal systemic availability of the extensively metabolized drug lidocaine is 65% as compared to 30% after oral administration.

Rectal absorption of drugs from aqueous or alcoholic solutions is generally much faster than from suppositories. Non-surfactant adjuvants, such as salicylates, increase rectal absorption of water-soluble drugs and also of high molecular weight compounds, such as insulin, heparin, and gastrin.

ORAL DOSING OF ABSORPTION ENHANCERS

Although oral dosing is generally more convenient than other dosage routes, oral absorption of many drugs is poor. As molecules become larger, more complex, and generally more lipophilic in the quest for new or improved efficacy, their absorption tends to decline.

To address this problem, absorption enhancers continue to be examined, so far with variable success. Some compounds that have been shown, largely in animal studies, to increase absorption of drugs are shown in Table Little or no information is available in humans for most of these compounds.

Nonsteroidal antiinflammatory agents (NSAIDs) NSAIDs, in particular indomethacin, diclofenac, mepirazole, phenylbutazone, and salicylate, can promote absorption of other drugs, including insulin, ampicillin, cephalothin, cefoxitin, and cefmetazole.

Most of these observations were made in the rat and frequently after rectal administration. Several mechanisms by which NSAIDs promote drug absorption have been postulated, but exact mechanisms are not known.

As NSAIDs are often irritating to the GI mucosa, and this may well relate to their absorption enhancing ability, the feasibility of their use to promote drug absorption is uncertain.

Table. Some Types of Oral Drug Absorption Enhancers.

Non-steroidal anti-inflammatory agents
surfactantants
Bile salts
Medium chain fatty acids
Mixed micelles
Liposomes
Azone
Cell permeation enhancers
Nanoparticles

Surfactants Effects

In view of their solubilizing effects and also their potential to change membrane permeability, surfactants have been considered as absorption enhancers, again mostly in animals. Polyoxyethylene ethers have been shown to enhance gastric or rectal absorption of lincomycin, penicillin, cephalosporins, and fosfomycin in rats and rabbits.

In rats, colonic absorption of interferon– alpha is increased from 3 to 8% by polyoxyethylene esters of oleic acid and oleic acid glycerides. Some studies have examined the effects of surfactants on intestinal absorption of insulin, with variable results. Both rectal and jejunal absorption of insulin was increased by anionic and cationic surfactants.

However, in humans, oral polyoxyethylene-20-oleyl ether resulted in poor and variable insulin absorption. Any enhancing effect of surfactants on drug absorption appears to be related to increased drug solubilization, modification of mucosal permeability, or reduction of resistance of the unstirred water layer at the GI membrane surface.

In general, unionic surfactants have little effect on membrane structure but cationic surfactants have been associated with reversible cell loss and loss of goblet cells. These effects must limit consideration of surfactants as absorption promoters, particularly for long term treatment.

Bile Salts

Bile contains conjugates of cholic acid and chenodeoxycholic acid, which emulsify dietary fat, facilitate lipolysis, and transport lipid molecules through the unstirred layer of the intestinal mucosa by micellar solubilization. The ability of bile salts to promote lipid absorption has prompted their investigation as absorption enhancers for drugs, with modest success.

Studies in animals have demonstrated increased intestinal absorption of heparin and interferon-alpha. Absorption of insulin can be increased by bile salts, both in experimental animals and in humans. The effect on drug absorption appears to correlate with mucosal damage. This, together with possible cocarcinogenic and comutagenic properties of secondary bile salts, reduces the attractiveness of bile salts as absorption enhancers.

Medium-chain Fatty Acids and Glycerides

The presence of medium chain fatty acids and glycerides in food products has stimulated interest in their potential utility as absorption enhancers.

Some fatty acids and glycerides have been shown to increase drug absorption under a variety of conditions, almost always in animals and in most cases after rectal dosing. However, some studies have yielded positive results after oral dosing.

Oral insulin bioavailability was increased to 9–13% relative to IM administration by a mixture of sodium dodecanoate and cetyl alcohol. Aftiraxone absorption was enhanced by glyceryl- 1-monooctanoate after oral, duodenal, and rectal administration to animals.

Despite the potential of these classes of compounds as absorption enhancers, they have been shown to have negative effects on mucosal membrane integrity. Additional research is needed to evaluate risks and benefits.

Consist of Fatty Acids Solubilized in Mixed Micelles

Mixed micelles consist of fatty acids solubilized by surfactants or bile salts. The effects of mixed micelles on drug absorption were reviewed by Muranishi Mixed micelles are effective absorption enhancers for compounds such as heparin, streptomycin, gentamycin, and insulin. The effect of mixed micelles on drug absorption tends to be greater at the distal region of the GI tract. The mechanism for increased absorption is not known. Some publications claim that they are safe to use. Others report a disordering effect on intestinal epithelial cells.

Consist of Vesicles in Liposomes

Liposomes consist of vesicles composed of bilayers or multilayers that contain phospholipids and cholesterol surrounding an aqueous compartment. Drug is entrapped within the liposome and is released from the liposome for absorption at the intestinal membrane surface.

This dosage form received considerable attention during the 1970s and 1980s, and several animal studies demonstrated potential for absorption enhancement. However, lack of effect in other studies, and also stability problems, have resulted in reduced interest in liposomes as absorption enhancers.

Azone Compounds

Azone (1-Dodecylazacycloheptan-2-one) and related compounds have been studied as transdermal penetration and oral absorption enhancers. Although some efficacy has been shown, an emulsifying agent appears to be necessary for azone to penetrate the intestinal mucosal membrane in order to promote drug absorption.

One study reported the absence of gross morphological damage after exposure of mucosa to azone but additional information on the effect of azone on overall mucosa structure is not avalable.

Cell Permeation Enhancers

Although the objective of most absorption enhancers is to avoid direct interaction with the mucosal membrane, cell permeation enhancers use this as a means to increase drug absorption. One form of enhancer currently of interest consists of glycosylated molecules, or facial amphiphiles.

It is claimed that these compounds temporarily increase membrane permeability. Molecules are designed to self-assemble in membranes to form transient pores that permit hydrophilic compounds to cross the membrane. This technology has considerable potential for absorption enhancement. No adverse effects have been reported to date.

Surface Area and Dissolution of Nanoparticles

From known relationships between surface area and dissolution, it is reasonable to predict that ultrafine particles may increase the dissolution rate of relatively insoluble compounds.If these particles are then stabilized to avoid aggregation and agglomeration, and yet retain fluidity, then a useful drug product could be obtained. This concept has found expression in a proprietary nanoparticle technology in which a drug is reduced to nanometer-size particles in the presence of stabilizers. Originally developed for IV computer imaging, this technology shows considerable promise to increase absorption of poorly water soluble compounds.

As the nanoparticle system is purely "formulation" in nature, it is unlikely to affect GI mucosal integrity. Nanoparticles are sufficiently small that they can be used parenterally, apparently without ill effects. More is likely to be heard about this novel absorption enhancer technology.

THE CELL MEMBRANE

The lipid is composed mainly of the phospholipids phosphatidylserine and phosphatidylinositol, and contains saturated and unsaturated fatty acids and sterols. The bilayer exhibits high permeability to hydrophobic molecules and low permeability to hydrophilic molecules.

The cell membrane is associated with intrinsic and extrinsic proteins. Intrinsic proteins are globular proteins that generally span the bilayer and are held within the membrane by hydrophobic and electrostatic interactions. The proteins can form channels, carriers, or pumps that enable polar molecules to cross the membrane.

Membrane Transport

Several mechanisms have been identified for drug transport across membranes. One of them is passive. The remainder utilize some type of carrier mechanism. Simple or Passive Membrane Transport This mechanism, based primarily on lipid solubility and concentration gradient, is responsible for membrane transport of the great majority of drugs.

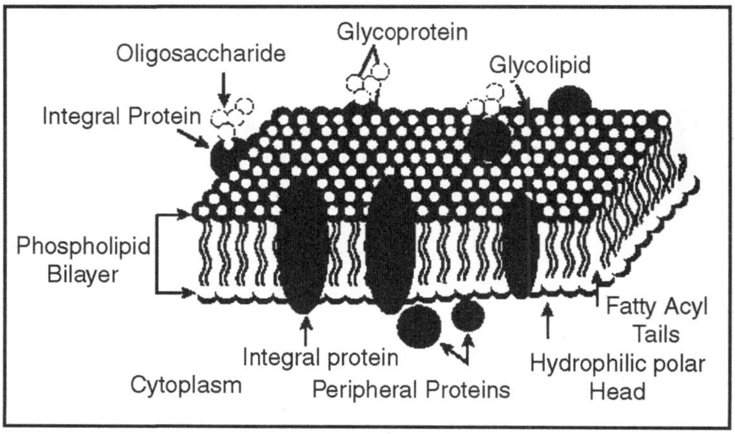

Fig. Structural Model of the Cell Membrane.

The membrane is composed of a bimolecular leaflet of phospholipid with the polar head groups facing the extracellular and cytosolic compartments and the acyl groups in the middle of the bilayer. Integral membrane proteins are embedded in the lipid bilayer. Integral proteins are glycosylated on the exterior surface and may be phosphorylated on the cytoplasmic surface. Extrinsic membrane proteins, peripheral proteins, are linked to the cytosolic surface of the intrinsic proteins by electrostatic interactions.

The range of membrane permeabilities is very high. Typically, hydrophobic molecules have high partition coefficients, while hydrophilic molecules have low partition coefficients. The direction and rate of transport are determined by the concentration gradient of a substance across the membrane. Molecules that are weak acids or bases cross membranes more readily when they are in the non-ionized form. However, aqueous solubility is favoured for the ionized form. In order to be available to cross any membrane, a drug must be in solution. This paradoxical requirement of both aqueous and lipid solubility is of particular concern in the area of drug absorption and presents a constant challenge in pharmaceutical formulation.

Transport Proteins

Many substances, particularly polar molecules, cross membranes at rates greater than those predicted from solubility and permeability data. Some can cross membranes against a concentration gradient. Unexpectedly high membrane permeability is related to transport proteins. Many transport proteins have been identified, cloned, and sequenced. Current knowledge has permitted an operational definition of carrier proteins as channels, carriers, and pumps. The current state of the art in identification and characterization of these systems has been described by Wright.

Channels

Initial speculation on the existence of small aqueous pores in membranes was based on high membrane permeability of small polar molecules. For example, the permeability of water is 1000-fold more, and that of urea 10- to 100-fold more, than predicted. These types of observations led to the prediction of aqueous channels with radii of approximately 4A°.

Water Channels

The presence of water channels has been demonstrated by successful cloning of proteins that increase membrane water permeability. These have been expressed in erythrocytes and in cells of the renal tubule.

Ion Channels

Evidence for ion channels in biological membranes was introduced in the 1970s. Possibly the most significant experiment was by Neher and Sakmann, who recorded single ion channel currents in muscle fibres.

Many types of channels for sodium, potassium, calcium, and chloride ions have been described, and each has a specific conductance, ion selectivity, and probability of opening. Ion channel opening may be controlled by voltage or by ligand, and channels are thus designated voltage-gated or ligand-gated. Each of the ion channels has a specific pharmacology. Sodium channels are 12 times more selective for sodium than for other cations. Calcium channels are 1000 times more selective for calcium than for other cations.

Facilitated Diffusion

Facilitated diffusion is a simple mechanism proposed to explain transport of water soluble compounds. The main characteristics of this transport system are that membrane permeability exceeds that predicted from partition coefficients, transport occurs down a concentration gradient, transport is saturable, and competition occurs between isomers. Facilitated diffusion has been used to explain cellular uptake of sugars and amino acids. Six human sugar transporters with different tissue distributions, substrate kinetics, and specificities have been identified. A number of facilitated amino acid transporters have also been identified in mammalian cells. System L, which transports neutral amino acids, such as leucine and phenylalanine, is probably the best known of these.

Pumps

Pumps are proteins that can transport ions against electrochemical potential gradients using adenosine- 5-triphosphate (ATP) as an energy source. Sodium– potassium pumps maintain intracellular sodium and potassium concentrations in animal cells and also control salt and water absorption by the epithelial cells in the intestine and kidney.

The sodium–potassium pump transports three sodium ions out of the cell and two potassium ions into the cell at the cost of one molecule of ATP. The 3: 2 coupling ratio results in net loss of sodium ions into the cell down an electrochemical gradient and maintains cell volume. Currently, considerable research is attempting to elucidate the structures of the various isoforms and subunits of sodium potassium pumps.

Cotransporters and Exchangers

There are many other examples of ions and ionized molecules accumulating in cells against their concentration gradient, such as uptake of iodine by the thyroid gland, accumulation of acids in liver cells, and absorption of sugars and phosphate by the small intestine. Recent studies have shown that these are governed by cotransport mechanisms. In all cases tested to date, sodium or hydrogen ion gradients are used to drive cotransporters, and these gradients are maintained by ion pumps. Glucose transport across the brush border of the small intestine is coupled with sodium transport, and uphill sugar transport is driven by the sodium gradient.

Cotransporters

Cotransporters use the sodium or hydrogen ion gradient to drive transport of a substrate. Many cotransporters have been described, cloned, sequenced, and expressed. The sodium–glucose cotransporter just described is one of these. Other cotransporters facilitate the transport of other sugars, osmolytes, and amino acids. In humans, a disorder of intestinal glucose and galactose absorption is due to a defective sodium–glucose transporter.

Antiporters

The best characterized antiporters, or exchangers, are the chloride–bicarbonate, sodium–hydrogen ion, and sodium–calcium exchangers. The cellular sodiumhydrogen ion exchanger controls cell volume, pH, growth, and sodium transport.

Mammalian isoforms of these exchangers have been cloned and sequenced. Sodium-hydrogen exchangers play a dominant role in the regulation of intracellular calcium, and thus the force of contraction of the heart. The therapeutic effect of cardiac glycosides is probably related to decreased sodium-calcium exchange in the heart caused by a decreased sodium–hydrogen gradient across the cell membrane.

Permeability Glycoprotein

Permeability glycoprotein (p-glycoprotein) is an ATPdependent efflux pump responsible for pumping substances out of cells. It is implicated in development of drug resistance in tumor cells. Localization of p-glycoprotein in the apical membranes of intestinal, liver, and kidney cells, and also at the blood brain barrier, provides potential for this pump to have a profound effect on drug absorption, distribution, and elimination, as well as in drug–drug interactions.

Pinocytosis and Endocytosis

Pinocytosis is a non-specific process whereby a substrate enters a cell by invagination to form an intracellular vesicle. Receptor-mediated endocytosis occurs when substrate binds to a specific membrane receptor.

Substrates ingested by cells in this way are stored in vesicles or degraded. Receptor-mediated endocytosis is involved in cellular uptake of immunoglobulin and low density lipoprotein.

6

Pharmacodynamics

After a single dose of a psychotropic medication, its plasma concentrations and clinical effects will increase, reach a maximum, and then decline with time. Pharmacokinetics deals with the time-course of drug concentration, whereas the discipline of pharmacodynamics refers to the time-course and intensity of drug action or response. The capacity to understand and predict individual differences in drug response is of critical importance in psychopharmacology, since the objective of drug treatment is to produce therapeutic benefit while minimizing side-effects. Advances in the clinical applications of pharmacodynamics are due to improved precision and objectivity in methods for measuring human drug response, and the emergence of the discipline of kinetic–dynamic modelling, in which drug concentration is used as a direct predictor of response.

METHODS OF MEASUREMENT

Each class of psychotropic drugs produces unique pharmacodynamic effects, as well as problems with measurement. Pharmacodynamic studies of sedative–anxiolytic drugs serve to illustrate available approaches to quantitation, and the limitations associated with each. Benzodiazepine derivatives, and other aminobutyric acid (GABA)-benzodiazepine receptor agonists, produce clinical sedation as a final common pathway of action. However, the clinical consequences of benzodiazepine agonist action depend on the therapeutic objective and the setting of drug administration.

Antianxiety– antipanic effects, enhancement of sleep, and reduction of seizure activity are usually described as primary therapeutic actions. Typical side-effects include excessive or persistent sedation, drowsiness, ataxia, incoordination, slowed reaction time, or slowed psychomotor performance. Some clinical actions are ambiguous; impaired memory, for example, may be desirable in the context of premedication before general anesthesia or sedation prior to endoscopy or cardioversion, but undesirable in other circumstances. Benzodiazepines also produce some effects whose direct link to clinical efficacy or side-effects are not directly established. These include: increased beta activity on the electroence-phalogram (EEG), reduced saccadic eye movement velocity, increased postural sway, elevated concentrations of plasma growth hormone, and reduced plasma concentrations of cortisol and ACTH. Measurement of the pharmacodynamic effects of benzodiazepine agonists becomes more reliable and objective as the effects become more removed from the primary therapeutic action. The time-course and intensity of primary therapeutic effects are the most difficult to measure. Clinical anxiolytic or sedative properties can be quantitated using global assessments by patients or trained observers, or by rating instruments targeted to specific symptoms or symptom groups.

These measurement approaches have been extremely useful in clinical psychopharmacology, but have limitations. They are subjective ratings, and can be influenced by the outlook, expectation, and experience of the patient as well as the observer. Quantitative drug effects must be evaluated as change scores relative to both the pretreatment baseline condition and, to inactive placebo. If special populations (such as the elderly) have a unique way of interpreting rating scale items or describing drug effects, this further complicates the methodology. Procedures quantitating secondary effects of benzodiazepine agonists generally focus on aspects of psychomotor performance, such as measures of reaction time, speed, and accuracy of task performance, and the capacity for information acquisition and recall. These measures are only partly objective, although they provide numeric results.

Again, change scores are the most meaningful; drug effects must be compared to pre-dose baseline performance, and to effects of placebo. Psychomotor performance procedures are influenced by practice—subjects' performance will progressively improve if they repeatedly perform a task in the absence of medication.

This can complicate interpretation of drug-associated changes, and must be accounted for in the experimental design. The study of special populations (such as the elderly) is also affected by baseline performances that differ from that of the control group. It is not clear how best to incorporate differences in baseline performance when comparing drug effects. The relation of drug effects on laboratory performance tasks to effects on real life tasks is not clearly established.

Even for complex and seemingly realistic laboratory tasks such as simulated automobile operation, the consequences of error are negligible in the laboratory, but potentially lethal on the road. Fully objective measures of pharmacodynamic effects of benzodiazepines are of increasing interest. Computerized analysis of the EEG can directly quantitate benzodiazepine effect, either by evaluation of activity in the beta frequency band by fast-Fourier transform, or by aperiodic analysis. In addition to being completely objective, these measures are unresponsive to placebo, and are not altered by practice or experience. No theoretical basis is yet established to link EEG changes to changes in therapeutic effects or psychomotor side effects.

Empirical data, however, demonstrate that the pharmaco-dynamic profile of EEG changes following benzodiazepine agonist administration closely intercorrelates with other measures of drug activity.

KINETIC–DYNAMIC MODELLING

Kinetic–dynamic modelling procedures directly evaluate the relation of drug concentrations (*C*) to drug effect (*E*), with the objective of determining how much variability in drug effect is attributable to measurable drug concentrations. The Sigmoid-*E*max relationship is commonly used in kinetic–dynamic modelling. When this model fits the data, investigators are reassured about the concentration–effect relationship, since many drug-receptor interactions fit the same model. The linkage is described by the following equation:

$$E = \frac{E_{max} \cdot C^A}{C^A + EC_{50}^A}$$

*E*max is the maximum drug effect (relative to baseline), which cannot be exceeded no matter how high the drug concentration; EC50, is the concentration at which the drug effect is 50 per cent of *E*max; *A* is an exponent of unknown biological significance related to the "steepness" of the curve in its linear phase.

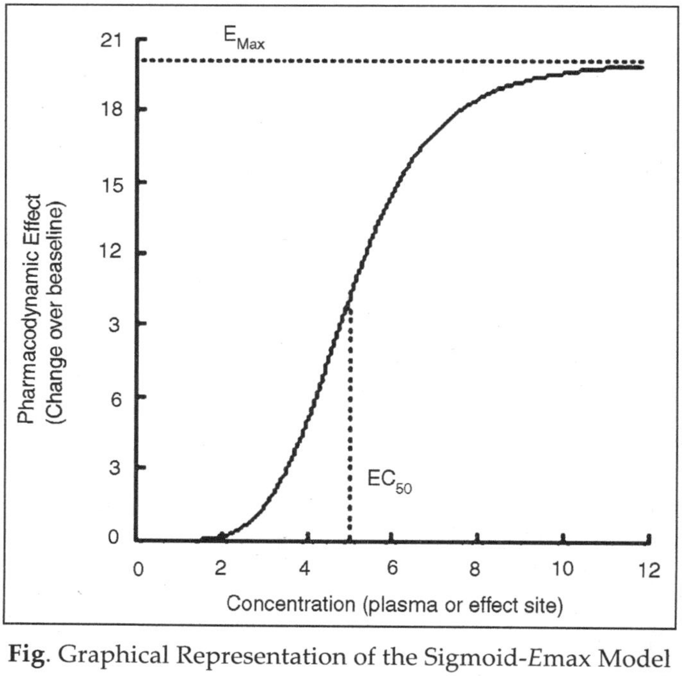

Fig. Graphical Representation of the Sigmoid-*E*max Model

When applicable, this model allows quantitation of individual differences in sensitivity to a given drug, relative potencies of drugs having the same mechanism of action, and the quantitative and qualitative effects of pharmacologic potentiators or antagonists. However, not all data sets are consistent with the Sigmoid-*E*max model, and it may be inappropriate or even misleading to "force" this model on data which it does not fit. In many cases, data from kinetic– dynamic studies can be described by an exponential function such as:

$$E = B \cdot CA$$

or by a linear function such as:

$$E = m \cdot C + b$$

Plasma Concentration Versus Effect-Site Concentration

The use of plasma drug concentration as the independent variable in kinetic–dynamic modelling procedures assumes that equilibration of drug between plasma and the site of action (effect-site) in brain is very rapid, such that plasma concentrations are proportional to concentrations at the effect site. This assumption may need modification in some circumstances. For the subject described in Figure maximum change over baseline in EEG beta activity is delayed following completion of the infusion of midazolam.

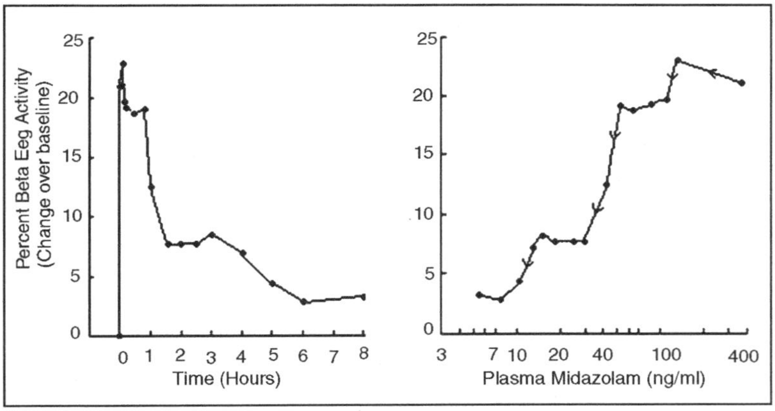

Fig. EEG was Monitored Prior to and for 8 hr After the Intravenous Dose of Midazolam.

The EEG signal was analysed by computer to determine the relative activity in the 13–30 cps range, also termed the beta range. The maximum EEG change was not attained immediately after the end of the infusion, but rather at *five* minutes after completion of the infusion. *Right*: A plot of plasma midazolam concentrations (*x* axis, logarithmic scale) versus EEG change (*y* axis) shows that the maximum EEG change does not correspond to the maximum plasma concentration. Arrows indicate the direction of increasing time. A kinetic–dynamic plot of EEG effect versus plasma concentration reveals "counterclockwise hysteresis," in which the maximum effect does not correspond to the maximum plasma concentration. This may be explained by the time necessary for midazolam to equilibrate between plasma and effect site. To account for this delay, the model is modified to include the hypothetical effect site.

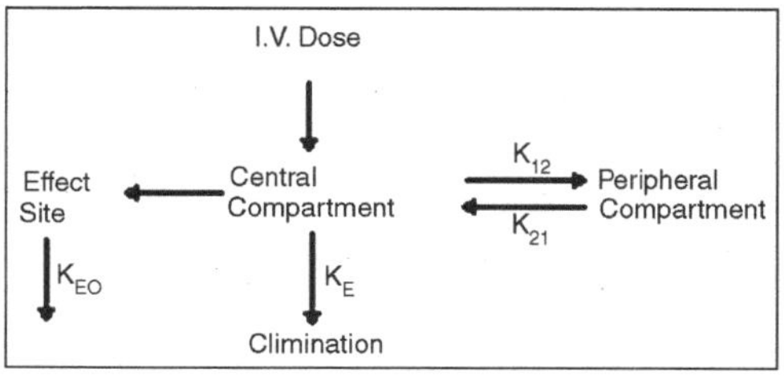

Fig. Modification of the Kinetic–Dynamic Model to Incorporate an effect-site, Presumed to Correspond to the Site of Pharmacologic Activity.

The first-order rate constant for drug exit from the effect-site (*K*EO) is effectively the rate constant determining the time necessary for drug equilibration between plasma and effect-site. Kinetic–dynamic linkage models are then applied to hypothetical effect-site concentrations rather than plasma concentrations. Effect-site concentrations, as opposed to plasma concentrations, are presumed to determine pharmacodynamic effects.

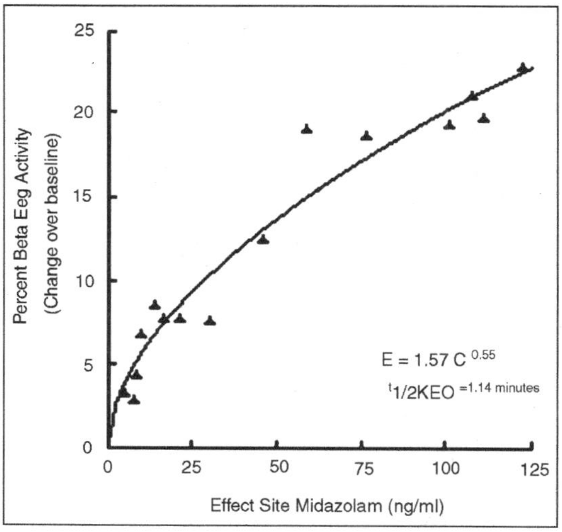

Fig. The Relation between Midazolam Concentrations at the Presumed Effect Site Versus EEG Change now Fits a Kinetic–Dynamic Model Consistent.

The first-order rate constant *K*EO governs the exit of drug from the hypothetical effect site, and also determines the observed equilibration of drug between plasma and effect-site. *K*EO can be used to calculate a half-life of equilibration (*t*1/2KEO). If *K*EO is incorporated as a variable in the iterative analysis, the resulting plot of effect-site concentration versus EEG effect eliminates the hysteresis and yields data points that are consistent with Equation.

ANTIMICROBIAL PHARMACODYNAMICS

The assessment of the activity of an antibiotic is crucial to the successful outcome of antimicrobial therapy. Non-microbiological factors such as host defence mechanisms, the location of an infection, underlying disease as well as the intrinsic pharmacokinetic and pharmacodynamic properties of the antibiotic.

Fundamentally, antibiotics are classified as either having lethal or bactericidal action against bacteria or are bacteriostatic, preventing bacterial growth. The bactericidal activity of antibiotics may be growth phase dependent and in most but not all cases the action of many bactericidal antibiotics requires ongoing cell activity and cell division for the drugs' killing activity. These classifications are based on laboratory behaviour; in practice, both of these are capable of ending a bacterial infection.

'In vitro' characterisation of the action of antibiotics to evaluate activity measure the minimum inhibitory concentration and minimum bactericidal concentration of an antimicrobial and are excellent indicators of antimicrobial potency. However, in clinical practice, these measurements alone are insufficient to predict clinical outcome. By combining the pharmacokinetic profile of an antibiotic with the antimicrobial activity, several pharmacological parameters appear to be significant markers of drug efficacy.

The activity of antibiotics may be concentration-dependent and their characteristic antimicrobial activity increases with progressively higher antibiotic concentrations. They may also be time-dependent, where their antimicrobial activity does not increase with increasing antibiotic concentrations; however, it is critical that a minimum inhibitory serum concentration is maintained for a certain length of time.

A laboratory evaluation of the killing kinetics of the antibiotic using kill curves is useful to determine the time- or concentration-dependence of actimicrobial activity.

Antibiotic classes

Antibiotics are commonly classified based on their mechanism of action, chemical structure or spectrum of activity. Most antibiotics target bacterial functions or growth processes.

Antibiotics which target the bacterial cell wall (penicillins, cephalosporins), or cell membrane (polymixins), or interfere with essential bacterial enzymes (quinolones, sulfonamides) usually are bactericidal in nature. Those which target protein synthesis, such as the aminoglycosides, macrolides and tetracyclines, are usually bacteriostatic.

Further categorization is based on their target specificity: "narrow-spectrum" antibiotics target particular types of bacteria, such as Gram-negative or Gram-positive bacteria, while broad-spectrum antibiotics affect a wide range of bacteria. In the last few years three new classes of antibiotics have been brought into clinical use. This follows a 40-year hiatus in discovering new classes of antibiotic compounds. These new antibiotics are of the following three classes: cyclic lipopeptides (daptomycin), glycylcyclines (tigecycline), and oxazolidinones (linezolid). Tigecycline is a broad-spectrum antibiotic, while the two others are used for Gram-positive infections. These developments show promise as a means to counteract the bacterial resistance to existing antibiotics.

Production

Since the first pioneering efforts of Florey and Chain in 1939, the importance of antibiotics to medicine has led to much research into discovering and producing them. The process of production usually involves the screening of wide ranges of microorganisms, and their testing and modification. Production is carried out using fermentation, usually in strongly aerobic form.

Administration

Oral antibiotics are simply ingested, while intravenous antibiotics are used in more serious cases, such as deep-seated systemic infections. Antibiotics may also sometimes be administered topically, as with eye drops or ointments.

Side Effects

Although antibiotics are generally considered safe and well tolerated, they have been associated with a wide range of adverse effects. Side effects are many, varied and can be very serious depending on the antibiotics used and the microbial organisms targeted.

The safety profiles of newer medications may not be as well established as those that have been in use for many years. Adverse effects can range from fever and nausea to major allergic reactions including photodermatitis and anaphylaxis. One of the more common side effects is diarrhea, sometimes caused by the anaerobic bacterium *Clostridium difficile,* which results from the antibiotic disrupting the normal balance of the intestinal flora, Such overgrowth of pathogenic bacteria may be alleviated by ingesting probiotics during a course of antibiotics. An antibiotic-induced disruption of the population of the bacteria normally present as constituents of the normal vaginal flora may also occur, and may lead to overgrowth of yeast species of the genus *Candida* in the vulvo-vaginal area. Other side effects can result from interaction with other drugs, such as elevated risk of tendon damage from administration of a quinolone antibiotic with a systemic corticosteroid.

Drug-drug Interactions

Contraceptive Pills

Hypothetically, interference of some antibiotics with the efficiency of birth control pills is thought to occur in two ways. Modification of the intestinal flora may result in reduced absorption of estrogens. Secondly, induction of hepatic liver enzymes causing them to metabolize the pill's active ingredients faster may affect the pill's usefulness.

However, the majority of studies indicate that antibiotics do not interfere with contraception. Even though a small percentage of women may experience decreased effectiveness of birth control pills while taking an antibiotic, the failure rate is comparable to those taking the pill. Moreover, there have been no studies that have conclusively demonstrated that disruption of the gut flora affects contraception. Interaction with the combined oral contraceptive pill through induction of hepatic enzymes by the antifungal medication griseofulvin and the broad-spectrum antibiotic rifampicin has been shown to occur. It is recommended that extra contraceptive measures are applied during antimicrobial therapy using these antimicrobials.

Alcohol

Alcohol can interfere with the activity or metabolization of antibiotics. It may affect the activity of liver enzymes, which break down the antibiotics.

Moreover, certain antibiotics, including metronidazole, tinidazole, cephamandole, ketoconazole, latamoxef, cefope-razone, cefmenoxime, and furazolidone, chemically react with alcohol, leading to serious side effects, which include severe vomiting, nausea, and shortness of breath.

Alcohol consumption while taking such antibiotics is therefore not recommended. Additionally, serum levels of doxycycline and erythromycin succinate may, in certain circumstances, be significantly reduced by alcohol consumption.

7

Biotransformation of Drugs

Most xenobiotics contain lipophilic functional groups to facilitate penetration of membrane barriers, thereby expediting intestinal absorption. Excretion of these compounds would occur slowly, owing to a high degree of protein binding and renal tubular reabsorption. Thus, a function of drug biotransformation is to render molecules more polar so that the products can be efficiently eliminated. Drug metabolism is primarily a hepatic event, although it also occurs in the kidney and, to a lesser extent, in the intestine, lung, and skin. By converting lipophilic substances to excretable metabolites, hepatic enzymes detoxify drugs by terminating their pharmacologic activity. Unfortunately, reactive intermediates, which themselves are responsible for some toxic effects, may be formed during metabolism.

PATHWAYS OF BIOTRANSFORMATION

Biotransformation pathways have been classified under the headings of Phase I and Phase II reactions. The former includes processes involving oxidation, reduction, and hydrolysis that result in the formation of new functional groups in the drug molecule. Many of the metabolites formed are then able to participate in a subsequent biotransformation process (Phase II) in which a polar hydrophilic moiety such as glucuronic acid, sulfate, or an amino acid is covalently linked (conjugated) to the functional group introduced in the Phase I reaction.

The usual overall result of these processes is the production of metabolites with progressively greater water solubility that facilitates their removal from the body. Biotransformations are generally catalyzed by enzymes present in the liver and other tissues. Some researchers consider that the biotransformation by the gut flora of drug metabolites excreted in the bile represents a third phase. Many drugs undergo a series of biotransformations, and the combinations of consecutive Phase I and Phase II reactions can produce a complex array of metabolites. Drugs already containing appropriate functional groups can undergo Phase II reactions without the need for Phase I biotransformation; examples are the arylacetic and arylpropionic acid antiinflammatory drugs (e.g., indomethacin and ibuprofen respectively), in which the carboxylic acid function can conjugate with glucuronic acid or amino acids such as glycine. In some cases, Phase I metabolites may not be detected, owing to their instability or high chemical reactivity. The latter type are often electrophilic substances, called reactive intermediates, which frequently react non-enzymically as well as enzymically with conjugating nucleophiles to produce a Phase II metabolite. A common example of this type is the oxidative biotransformation of an aromatic ring and conjugation of the resulting arene oxide (epoxide) with the tripeptide glutathione. Detection of metabolites derived from this pathway often points to the formation of precursor reactive electrophilic Phase I metabolites, whose existence is nonetheless only inferred.

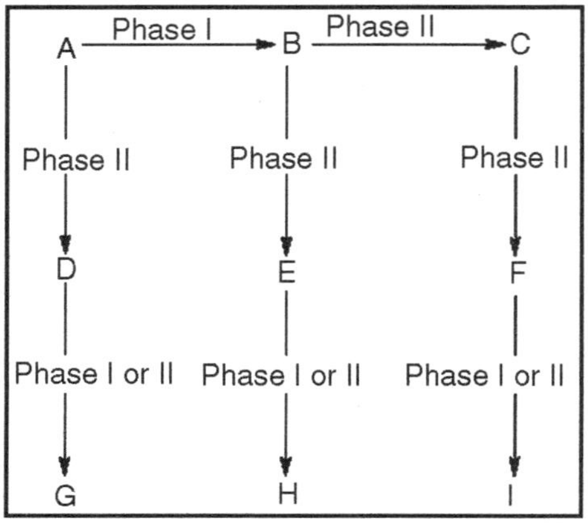

Fig. Biotransformation of Drug A by Phase I and Phase II Reactions to Several Types of Metabolite (B through I).

Certain biotransformation processes are reversible, and formation of an inactive metabolite that can be converted back to the active drug delays the removal of the drug from the body and probably prolongs the duration of exposure of the target tissues to the drug. The common processes that can contribute to this phenomenon are oxidation/reduction of secondary alcohols/ketones, sulfides/sulfoxides, and tertiary amines/N-oxides, all of which are reversible processes. The pathways of biotransformation that a given drug follows are determined by many factors, including lipophilicity, molecular size and shape, steric and electronic characteristics, stereochemistry, acidity, and basicity.

The interrelationships between these factors are not well understood, as judged by the present inability to predict biotransformations accurately except perhaps within a narrow group of closely related compounds. This inability is primarily due to the limited depth of available background information and insufficient knowledge of the factors affecting the interaction of compounds with the active sites of enzymes responsible for biotransformation. The knowledge of biotransformation processes is continually increasing, and new pathways are still being discovered.

DRUG BIOTRANSFORMATION REACTIONS

HYDROLYSIS BIOTRANSFORMATION REACTIONS

Other important Phase I drug biotransformation reactions are those involving hydrolytic processes. The types of groups hydrolyzed include esters, amides, thioesters, hydroxamates, oximes, carbamates, hydrazides, organophosphates, and perhaps even nitriles.

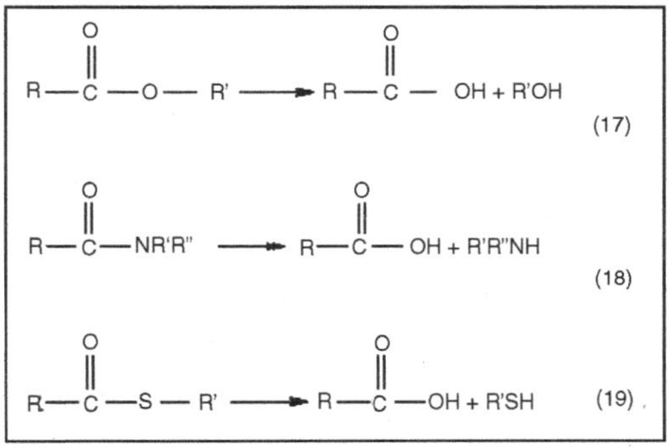

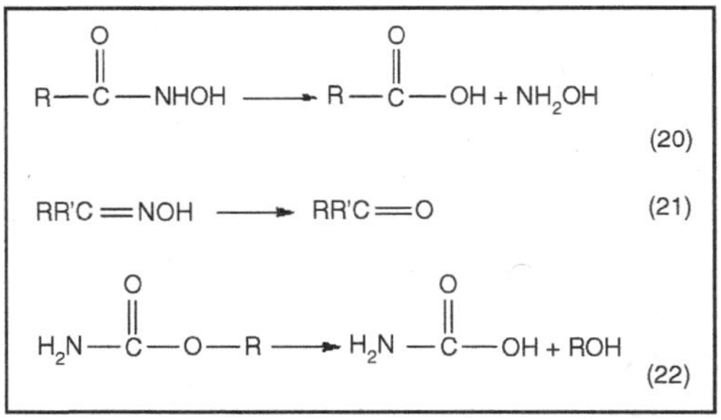

The hydrolytic reactions are mostly catalyzed by carboxylesterases although other names may be used for the enzymes involved depending on the substrates concerned, for example, amidases, or phosphatases. With the increasing number of peptides being synthesized as potential drugs, mention should be made of peptidases that cleave (by hydrolysis) the amide (peptide) linkage between adjacent amino acids.

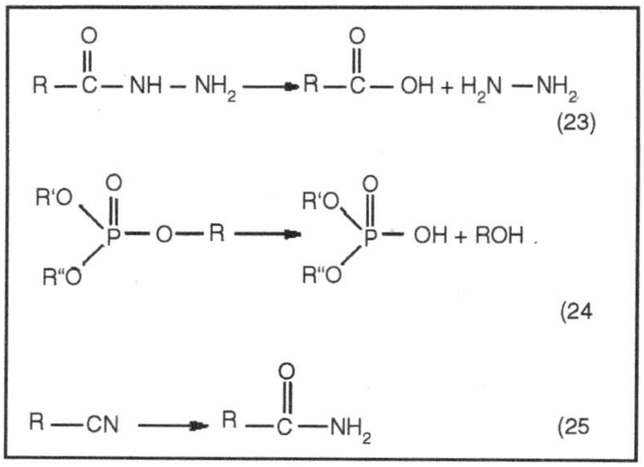

The carboxylesterases are a family of somewhat non-specific enzymes that are not easily classified. They have an endogenous role in, for example, transport. B-esterases (serine hydrolases) are those sensitive to inhibition by organophosphates such as paraoxon that esterify a serine residue at the enzymes' active site; they are present in most mammalian tissues and biofluids such as the plasma. Other esterases are A-esterases (arylesterases) and the C-esterases (acetylesterases).

Esterase activity is particularly high in the endoplasmic reticulum of the liver. Some esterase activity appears to be tissue and species specific. The latter can lead to marked species differences in ester half-lives in the body, ranging, for example, from only a few minutes in the rat to several hours in the dog.

Because both these species are commonly used in drug evaluation, their comparison with humans becomes of paramount importance when ester-containing drugs are being assessed. The specificity of carboxylesterases appears to depend on the nature of the substituents (R,R0,R00) rather than on the type of heteroatom (O, S, or N) attached to the carbonyl group.

Usually esters are hydrolyzed more rapidly than the corresponding amides, and this difference can be exploited in drug design; the duration of action of ester-containing drugs eliminated by hydrolysis could be prolonged by replacement of the ester with the amide link.

The hydrolysis of an ester or amide link is usually associated with detoxication of the drug because of the greater water solubility and thus more facile excretion of the resultant hydrolysis products. However, exceptions exist; for example, salicylic acid produced by the hydrolysis of acetylsalicylic acid is still pharmacologically active and has a longer half-life than the parent drug.

The hydration of epoxides can be viewed as a special form of hydrolysis in which the elements of water are added to the epoxide without the resultant formation of more than one product, such as occurs in most hydrolytic reactions. This reaction is catalyzed by epoxide hydrolase.

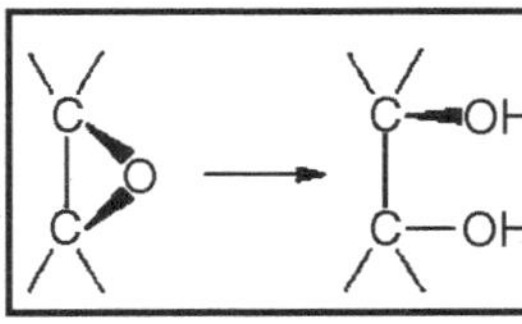

(26)

Epoxides are usually chemically quite reactive as a consequence of the strain inherent in the 3-membered cyclic ether ring and the electrophilic nature of the ring's carbon atoms.

Through the action of epoxide hydrolase (epoxide hydrase, epoxide hydratase), they react stereoselectively with water to produce vicinal diols that are predominantly in the trans configuration because nucleophilic attack by OH^-, at the least hindered carbon atom, occurs from the opposite side of the molecule to where the epoxide ring is situated. Although the diol is less reactive than the parent epoxide, it may serve as a substrate for further epoxidation elsewhere, particularly in large hydrophobic molecules such as polycyclic aromatic hydrocarbons in which it leads, for example, to the formation of highly toxic dihydrodiol epoxides resistant to epoxide hydrolase.

Epoxide hydrolase activity is present in most mammalian tissues and is high in the endoplasmic reticulum and cytosol of the liver. Humans appear to have considerable intersubject variability in the activity of epoxide hydrolase that may impact on individual susceptibility to the toxic action of certain epoxides: This variability argues for the existence of several epoxide hydrolases.

Some epoxides are resistant to epoxide hydrolase: Carbamazepine, for example, ismetabolizedto an epoxide that survives biotransformation and is excreted in urine.

REDUCTIVE BIOTRANSFORMATION PROCESSES

Reductive biotransformation processes are less common than those involving oxidation, but a number of functional groups are susceptible to reduction. Some may occur non-enzymically through the action of biologic reducing agents. Certain processes are part of a reversible oxidoreductase system, and the direction in which the equilibrium lies depends on the properties of the oxidized and reduced products.

Reductive processes are favored by anaerobic conditions, and besides the action of reductases in tissues such as liver, kidney, and lung, micro-organisms in the anaerobic environment of the intestinal tract may produce reduced metabolites that are subsequently absorbed.

Because reduction may often lead to metabolites of equivalent or greater lipophilicity than the precursor, it is seldom the only process occurring before a metabolite is excreted.

Carbon

Numerous ketones serve as substrates for alcohol dehydrogenases. These include alicyclic and arylalkyl ketones.

Unlike aldehydes, ketones are essentially unreactive, apart from being reduced when the equilibrium shifts predominantly to the secondary alcohol that is readily conjugated, thereby encouraging further reduction of the ketone. If ketone reduction results in the formation of an asymmetric carbon, a high degree of stereospecificity often occurs, even though the enzyme systems demonstrate a low degree of substrate specificity.

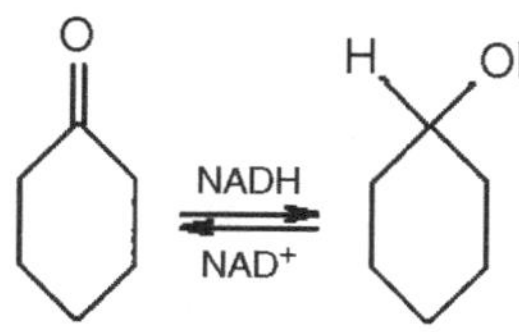

(14)

Alcohol dehydrogenase catalyzes the oxidation of primary alcohols to aldehydes, as well as the reverse reaction. However, due to the facile oxidation of aldehydes to acids, the reaction equilibrium is driven to the right.

$$RCH_2OH + NAD^+ \rightleftharpoons RCHO + NADH + H^+ \quad (15)$$

Thus administration of an aldehyde invariably results in the appearance of the corresponding acid as an excreted metabolite, although in some instances both the acid and alcohol are formed. Thus the aldehyde formed by deamination of the antihistamine chlorpheniramine is metabolized to a mixture of the corresponding acid and alcohol.

Reduction of carbon–carbon double bonds that may be instigated by gut flora has been observed.

Nitrogen

The types of nitrogen-containing compounds that are most frequently involved in reductive biotransformation are those containing nitro, azo, and N-oxide functional groups. Similar enzymes are involved that are generally located in the endoplasmic reticulum or cytosol of the liver or in the intestinal microflora. Complete reduction of a nitro compound to the primary amine involves a six-electron transfer and proceeds through nitroso and hydroxylamine intermediates.

$$R-NO_2 \longrightarrow R-NO \longrightarrow R-NHOH \quad (16)$$
$$\longrightarrow R-NH_2$$

Additional free-radical intermediates, such as the nitro anion radical and hydronitroxide radical, involving one-electron transfers have also been proposed. The benzodiazepine nitrazepam is reduced by liver microsomes to the corresponding amine.

However, reductions involving NADPH-cytochrome c reductase may form only the hydroxylamine. A high level of nitroreductase activity is associated with anaerobic bacteria in the gut, that are capable of reducing aromatic nitro compounds to the corresponding amines.

When the nitro group is located adjacent to a heteroatom in a heterocyclic ring, reduction to the nitroso or hydroxylamine intermediates often leads to cleavage of the ring. The hydroxylamine function in hydroxamic acids can also be reduced to the corresponding amides.

Aliphatic and aromatic tertiary amine N-oxides can be reduced to the corresponding amines. Reduction can occur either in liver cytosol, endoplasmic reticulum, or mitochondria, in extrahepatic tissues such as the kidney and lung, or by the gut flora.

Azoreductase activity is present in both the endoplasmic reticulum and the cytosol of the liver. It is thought to produce the hydrazo intermediate that is cleaved to give two primary amines.

The best known example is that of the azo dye Prontosil that is cleaved to form sulfanilamide, an active antibacterial compound, a reaction that involves a four-electron transfer. Certain halogenated hydrocarbons such as halothane, undergo reductive dehalogenation.

Sulfur

Most functional groups bearing sulfur, other than thiols, are primarily metabolized by oxidative processes. However, sulfoxides can be reduced to sulfides. The anti-inflammatory drug sulindac is reduced to the sulfide, which is an active metabolite, and it can be oxidized back to the sulfoxide. Disulfides are reduced to the corresponding sulfides.

OXIDATION BIOTRANSFORMATION PROCESS

Oxidation is the most commonly encountered biotransformation process, and few drugs undergo a Phase I reaction that does not involve it. Oxidation occurs most frequently at saturated (sp^3) or unsaturated (sp^2) carbon and with different valency states of nitrogen and sulfur present in either acyclic or cyclic structural environments.

Oxidation tends to decrease lipophilicity by the introduction of hydrophilic functional groups forming metabolites that may be more readily excreted. In addition, many of these functionalized metabolites are also substrates for Phase II conjugation reactions.

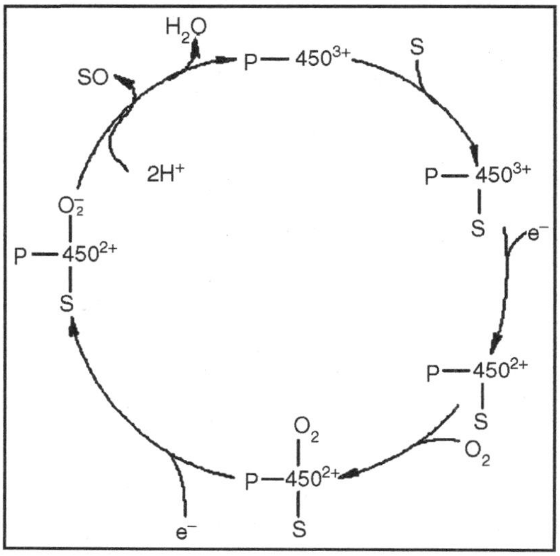

Fig. Postulated Mechanism of Cytochrome P450 Catalysis Net Reaction: Substrate + $2e^-$ + $2H^+$ O_2 $\rightarrow$ Substrate — O + H_2O.

The non-specific oxidation of drugs comprising many different chemical classes is catalyzed by a multienzyme mixed-function oxygenase system involving cytochrome P450 as the terminal oxidase. It is located in the endoplasmic reticulum of cells. Cytochrome P450 is a multigenic family of heme proteins, that can be distinguished from each other by their chromatographic and immunologic behaviour, as well as by substrate specificity. The mechanisms of cytochrome P450 catalysis has not been established unequivocally, but a generally accepted scheme is shown in Figure. The cytochromes P450 are the most versatile and extensively studied of the enzyme systems responsible for drug biotransformation.

After binding of the substrate to oxidized cytochrome P450, a one-electron transfer from NADPH is catalyzed by a flavoprotein reductase. Molecular oxygen combines with the reduced enzyme-substrate complex to form a ternary complex that accepts a second electron, possibly derived from NADH via cytochrome b5. One oxygen atom is transferred from the activated oxygen to the substrate to form the oxidized product. Water and oxidized cytochrome P450 also result. Other mono-oxygenases are not cytochrome P450 dependent, such as flavoproteins located in the endoplasmic reticulum that are involved in the oxidation of tertiary amines to N-oxides and of various sulfur compounds. Yet other oxidative enzymes, including alcohol and aldehyde dehydrogenases and monoamine oxidases, are located in the mitochondria or cytosol.

Aliphatic Carbon

Oxidation of primary equation, secondary equation, or tertiary equation carbon atoms all occur to give the corresponding alcohols. Further oxidation of a primary alcohol yields the aldehyde, which is usually rapidly converted to the carboxylic acid. The oxidation of secondary alcohols to ketones generally leads to less hydrophilic metabolites and is less common.

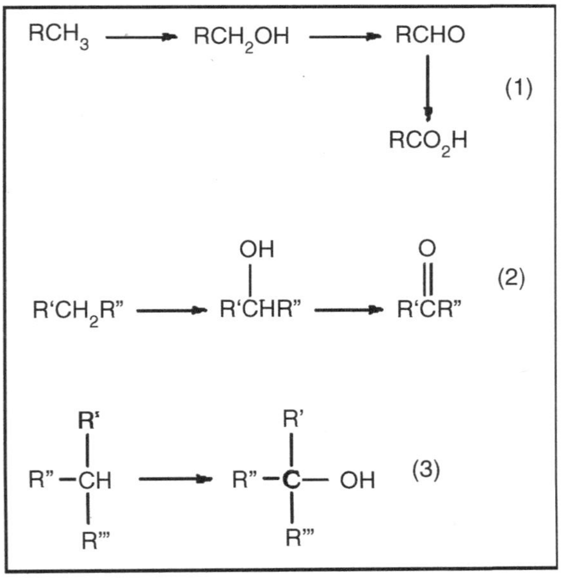

Metabolites derived by loss of an alkyl or arylalkyl group from ethers, thioethers, amines, and amides represent common biotransformation pathways (R0, R00 = H, alkyl or aryl). These processes involve oxidation on carbon adjacent to the heteroatom. The intermediates are generally unstable and readily decompose to the corresponding alcohol, thiol, amine, or amide and an aldehyde. Intermediates formed from amides are more stable and may be detected as excreted metabolites. If a secondary carbon atom is adjacent to the heteroatom, then this portion of the molecule is released as a ketone.

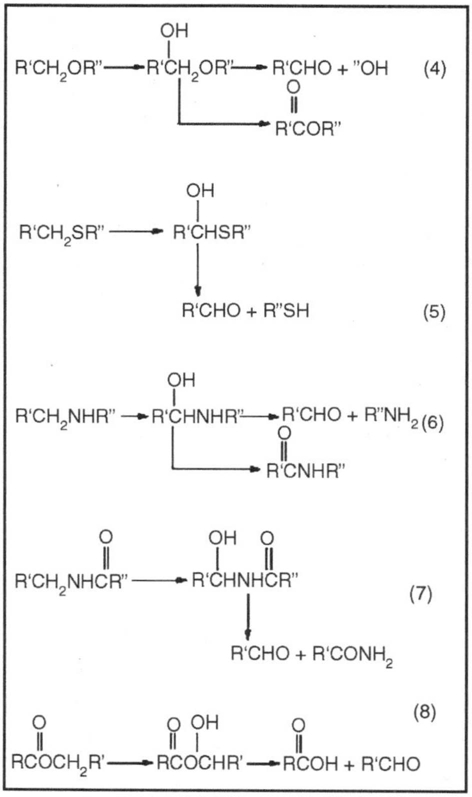

The heteroatom may also be located in a cyclic structure (e.g., morpholine, piperazine). Two processes have been adopted for amines, namely, N-dealkylation or deamination, that are essentially the same event. In general, which of the two terms applies depends on the side of the C–N bond that is of greatest concern, that is, whether interest is focused on the formation of the carbonyl compound or the amine.

For a primary amine, therefore, the process is termed deamination because the fate of the ammonia generated is of no interest. For amines, the intermediate hydroxylated products may be oxidized further to a carbonyl function, resulting in formation of amides. Cleavage of esters can also occur by an oxidative process of this type. The only difference compared to hydrolysis is that instead of forming an alcohol, this part of the molecule is released as an aldehyde.

Aromatic and Unsaturated Hydrocarbons

Double bonds in unsaturated hydrocarbons are oxidized by epoxidation. The resultant epoxides are generally reactive substrates that are biotransformed further to alcohols or glycols through the action of epoxide hydrolases. Hydroxylation of aromatic compounds to phenols has long been considered an important detoxication process, particularly as phenols can conjugate readily with glucuronic acid or sulfate to give highly watersoluble metabolites, that are excreted. The oxidation process can involve formation of arene oxide intermediates in which an aromatic double bond has undergone epoxidation. These epoxides are generally unstable and considered to be reactive intermediates. Phenols, nominally the major product of aromatic hydroxylation, do not necessarily require formation of intermediate epoxides. Epoxides can give rise to several different types of metabolites by spontaneous isomerization to phenols, enzymic hydration to dihydrodiols and conjugation with glutathione (GSH). The pattern of metabolites depends in part on the intrinsic reactivity of the epoxide. Epoxides can also react with a variety of other biologic nucleophilic centres in cellular macromolecules, such as DNA, RNA, and proteins; this may ultimately produce toxic effects. Other aromatic ring systems are also hydroxylated via postulated epoxide intermediates, the best known being furans.

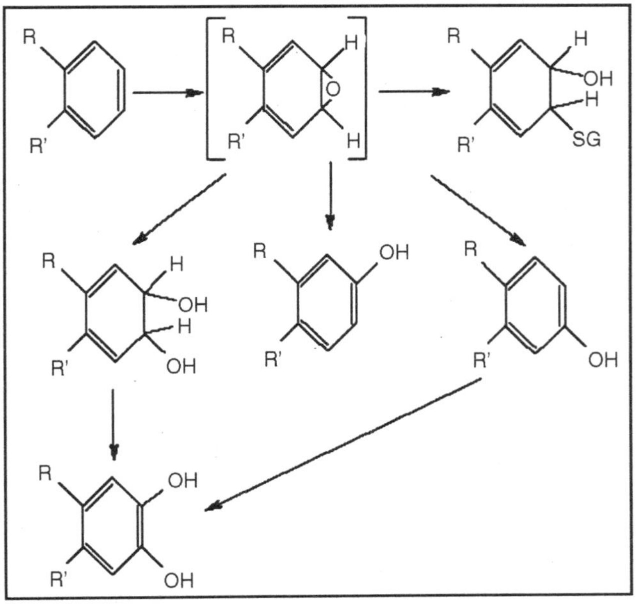

Fig. Biotransformation of an Aromatic Hydrocarbon to an Epoxide that Undergoes Various other Reactions.

The site of oxidation in an aromatic ring is dependent on the type of cytochrome P450 involved and on electronic and steric factors associated with the substrate structure. For substituted aromatic compounds, each epoxide can give rise to two isomeric phenols. Thus, methsuximide forms both a dihydrodiol and two isomeric phenols in animals and humans.

The dihydrodiols may be reduced to catechols through the action of dihydrodioldehydrogenase. Catechols can also be formed by two sequential hydroxylations. These catechols can be monomethylated by catechol O-methyltransferase, and are often the terminal metabolites formed from this pathway.

A rather novel biotransformation that can occur as part of an aromatic hydroxylation process has been called the NIH shift whereby a substituent such as bromo, chloro, and fluoro at the position of hydroxylation undergoes migration to the adjacent position.

Relevant examples are relatively few but this may be due to the lack of rigorous identification of ringhydroxylated metabolites.

Sulfur

Oxidation of divalent sulfur atoms in thioethers is a common biotransformation of sulfur-containing compounds. Oxidation proceeds in two stages, first to the sulfoxide and then to the sulfone.

Sulfoxides have increased polarity and are often observed as excreted metabolites, but they can also be reduced back to the sulfide. Formation of the sulfoxide creates an asymmetric centre, and stereospecific oxidation can occur. Sulfones tend to be terminal metabolites; no evidence for their reduction exists.

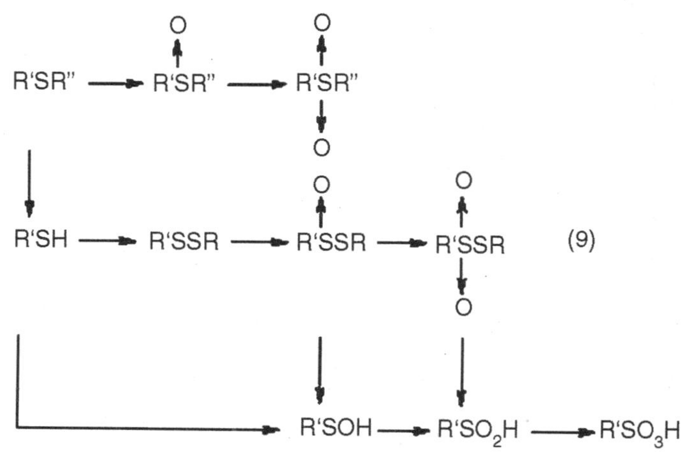

(9)

Thioethers can be cleaved to thiols that may be converted to stable oxygenated compounds, such as sulfinic and sulfonic acids, that are highly water-soluble and readily excreted. Thiols may be oxidized first to disulfides, in which further oxidation of sulfur then occurs before cleavage to sulfenic and sulfinic acids.

The thiocarbonyl group in compounds such as thioamides is biotransformed by oxidative attack on sulfur. Sulfoxidation of the thioamide occurs first and is followed by further S-oxidation to give an unstable intermediate that is readily hydrolyzed to the amide and thiosulfate.

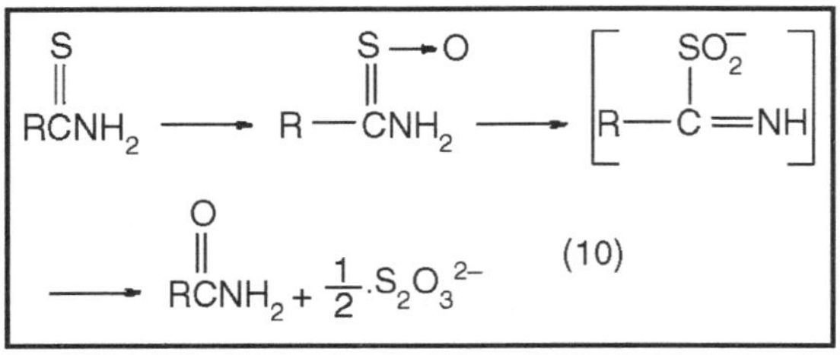

Nitrogen

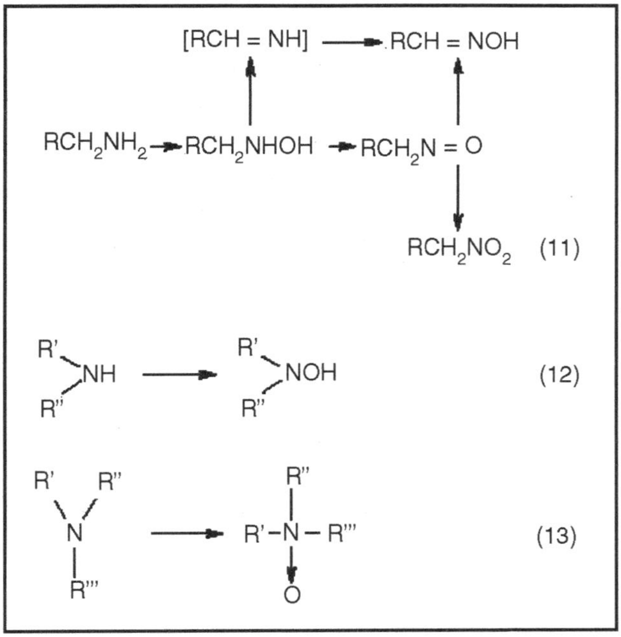

N-oxidation can occur in a number of ways to give either hydroxylamines from primary and secondary amines, hydroxamic acids from amides, or N-oxides from tertiary amines. The enzyme systems involved are either cytochrome P450 or a flavoprotein oxygenase. Hydroxylamines may be further oxidized to a nitro compound via a nitroso intermediate.

Oximes can be formed by rearrangement of the nitroso intermediate or N-hydroxylation of an imine, that could in turn be derived by dehydration of a hydroxylamine. N-Oxides may be formed from both tertiary arylamines and alkylamines and from nitrogen in heterocyclic aromatic systems, such as a pyridine ring.

NEED FOR DRUG BIOTRANSFORMATION

During their lives, humans take a wide, almost infinite variety of substances into their bodies, some of which are necessary for their well-being and provide the raw materials essential for the processes of intermediary metabolism. However, other substances which the body does not require are also ingested, inhaled, or absorbed; these substances have been described as exogenous compounds, foreign compounds, or xenobiotics (Greek xenos = foreign) and can be either naturally occurring or man-made.

The physicochemical properties (e.g., lipid solubility) that enable these compounds to be absorbed readily into the body would tend to preclude their facile excretion. Consequently, they would accumulate in the body and eventually cause deleterious effects. As a counter, metabolic processes have been adapted during evolution to deal with foreign compounds by converting (biotransforming) them into derivatives of great polarity (i.e., water solubility) that are consequently more easily excreted.

Drugs are one type of foreign compound that the body usually has to biotransform in order to effect their removal. The action of many drugs is curtailed by biotransformation. With some drugs, however, biotransformation results in metabolites that share the pharmacologic action of the parent drug or may even exhibit other properties.

The polynitrated organic nitrate vasodilators, such as isosorbide dinitrate, have metabolites that, as long as they contain nitrate groups, can exert the antianginal action of the parent drug, albeit at lower potency. Metabolites of benzodiazepines, such as diazepam, also possess its anxiolytic properties, and one of its metabolites, temazepam, is used as a hypnotic. Yet others (prodrugs) rely on biotransformation to generate the pharmacologically active agent. The depot antipsychotics are of this type; haloperidol decanoate relies on ester hydrolysis in the body to release the active drug, haloperidol.

CHEMICAL CONVERSION OF DRUGS

The term biotransformation of drugs can be defined as the chemical conversion of drugs to other compounds in the body. However, this definition excludes degradation due to any inherent chemical instability of drugs in biological media. These conversions are usually mediated by enzymes in the body's organs, tissues, or biofluids, but occasionally they may occur by nonenzymic reactions and even by a combination of both enzymic and non-enzymic processes. Synonymous with biotransformation in this context is the word metabolism, and indeed the terms drug biotransformation and drug metabolism are used to describe the same events and are often interchanged, seemingly somewhat randomly. If anything, the latter is the more popular term. Compared to the established scientific disciplines, the study of the biotransformation of drugs constitutes a relatively new subject area: the earliest scientific papers are little more than a century old. Regarded by many as a branch of pharmacology, drug biotransfo-rmation more properly belongs to biochemistry.

However, knowledge of the biotransformation of drugs often contributes, in no small way, to a better understanding of pharmacologic, toxicologic, and clinical phenomena, and papers on drug biotransformation are to be found in journals specializing in these other subjects as well as those of biochemistry and chemical analysis. In the last three decades, exponents of drug biotransformation have created their own journals and review series.

A leading pioneer in drug biotransformation research was the late R.T. Williams, who wrote Detoxication Mechanisms, the first major text on the subject published over 40 years ago, and a classic to this day. Subsequently, the literature on drug biotransformation grew steadily until recent years, when it seems to have expanded almost exponentially. The major driving forces for this proliferation include the realization of the importance of drug biotransformation knowledge in promoting a better understanding of certain pharmacologic, toxicologic, and/or clinical findings; governmental legislation requiring drug biotransformation studies as part of new chemical entity safety evaluation programs; major advances in the necessary scientific instrumentation; and the commercial availability of isotopically labeled compounds whose use can greatly facilitate drug biotransformation studies.

METHODS FOR THE STUDY OF DRUG BIOTRANSFORMATION

In Vivo

Investigations of biotransformation pathways in vivo require the collection and analysis of appropriate biologic samples. The types of sample collected include urine, feces, expired air, blood and/or plasma, bile, milk, saliva, synovial fluid, and tissues.

These samples can be divided into two groups:

- Those requiring complete collection (e.g., urine and feces) in order to provide quantitative as well as qualitative information on the excretion of the drug and its metabolites
- Those such as blood and milk that are subsampled at specific times to yield information on the identity and time-related concentrations of the drug and its metabolites that contribute to systemic exposure.

Samples of the above types can be obtained from all the common laboratory animals used in biomedical research. In addition, with the exception of bile and tissues, similar samples can usually be obtained from humans without great difficulty. One approach to the identification of metabolites is to investigate the presence of specific compounds in biologic samples when reference compounds are available.

Quantitative data will require validated bioanalytic methods for each specific compound and may not be readily obtained from certain biological matrices (e.g., feces). In addition, the possible presence of, and interference from, unknown metabolites would frequently create uncertainty.

Another approach is to attempt isolation and identification of all major metabolites, regardless of the availability of reference compounds. This approach requires an unambiguous method of detecting metabolites, such as use of a radiolabeled form of the drug.

Provided that the drug is radiolabeled in a metabolically stable position, this approach allows accurate measurement of metabolites in terms of percentage of administered dose and thence in concentration or quantity units. Procedures are well established for the detection and measurement of radiolabeled metabolites after chromatographic separation mainly by high-performance liquid chromatography and thinlayer chromatography.

Although chromatographic comparison with reference compounds may provide reasonable evidence for the identity of metabolites, more rigorous characterization by application of spectroscopic techniques to isolated samples is highly desirable.

Isolation and purification of metabolites are an important prelude to successful structural elucidation. The higher the state of chemical purity, the greater the chance of success in identification. Skillful selection of procedures according to the physicochemical properties of the metabolite can greatly facilitate generation of a viable sample.

The most common technique for structure elucidation is mass spectrometry due to the very high sensitivity and versatility of this technique for the introduction of samples into the mass spectrometer.

Further developments to increase sensitivity and specificity, new ionization techniques, and direct combination with high-performance liquid chromatography are ensuring the prime role of mass spectrometry.

Other techniques that complement mass spectrometry are nuclear magnetic resonance spectroscopy (particulary the emerging technique of LC-NMR), that generally requires larger samples of greater purity, and less commonly, infrared and ultraviolet spectroscopy.

In Vitro

Several biologic systems can be utilized for the study of drug biotransformation in vitro. These range in remoteness from the living animal in the following order: perfused organs, tissue slices, isolated cells, subcellular preparations, crude enzyme preparations, and purified enzymes.

In complexity, these systems range from the relatively simple to the exceedingly awkward; each system has its advantages and disadvantages, and none is ideal. Undoubtedly the most popular system has been the subcellular preparation because of its ease of production and convenience of use, and because expensive equipment is not always necessary.

Indeed, much of the earlier biotransformation knowledge in the literature has been gleaned from the use of subcellular preparations.

However, for a better understanding of catalytic mecha-nism, substrate specificity, and enzyme kinetics, purified enzymes are necessary. In recent years, the increasing availability of cDNA-expressed human enzymes responsible for drug biotransformation is greatly facilitating drug discovery and development. Several of the human cytochrome P450 enzymes and some others are commercially available in this form.

Perfused organs and isolated cells are often regarded as providing a better reflection of what might occur in the intact animal in vivo when the overall biotransformation of a compound is being considered, the latter are the much more popular.

STEREOSELECTIVE AND STEREOSPECIFIC BIOTRANSFORMATION

Many drugs contain one or more asymmetric centers, and thus consist of a mixture of enantiomers or diastereoisomers. For each asymmetric centre, there is a pair of enantiomers in a mirror-image relationship. Both pharmacologic activity and biotransformation usually require interaction of the drug with specific receptors or enzymes that can be highly stereoselective according to the steric configuration at the site of interaction.

Hence, a pair of enantiomers may differ in biologic activity and rates of biotransformation. The latter results in stereoselective pharmacokinetics, thereby also influencing biologic activity. Due to this fact, there has been a drive toward the development of single enantiomers as new drugs that provides the opportunity to optimise the therapeutic ratio (index).

In fact, single enantiomers of existing racemic drugs such as ibuprofen and bupivacaine have been developed as new products. Development of racemic drugs now requires detailed investigations of the pharmacokinetics and biotransformation of the individual enantiomers.

The cardiovascular drug propranolol is used as a racemate (i.e., an equal mixture of two enantiomers), with most of the therapeutic activity mediated by the (–)isomer. One of its major biotransformation pathways in the dog is glucuronidation, and urinary excretion of the glucuronic acid conjugate of (–) propranolol is about fourfold greater than that of the (+)isomer; thus, plasma concentrations of the latter are greater than those of the former.

Ring hydroxylation is another important pathway, and for this process in humans, there is a preference for the (+)isomer, leading to greater plasma concentrations of the active (–)isomer. Some biotransformations introduce an asymmetric centre into a drug and these often proceed stereospecifically. The most common examples are hydroxylation of a secondary carbon and the reduction of ketones to secondary alcohols. Ibuprofen undergoes both ω and ω -1 oxidation of the isobutyl side chain, and formation of the resulting carboxylic acid metabolite introduces a second asymmetric centre into the molecule. Both ibuprofen enantiomers have been shown to undergo stereospecific oxidation to give a metabolite with the same configuration at the new asymmetric centre.

Stereochemical inversion is another aspect of stereospecific biotransformation. The 2-arylpropionic acids containing an asymmetric centre represent a major class of the non-stero-idal anti-inflammatory drugs.

A notable aspect of their biotransform-ation is the conversion of one enantiomer to the other. Although most of these drugs were originally developed and used as a racemic mixture, only one of the isomers is usually regarded as being biologically active. The inactive (R) isomer of ibuprofen is converted to the active (S) isomer in vivo. A possible mechanism is dehydration to an *a*-methylene intermediate; this stage, involving a flavin-dependent dehydrogenase (FAD), is stereospecific.

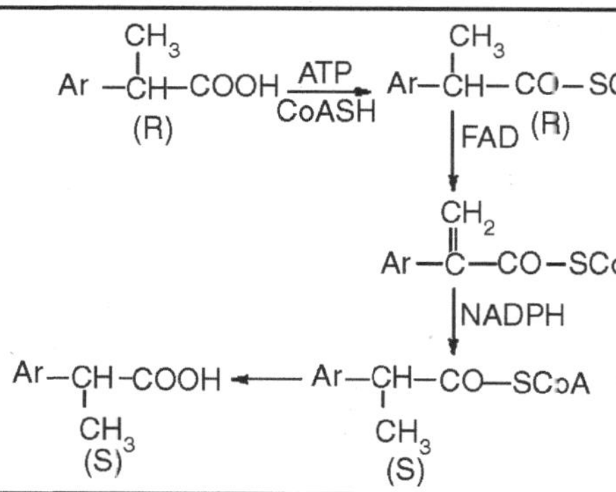

(34)

FACTORS AFFECTING DRUG BIOTRANSFORMATION

The preceding sections have illustrated that drugs can be biotransformed by a variety of reactions that are generally enzyme catalyzed. The rates of these reactions and their relative importance may be modulated by a number of potential interacting factors that can be broadly classified as physiologic or environmental. The influence of most of the factors mentioned has been revealed by laboratory animal studies, often involving high dose levels and exaggerated systemic exposure. In humans, where a single factor may not be seen to exert any notable effect, the interplay of several factors could contribute to important alterations in drug biotransformation and thereby to drug action.

Physiologic

Age

The activities of the enzymes responsible for drug biotransformation generally seem relatively low in the fetus, then increase rapidly after birth to reach adult levels, and then eventually decline in old age.

However, the developmental profiles of the various enzymes probably depend on the species and the enzyme system. The decline of drug biotransformation capability in old age may have important clinical consequences because the elderly comprise much of the patient population, whereas the pharmacokinetics and biotransformation of new drugs are mostly studied and first defined in young, healthy subjects.

Similarly, limited drug biotransformation capability in the very young requires careful selection of drug dose regimens in pediatrics.

Disease

In modern drug development, a knowledge of drug biotransformation and pharmacokinetics in certain disease states is essential, especially disease states of the liver and kidney because of their key role in drug elimination from the body. The impairment of these organs by disease could result in altered pharmacokinetics and drug and/or metabolite accumulation in the body. Cardiovascular disease is also of importance because of possible alterations in blood flow that could affect drug transport to the eliminating organs.

Hormones

A deficiency or surfeit of certain hormones can influence the activity of the enzymes involved in drug biotransformation. Apart from the sex hormones, those of the adrenal, pancreas, pituitary, and thyroid also exert effects on the regulation of certain enzyme activities.

Some hormonal disease states, such as diabetes and hyperthyroidism, could have important implications for drug biotransformation. Alterations in drug biotransformation can occur in pregnancy, and this subject area requires much more study.

Pharmacogenetics

Intersubject variability in drug biotransformation can be very large primarily due to genetic factors modulated by other factors such as those mentioned elsewhere in this section. Thus the ability of certain individuals to biotransform particular drugs is greatly diminished owing to a genetically inherited deficiency in the enzyme(s) mainly responsible for the biotransformation(s) of those drugs.

Although enzyme systems, such as the cytochromes P450 (CYP), have broad and overlapping substrate specificities, one enzyme alone (because of high enzyme-substrate affinity) may be dominant in the biotransformation of a particular drug. Thus a deficiency in this enzyme (or its inhibition by concomitant medication) can lead to exaggerated pharmacologic or toxicologic effects. For example, about 10% of the Caucasian population is deficient in CYP2D6, whereas this deficiency is less than 1% in the Japanese population where the major known CYP deficiency of about 20% is in CYP2C19 (about 5% in Caucasians).

Drugs that have been shown to be subject to pharmacogenetic differences in biotransformation include debrisoquine, perhexiline, phenformin, mephenytoin, tolbutamide, dapsone, isoniazid and sulfadimidine. The latter three drugs are primarily biotransformed by N-acetylation, a pathway for which about 50% of the United Kingdom population can be classified as slow acetylators and the rest as fast acetylators.

In modern drug development, it is prudent to avoid developing those compounds likely to be exclusively biotransformed by an enzyme known to be deficient in a notable proportion of the population.

Sex (gender)

The biotransformation of certain drugs may differ in males and females. Such sex differences are fairly common in the rat, which is probably the most widely utilized laboratory animal in drug safety evaluation, but have been infrequently reported for humans or even for other laboratory animal species.

When encountered, these sex differences appear to arise as a result of the regulation of drug biotransformation by hormones such as growth hormone.

ENVIRONMENTAL

Chemical Exposure

Exposure to chemicals can be intentional or unintentional. The former includes the ingestion of drugs and food additives, alcohol consumption, and tobacco smoking; the latter includes ingestion of agrochemical residues present in food and chemical exposure in the workplace or environment.

The extent of exposure may be sufficient to affect the activities of the enzymes involved in drug biotransformation to cause either an increase (induction) or a decrease (inhibition) of these activities.

Barbiturates, for example, can cause an increase, and cimetidine can cause a decrease. Induction caused by the organochlorine pesticides DDT and dieldrin and the consequences thereof have been well publicized.

The consequences of inhibition of biotransformation are usually more serious than those of induction, and drugs, such as terfenadine and mibefradil, have been withdrawn from clinical use owing to unacceptable drug–drug interactions arising from enzyme inhibition.

Paradoxically, others, such as ketoconazole and quinidine, that are potent inhibitors of cytochrome P450 enzymes, remain in clinical use.

Diet

The diet contains an almost infinite number of foreign chemicals, and the enzymes of drug biotransformation probably evolved to cope with these chemicals. Consequently they were already available to deal with subsequently developed drugs.

The activities of these enzymes can be affected by dietary constituents that may serve to increase or decrease them or cause both effects; in alcohol ingestion, enzyme activities show a short-term decrease (inhibition) and then increase (induction). They can also be affected by nutritional status and reflect protein, fat, carbohydrate, mineral, and vitamin intake.

Stress

Stress can perturb the well being of the body and result in alterations in the rate of drug biotransformation. Such stress factors include abnormal temperature, violent exercise, and disease states.

SITES OF DRUG BIOTRANSFORMATION

The liver is usually considered to be the main site of drug biotransformation. Other organs and tissues, including those located at portals of entry or exit of drugs into or from the body, such as the intestinal tract, lungs, skin, and kidneys, also make major contributions.

Certain other organs and tissues may be important to the biotransformation of particular drugs; for example, nitroglycerin has been shown to be biotransformed in vascular tissue. Within an individual organ or tissue, the enzyme systems responsible for drug biotransformation are not uniformly distributed but vary in activity according to cell type and differ in subcellular localization.

In the lung, for example, cytochrome P450 activity is relatively high in Clara cells. Although most types of organs and tissues contain the enzyme systems involved in drug biotransformation, the enzyme activities may be expressed by a family of enzymes, whose individual distribution may be organ or tissue specific. The profile of these enzymes can be altered, for example, during treatment with certain compounds.

The biotransformation of many drugs is often slower in humans than in the laboratory animal species used in drug development and safety evaluation.

Lower activities of the respective enzyme systems may be the cause, and indeed the rates of drug biotransformation and the bodyweights of animal species appear to be correlated to some extent.

IMPACT ON ROUTES OF EXCRETION

The two main routes of systemic excretion of drugs and their metabolites are renal (urinary) and hepatic (biliary). Which of these occurs or predominates depends mainly on the physicochemical properties of the metabolites.

Sulfate conjugates are more likely to be present in urine and glucuronic acid conjugates appear more often in bile. Glutathione conjugates are almost exclusively excreted in bile. However, this is both compound- and species-dependent due to molecular- weight thresholds for biliary excretion. Thus, many glucuronides are almost always actively excreted in the bile of rats but excreted primarily in urine of humans and non-human primates. Conjugation with sulfate, glucuronic acid, and glutathione increases the molecular-weight by as much as 80, 177, and 307 mass units, respectively.

The molecular-weight thresholds for active biliary excretion are about 330 for the rat, and dog; 450 for the rabbit and non-human primate; and 500 for humans.

8

Vitamins

ORGANIC COMPOUNDS OF THE VITAMINS

Vitamin, any of the organic (carbon-containing) compounds that the body requires in small amounts to maintain health and function properly. Children additionally need vitamins to grow. The body gets most of its vitamins from the foods we eat.

A healthy diet with plenty of fruits and vegetables should provide nearly all of the vitamins a person needs. Scientists have classified 13 compounds as vitamins. They have given most of these vitamins letter or letter plus number names, such as A, B12, and D. Most vitamins are produced by plants. Some, such as vitamin D, are produced only by animals. A few vitamins are made by the body itself. For example, bacteria in the digestive tract help produce vitamin K, and the skin uses sunlight to produce vitamin D.

Vitamins are also manufactured for sale as supplements for people who need additional vitamins to meet their body's requirements. For example, doctors often prescribe vitamin supplements for pregnant and nursing women to provide the additional nutrients needed by a rapidly growing fetus or infant. Older adults may not meet their vitamin requirements through food because the body's ability to absorb vitamins is impaired with age.

Expressed in milligrams (mg) or international units (IU) for adults and children of normal health, these recommendations provide useful guidelines for daily intake. Such guidelines are useful not only for professionals in nutrition but also for the growing number of families and individuals who eat irregular meals and rely on prepared foods, most of which are now required to carry nutritional labeling.

VITAMINS ARE VITAL

The word vitamine, later shortened to vitamin, was coined by Polish-American chemist Casimir Funk in the early 20th century. Funk was searching for the then-unknown substance in foods that prevents such diseases as beriberi, rickets, and scurvy. After experiments on pigeons, Funk guessed that an amine-a compound containing nitrogen and hydrogen-in foods was responsible. He called this substance a vital (necessary for life) amine, or vitamine. Scientists later discovered that not all vitamins were amines, however.

Vitamins help the body carry out essential biochemical processes. Vitamins generally combine with proteins to create enzymes that promote chemical reactions. These enzymes play an important role in metabolism-the reactions that break down the fats, carbohydrates, and proteins in food so that the body can use them for energy and cell repair.

The enzymes also promote reactions involved in the formation of bone, hormones, blood cells, nervous-system chemicals, and genetic material. Without vitamins, many of these reactions would slow down or cease. The different vitamins are not chemically related, and most differ in their actions in the body.

Deficiency of particular vitamins can lead to various diseases. Too little vitamin C, for example, can cause rickets, a disease in which the bones fail to develop properly. Bowlegs or"knock knees" are signs of rickets. Rickets and the other vitamin-deficiency diseases that interested Casimir Funk have largely disappeared in the developed world as the result of fortified foods and improved nutrition. But these diseases still occur in the developing world, especially where malnutrition is common.

Interest in vitamins grew during the 1990s, especially in those vitamins that act as antioxidants-vitamins A, C, and E. Antioxidants neutralize molecules known free radicals that cause cell damage. Free radicals have been linked to a number of disorders, including Alzheimer's disease, arteriosclerosis, cancer, diabetes, and Parkinson disease. Researchers theorized that increased doses of antioxidants might also prevent and even cure such diseases. These claims have so far not panned out, however, as scientific studies have failed to demonstrate the preventive abilities of antioxidants. Research on antioxidants and other vitamins continues.

VITAMINS AND MINERALS

Vitamins are organic molecules that function in a wide variety of capacities within the body. The most prominent function of the vitamins is to serve as cofactors (co-enzymes) for enzymatic reactions. The distinguishing feature of the vitamins is that they generally cannot be synthesized by mammalian cells and, therefore, must be supplied in the diet. The vitamins are of two distinct types, water soluble and fat soluble. The minerals that are considered of dietary significance are those that are necessary to support biochemical reactions by serving both functional and structural roles as well as those serving as electrolytes.

The use of the term dietary mineral is considered archaic since the intent of the term "mineral" is to describe ions not actual minerals. There are both quantity elements required by the body and trace elements. The quantity elements are sodium, magnesium, phosphorous, sulfur, chlorine, potassium and calcium. The essential trace elements are manganese, iron, cobalt, nickel, copper, zinc, selenium, molybdenum, and iodine. Additional trace elements (although not considered essential) are boron, chromium, fluoride, and silicon.

THIAMIN

Thiamin is also known as vitamin B_1. Thiamin is derived from a substituted pyrimidine and a thiazole which are coupled by a methylene bridge. Thiamin is rapidly converted to its active form, thiamin pyrophosphate, TPP, in the brain and liver by a specific enzyme, thiamin diphosphotransferase.

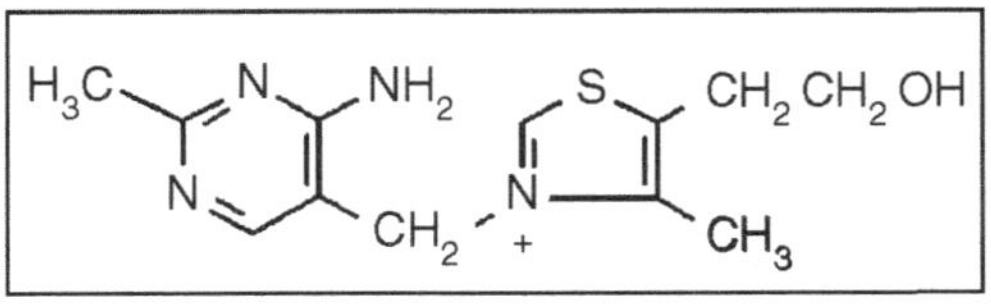

Fig. Thiamin Structure.

TPP is necessary as a cofactor for the pyruvate dehydrogenase and α-ketoglutarate dehydrogenase catalyzed reactions as well as the transketolase catalyzed reactions of the pentose phosphate pathway. A deficiency in thiamin intake leads to a severely reduced capacity of cells to generate energy as a result of its role in these reactions.

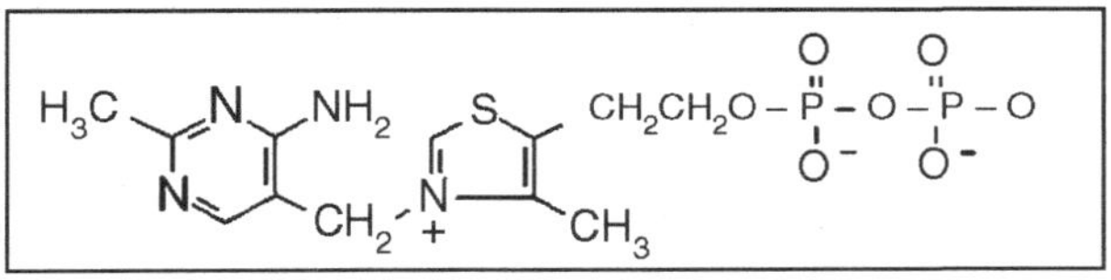

Fig. Thiamin Pyrophosphate.

The dietary requirement for thiamin is proportional to the caloric intake of the diet and ranges from 1.0–1.5 mg/day for normal adults. If the carbohydrate content of the diet is excessive then an increase in thiamin intake will be required.

CLINICAL SIGNIFICANCES OF THIAMIN DEFICIENCY

The earliest symptoms of thiamin deficiency include constipation, appetite suppression, nausea as well as mental depression, peripheral neuropathy and fatigue. Chronic thiamin deficiency leads to more severe neurological symptoms including ataxia, mental confusion and loss of eye coordination. Other clinical symptoms of prolonged thiamin deficiency are related to cardiovascular and musculature defects.

The severe thiamin deficiency disease known as Beriberi, is the result of a diet that is carbohydrate rich and thiamin deficient. An additional thiamin deficiency related disease is known as Wernicke-Korsakoff syndrome. This disease is most commonly found in chronic alcoholics due to their poor dietetic lifestyles.

RIBOFLAVIN

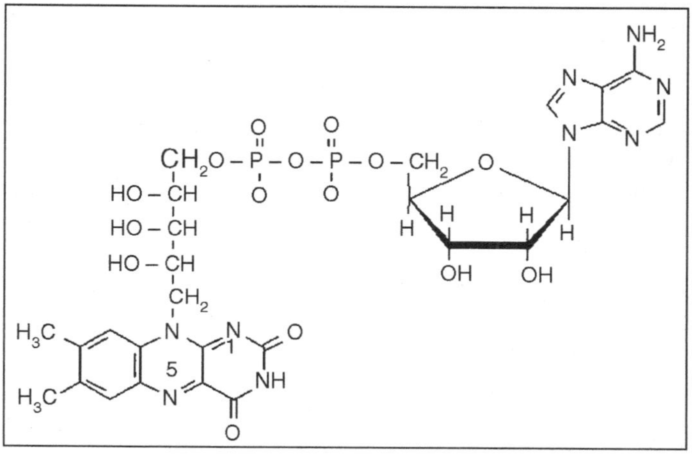

Fig. Structure of FAD.

Riboflavin is also known as vitamin B_2. Riboflavin is the precursor for the coenzymes, flavin mononucleotide (FMN) and flavin adenine dinucleotide (FAD).

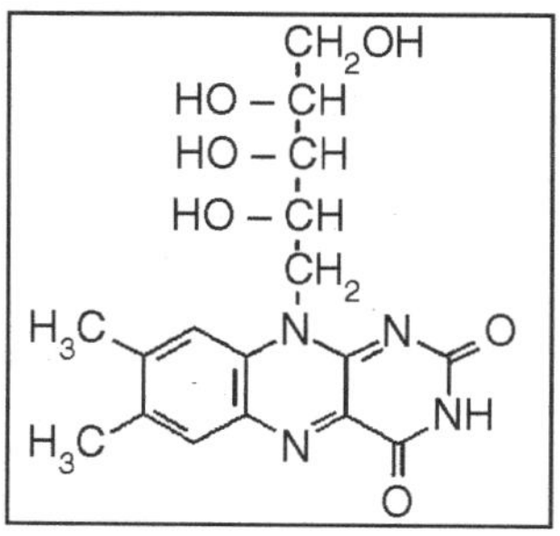

Fig. Riboflavin Structure.

The enzymes that require FMN or FAD as cofactors are termed flavoproteins. Several flavoproteins also contain metal ions and are termed metalloflavoproteins. Both classes of enzymes are involved in a wide range of redox reactions, e.g. succinate dehydrogenase and xanthine oxidase.

During the course of the enzymatic reactions involving the flavoproteins the reduced forms of FMN and FAD are formed, $FMNH_2$ and $FADH_2$, respectively. The hydrogens of $FADH_2$ are on nitrogens 1 and 5 as indicated in the figure. The normal daily requirement for riboflavin is 1.2–1.7 mg/day for normal adults.

CLINICAL SIGNIFICANCES OF FLAVIN DEFICIENCY

Riboflavin deficiencies are rare in the United States due to the presence of adequate amounts of the vitamin in eggs, milk, meat and cereals. Riboflavin deficiency is often seen in chronic alcoholics due to their poor dietetic habits.

Symptoms associated with riboflavin deficiency include itching and burning eyes, angular stomatitis and cheilosis (cracks and sores in the mouth and lips), bloodshot eyes, glossitis (inflammation of the tongue leading to purplish discoloration), seborrhea (dandruff, flaking skin on scalp and face), trembling, sluggishness, and photophobia (excessive light sensitivity). Riboflavin decomposes when exposed to visible light. This characteristic can lead to riboflavin deficiencies in newborns treated for hyperbilirubinemia by phototherapy.

NIACIN

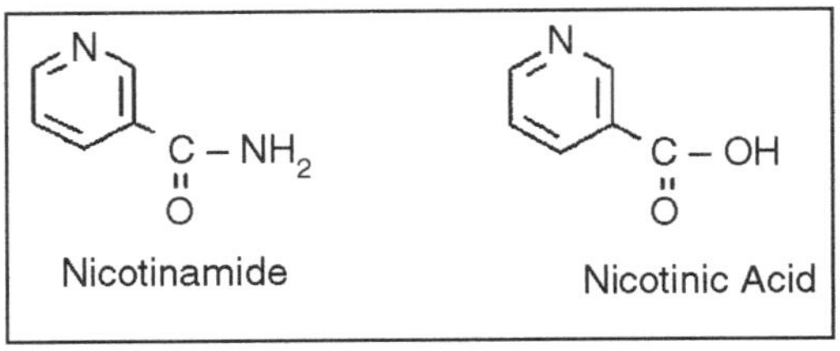

Niacin (nicotinic acid and nicotinamide) is also known as vitamin B_3. Both nicotinic acid and nicotinamide can serve as the dietary source of vitamin B_3. Niacin is required for the synthesis of the active forms of vitamin B_3, nicotinamide adenine dinucleotide (NAD^+) and nicotinamide adenine dinucleotide phosphate ($NADP^+$). Both NAD^+ and $NADP^+$ function as cofactors for numerous dehydrogenases, e.g., lactate dehydrogenase and malate dehydrogenase.

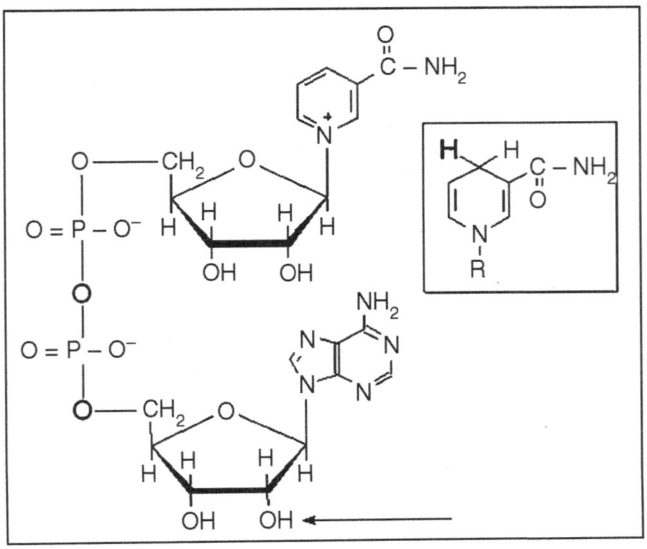

Structure of NAD^+NADH is shown in the box insert. The -OH phosphorylated in $NADP^+$ is indicated by the red arrow.

Niacin is not a true vitamin in the strictest definition since it can be derived from the amino acid tryptophan. However, the ability to utilize tryptophan for niacin synthesis is inefficient (60 mg of tryptophan are required to synthesize 1 mg of niacin).

Also, synthesis of niacin from tryptophan requires vitamins B_1, B_2 and B_6 which would be limiting in themselves on a marginal diet. The recommended daily requirement for niacin is 13–19 niacin equivalents (NE) per day for a normal adult. One NE is equivalent to 1 mg of free niacin).

CLINICAL SIGNIFICANCES OF NIACIN AND NICOTINIC ACID

A diet deficient in niacin (as well as tryptophan) leads to glossitis of the tongue (inflammation of the tongue leading to purplish discoloration), dermatitis, weight loss, diarrhea, depression and dementia. The severe symptoms, depression, dermatitis and diarrhea, are associated with the condition known as pellagra.

Several physiological conditions (e.g. Hartnup disease and malignant carcinoid syndrome) as well as certain drug therapies (e.g. isoniazid) can lead to niacin deficiency. In Hartnup disease tryptophan absorption is impaired and in malignant carcinoid syndrome tryptophan metabolism is altered resulting in excess serotonin synthesis. Isoniazid (the hydrazide derivative of isonicotinic acid) is a primary drug for chemotherapy of tuberculosis.

Nicotinic acid (but not nicotinamide) when administered in pharmacological doses of 2–4 g/day lowers plasma cholesterol levels and has been shown to be a useful therapeutic for hypercholesterolemia. The major action of nicotinic acid in this capacity is a reduction in fatty acid mobilization from adipose tissue.

Although nicotinic acid therapy lowers blood cholesterol it also causes a depletion of glycogen stores and fat reserves in skeletal and cardiac muscle. Additionally, there is an elevation in blood glucose and uric acid production. For these reasons nicotinic acid therapy is not recommended for diabetics or persons who suffer from gout.

PANTOTHENIC ACID

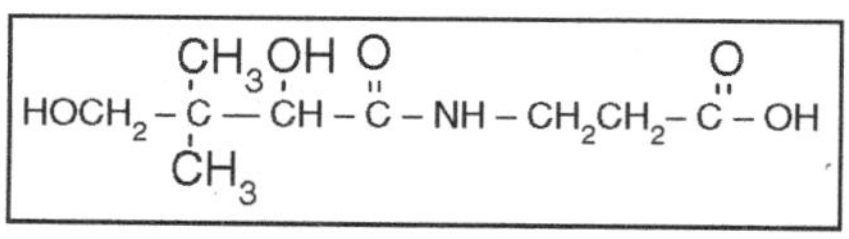

Fig. Pantothenic Acid.

Pantothenic acid is also known as vitamin B_5. Pantothenic acid is formed from β-alanine and pantoic acid. Pantothenate is required for synthesis of coenzyme A, CoA and is a component of the acyl carrier protein (ACP) domain of fatty acid synthase. Pantothenate is, therefore, required for the metabolism of carbohydrate via the TCA cycle and all fats and proteins. At least 70 enzymes have been identified as requiring CoA or ACP derivatives for their function. Deficiency of pantothenic acid is extremely rare due to its widespread distribution in whole grain cereals, legumes and meat. Symptoms specific to pantothenate deficiency are difficult to assess since they are subtle and resemble those of other B vitamin deficiencies.

These symptoms include painful and burning feet, skin abnormalities, retarded growth, dizzy spells, digestive disturbances, vomiting, restlessness, stomach stress, and muscle cramps.

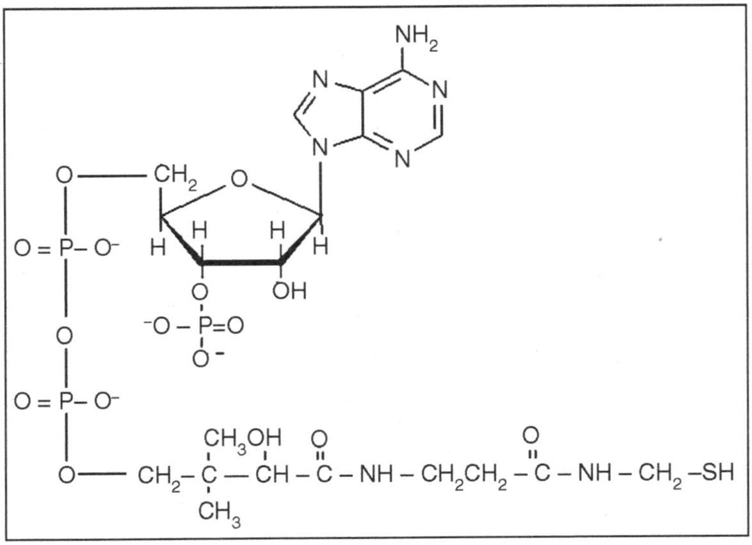

Fig. Coenzyme A.

VITAMIN B_6

Pyridoxal, pyridoxamine and pyridoxine are collectively known as vitamin B_6. All three compounds are efficiently converted to the biologically active form of vitamin B_6, pyridoxal phosphate (PLP). This conversion is catalyzed by the ATP requiring enzyme, pyridoxal kinase. Pyridoxal kinase requires zinc for full activity thus making it a metaloenzyme.

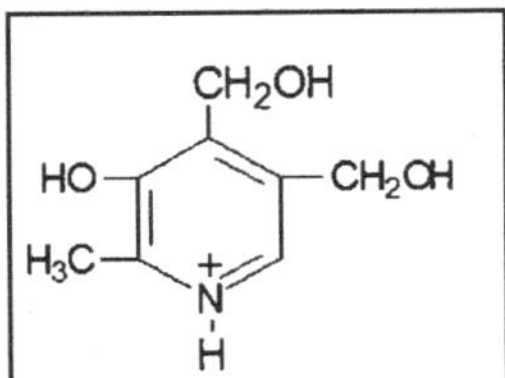

Pyridoxine

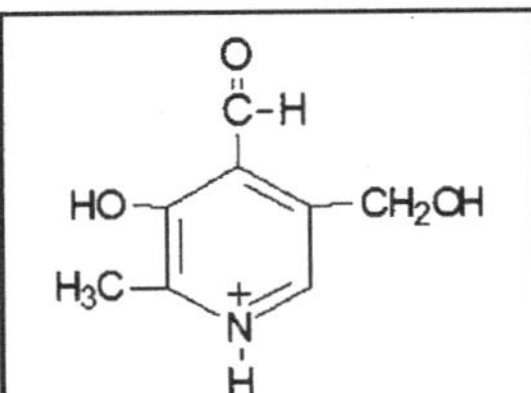

Pyidoxal

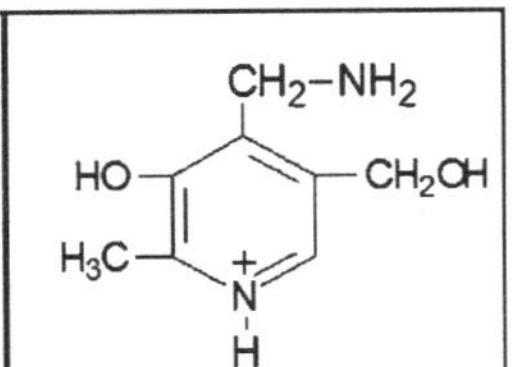

Pyridoxamine

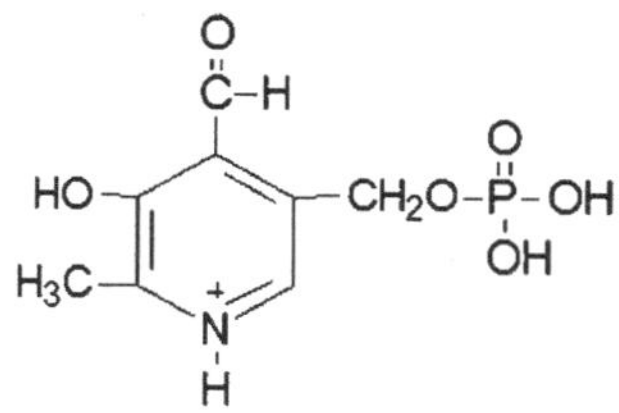

Fig. Pyridoxal Phosphate.

Pyridoxal phosphate functions as a cofactor in enzymes involved in transamination reactions required for the synthesis and catabolism of the amino acids as well as in glycogenolysis as a cofactor for glycogen phosphorylase and as a co-factor for the synthesis of the inhibitory neurotransmitter ã-aminobutyric acid (GABA). The requirement for vitamin B_6 in the diet is proportional to the level of protein consumption ranging from 1.4–2.0 mg/day for a normal adult. During pregnancy and lactation the requirement for vitamin B_6 increases approximately 0.6 mg/day.

Deficiencies of vitamin B_6 are rare and usually are related to an overall deficiency of all the B-complex vitamins. Isoniazid and penicillamine (used to treat rheumatoid arthritis and cystinurias) are two drugs that complex with pyridoxal and PLP resulting in a deficiency in this vitamin.

Deficiencies in pyridoxal kinase result in reduced synthesis of PLP and are associated with seizure disorders related to a reduction in the synthesis of GABA. Other symptoms that may appear with deficiency in vitamin B_6 include nervousness, insomnia, skin eruptions, loss of muscular control, anemia, mouth disorders, muscular weakness, dermatitis, arm and leg cramps, loss of hair, slow learning, and water retention.

BIOTIN

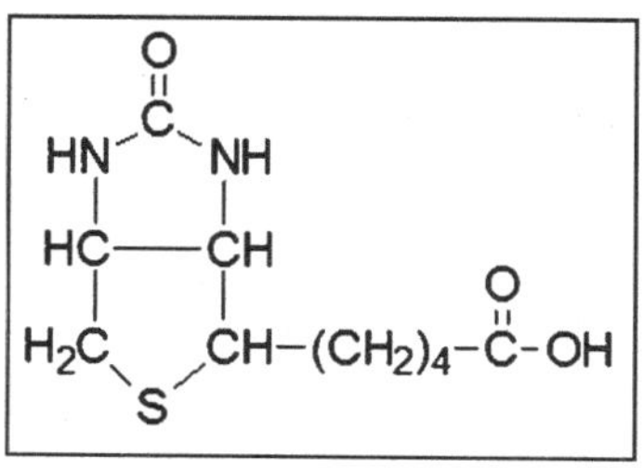

Fig. Biotin

Biotin is the cofactor required of enzymes that are involved in carboxylation reactions, e.g. acetyl-CoA carboxylase and pyruvate carboxylase. Biotin is found in numerous foods and also is synthesized by intestinal bacteria and as such deficiencies of the vitamin are rare. Deficiencies are generally seen only after long antibiotic therapies which deplete the intestinal fauna or following excessive consumption of raw eggs. The latter is due to the affinity of the egg white protein, avidin, for biotin preventing intestinal absorption of the biotin. Symptoms that may appear if biotin is deficient are extreme exhaustion, drowsiness, muscle pain, loss of appetite, depression, and grayish skin colour.

COBALAMIN

Cobalamin is more commonly known as vitamin B_{12}. Vitamin B_{12} is composed of a complex tetrapyrrol ring structure (corrin ring) and a cobalt ion in the centre. Vitamin B_{12} is synthesized exclusively by microorganisms and is found in the liver of animals bound to protein as methycobalamin or 5'-deoxyadenosylcobalamin. The vitamin must be hydrolyzed from protein in order to be active. Hydrolysis occurs in the stomach by gastric acids or the intestines by trypsin digestion following consumption of animal meat. The vitamin is then bound by intrinsic factor, a protein secreted by parietal cells of the stomach, and carried to the ileum where it is absorbed. Following absorption the vitamin is transported to the liver in the blood bound to transcobalamin II.

There are only two clinically significant reactions in the body that require vitamin B_{12} as a cofactor. During the catabolism of fatty acids with an odd number of carbon atoms and the amino acids valine, isoleucine and threonine the resultant propionyl-CoA is converted to succinyl-CoA for oxidation in the TCA cycle. One of the enzymes in this pathway, methylmalonyl-CoA mutase, requires vitamin B_{12} as a cofactor in the conversion of methylmalonyl-CoA to succinyl-CoA. The 5'-deoxyadenosine derivative of cobalamin is required for this reaction. The second reaction requiring vitamin B_{12} catalyzes the conversion of homocysteine to methionine and is catalyzed by methionine synthase. This reaction results in the transfer of the methyl group from N^5-methyltetrahydrofolate to hydroxycobalamin generating tetrahydrofolate (THF) and methylcobalamin during the process of the conversion.

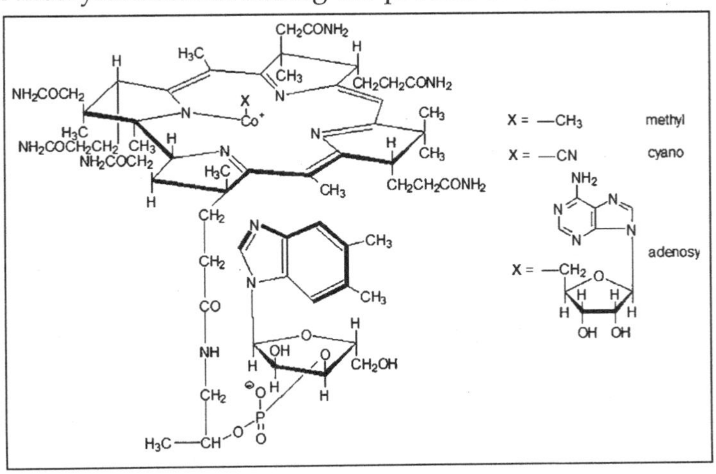

Fig. Vitamin B_{12}

CLINICAL SIGNIFICANCES OF B_{12} DEFICIENCY

The liver can store up to six years worth of vitamin B_{12}, hence deficiencies in this vitamin are rare. Pernicious anemia is a megaloblastic anemia resulting from vitamin B_{12} deficiency that develops as a result a lack of intrinsic factor in the stomach leading to malabsorption of the vitamin. The anemia results from impaired DNA synthesis due to a block in purine and thymidine biosynthesis. The block in nucleotide biosynthesis is a consequence of the effect of vitamin B_{12} on folate metabolism. When vitamin B_{12} is deficient essentially all of the folate becomes trapped as the N^5-methylTHF derivative as a result of the loss of functional methionine synthase. This trapping prevents the synthesis of other THF derivatives required for the purine and thymidine nucleotide biosynthesis pathways.

Neurological complications also are associated with vitamin B_{12} deficiency and result from a progressive demyelination of nerve cells. The demyelination is thought to result from the increase in methylmalonyl-CoA that result from vitamin B_{12} deficiency.

Methylmalonyl-CoA is a competitive inhibitor of malonyl-CoA in fatty acid biosynthesis as well as being able to substitute for malonyl-CoA in any fatty acid biosynthesis that may occur. Since the myelin sheath is in continual flux the methylmalonyl-CoA-induced inhibition of fatty acid synthesis results in the eventual destruction of the sheath.

The incorporation methylmalonyl-CoA into fatty acid biosynthesis results in branched-chain fatty acids being produced that may severely alter the architecture of the normal membrane structure of nerve cells.

Deficiencies in B_{12} can also lead to elevations in the level of circulating homocysteine. Elevated levels of homocysteine are known to lead to cardiovascular dysfunction. Due to its high reactivity to proteins, homocysteine is almost always bound to proteins, thus thiolating them leading to their degradation.

Homocysteine also binds to albumin and hemoglobin in the blood. The detrimental effects of homocysteine are thought to be due to its' binding to lysyl oxidase, an enzyme responsible for proper maturation of the extracellular matrix proteins collagen and elastin.

Production of defective collagen and elastin has a negative impact on arteries, bone and skin and the effects on arteries are believed to be the underlying cause for cardiac dysfunction associated with elevated serum homocysteine.

FOLIC ACID

Folic Acidpositions 7 and 8 carry hydrogens in dihydrofolate (DHF) positions 5–8 carry hydrogens in tetrahydrofolate (THF) Folic acid is a conjugated molecule consisting of a pteridine ring structure linked to para-aminobenzoic acid (PABA) that forms pteroic acid.

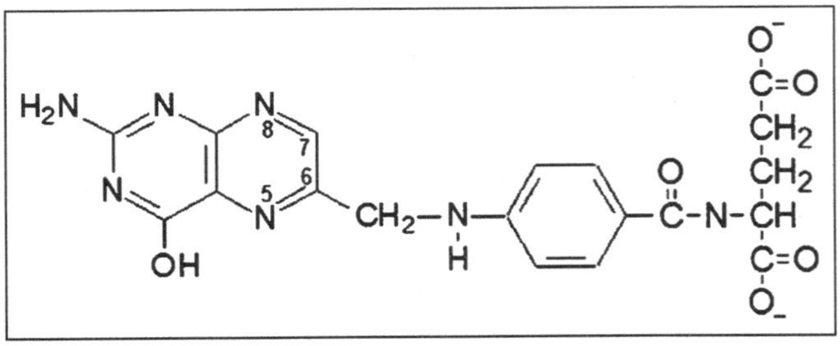

Fig. Folic Acid.

Folic acid itself is then generated through the conjugation of glutamic acid residues to pteroic acid. Folic acid is obtained primarily from yeasts and leafy vegetables as well as animal liver. Animal cannot synthesize PABA nor attach glutamate residues to pteroic acid, thus, requiring folate intake in the diet.

When stored in the liver or ingested folic acid exists in a polyglutamate form. Intestinal mucosal cells remove some of the glutamate residues through the action of the lysosomal enzyme, conjugase.

The removal of glutamate residues makes folate less negatively charged (from the polyglutamic acids) and therefore more capable of passing through the basal lamenal membrane of the epithelial cells of the intestine and into the bloodstream. Folic acid is reduced within cells (principally the liver where it is stored) to tetrahydrofolate (THF also H_4folate) through the action of dihydrofolate reductase (DHFR), an NADPH-requiring enzyme. The function of THF derivatives is to carry and transfer various forms of one carbon units during biosynthetic reactions. The one carbon units are either methyl, methylene, methenyl, formyl or formimino groups.

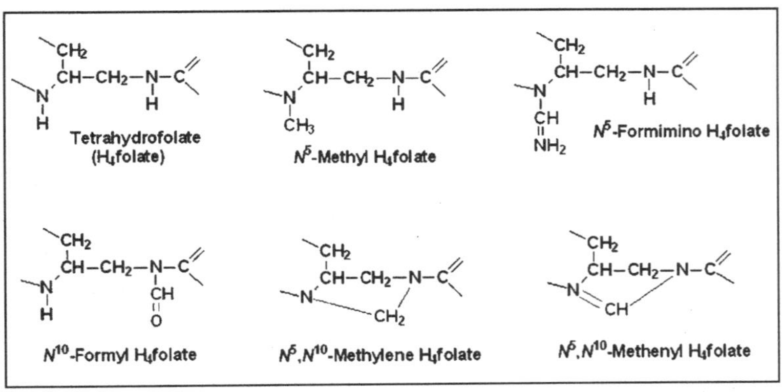

Active centre of tetrahydrofolate (THF).Note that the N^5 position is the site of attachment of methyl groups, the N^{10} the site for attachment of formyl and formimino groups and that both N^5 and N^{10} bridge the methylene and methenyl groups. These one carbon transfer reactions are required in the biosynthesis of serine, methionine, glycine, choline and the purine nucleotides and dTMP. The ability to acquire choline and amino acids from the diet and to salvage the purine nucleotides makes the role of N^5,N^{10}-methylene-THF in dTMP synthesis the most metabolically significant function for this vitamin. The role of vitamin B_{12} and N^5-methyl-THF in the conversion of homocysteine to methionine also can have a significant impact on the ability of cells to regenerate needed THF.

CLINICAL SIGNIFICANCES OF FOLATEDEFICIENCY

Folate deficiency results in complications nearly identical to those described for vitamin B_{12} deficiency. The most pronounced effect of folate deficiency on cellular processes is upon DNA synthesis. This is due to an impairment in dTMP synthesis which leads to cell cycle arrest in S-phase of rapidly proliferating cells, in particular hematopoietic cells.

The result is megaloblastic anemia as for vitamin B_{12} deficiency. The inability to synthesize DNA during erythrocyte maturation leads to abnormally large erythrocytes termed macrocytic anemia.

Folate deficiencies are rare due to the adequate presence of folate in food. Poor dietary habits as those of chronic alcoholics can lead to folate deficiency. The predominant causes of folate deficiency in non-alcoholics are impaired absorption or metabolism or an increased demand for the vitamin. The predominant condition requiring an increase in the daily intake of folate is pregnancy.

This is due to an increased number of rapidly proliferating cells present in the blood. The need for folate will nearly double by the third trimester of pregnancy. Certain drugs such as anticonvulsants and oral contraceptives can impair the absorption of folate. Anticonvulsants also increase the rate of folate metabolism.

ASCORBIC ACID

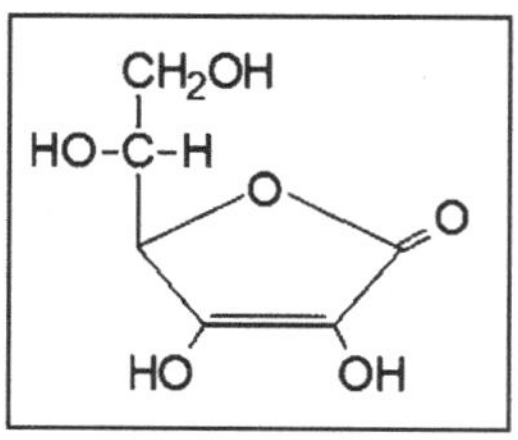

Fig. Ascorbic Acid.

Ascorbic acid is more commonly known as vitamin C. Ascorbic acid is derived from glucose via the uronic acid pathway. The enzyme L-gulonolactone oxidase responsible for the conversion of gulonolactone to ascorbic acid is absent in primates making ascorbic acid required in the diet.

The active form of vitamin C is ascorbate acid itself. The main function of ascorbate is as a reducing agent in a number of different reactions. Vitamin

C has the potential to reduce cytochromes a and c of the respiratory chain as well as molecular oxygen. The most important reaction requiring ascorbate as a cofactor is the hydroxylation of proline residues in collagen.

Vitamin C is, therefore, required for the maintenance of normal connective tissue as well as for wound healing since synthesis of connective tissue is the first event in wound tissue remodeling. Vitamin C also is necessary for bone remodeling due to the presence of collagen in the organic matrix of bones.

Several other metabolic reactions require vitamin C as a cofactor. These include the catabolism of tyrosine and the synthesis of epinephrine from tyrosine and the synthesis of the bile acids. It is also believed that vitamin C is involved in the process of steroidogenesis since the adrenal cortex contains high levels of vitamin C which are depleted upon adrenocorticotropic hormone (ACTH) stimulation of the gland.

Deficiency in vitamin C leads to the disease scurvy due to the role of the vitamin in the post-translational modification of collagens. Scurvy is characterized by easily bruised skin, muscle fatigue, soft swollen gums, decreased wound healing and hemorrhaging, osteoporosis, and anemia.

Vitamin C is readily absorbed and so the primary cause of vitamin C deficiency is poor diet and/or an increased requirement. The primary physiological state leading to an increased requirement for vitamin C is severe stress (or trauma).

This is due to a rapid depletion in the adrenal stores of the vitamin. The reason for the decrease in adrenal vitamin C levels is unclear but may be due either to redistribution of the vitamin to areas that need it or an overall increased utilization.

VITAMIN A

Vitamin A consists of three biologically active molecules, retinol, retinal (retinaldehyde) and retinoic acid.

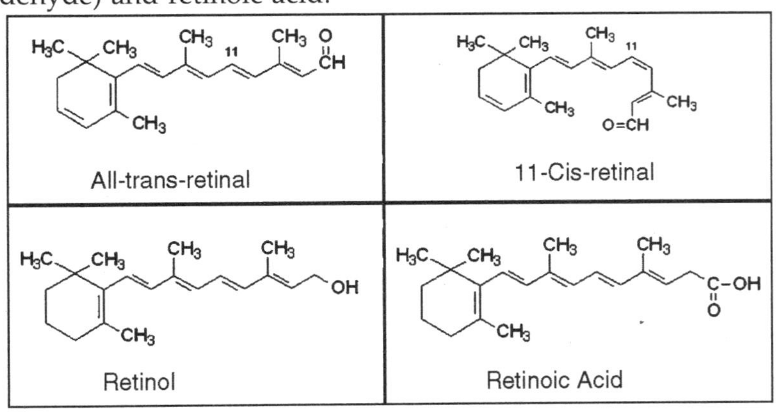

All-trans-retinal

11-Cis-retinal

Retinol

Retinoic Acid

Each of these compounds are derived from the plant precursor molecule, b-carotene (a member of a family of molecules known as carotenoids). Beta-carotene, which consists of two molecules of retinal linked at their aldehyde ends, is also referred to as the provitamin form of vitamin A.

Ingested β-carotene is cleaved in the lumen of the intestine by β-carotene dioxygenase to yield retinal. Retinal is reduced to retinol by retinaldehyde reductase, an NADPH requiring enzyme within the intestines. Retinol is esterified to palmitic acid and delivered to the blood via chylomicrons.

The uptake of chylomicron remnants by the liver results in delivery of retinol to this organ for storage as a lipid ester within lipocytes. Transport of retinol from the liver to extrahepatic tissues occurs by binding of hydrolyzed retinol to aporetinol binding protein (RBP).

The retinol-RBP complex is then transported to the cell surface within the Golgi and secreted. Within extrahepatic tissues retinol is bound to cellular retinol binding protein (CRBP). Plasma transport of retinoic acid is accomplished by binding to albumin.

GENE CONTROL EXERTED BY RETINOIC ACID

Within cells retinoic acid (as well as the related compound 9-*cis*-retinoic acid) bind to specific receptor proteins that are members of the nuclear receptors family of hormone receptors. Following binding, the receptor-retinoic acid complex interacts with specific sequences in several genes involved in growth and differentiation and affects expression of these genes. In this capacity retinoic acid is considered a hormone of the steroid/thyroid hormone superfamily of proteins.

The retinoic acid receptors are abbreviated RAR and the related retinoid X receptors (RXRs) bind 9-*cis*-retinoic acid. Several genes, whose patterns of expression are altered by retinoic acid, are involved in the earliest processes of embryogenesis including the differentiation of the three germ layers, organogenesis and limb development.

To date, there are at least 130 genes that have been shown to be directly regulated (either positive or negative) via interaction with RAR or RXR.

VISION AND THE ROLE OF VITAMIN A

Photoreception in the eye is the function of two specialized cell types located in the retina; the rod and cone cells. Both rod and cone cells contain a photoreceptor pigment in their membranes. The photosensitive compound of most mammalian eyes is a protein called opsin to which is covalently coupled an aldehyde of vitamin A.

The opsin of rod cells is called scotopsin. The photoreceptor of rod cells is specifically called rhodopsin or visual purple.

This compound is a complex between scotopsin and the 11-*cis*-retinal (also called 11-*cis*-retinene) form of vitamin A. Rhodopsin is a serpentine receptor imbedded in the membrane of the rod cell. Coupling of 11-*cis*-retinal occurs at three of the transmembrane domains of rhodopsin.

Intracellularly, rhodopsin is coupled to a specific G-protein called transducin. When the rhodopsin is exposed to light it is bleached releasing the 11-*cis*-retinal from opsin. Absorption of photons by 11-*cis*-retinal triggers a series of conformational changes on the way to conversion all-*trans*-retinal. One important conformational intermediate is metarhodopsin II. The release of opsin results in a conformational change in the photoreceptor.

This conformational change activates transducin, leading to an increased GTP-binding by the a-subunit of transducin. Binding of GTP releases the α-subunit from the inhibitory β- and ã-subunits. The GTP-activated α-subunit in turn activates an associated phosphodiesterase; an enzyme that hydrolyzes cyclic-GMP (cGMP) to GMP. Cyclic GMP is required to maintain the Na^+ channels of the rod cell in the open conformation.

The drop in cGMP concentration results in complete closure of the Na^+ channels. Metarhodopsin II appears to be responsible for initiating the closure of the channels. The closing of the channels leads to hyperpolarization of the rod cell with concomitant propagation of nerve impulses to the brain.

ADDITIONAL ROLE OF RETINOL

Retinol also functions in the synthesis of certain glycoproteins and mucopolysaccharides necessary for mucous production and normal growth regulation. This is accomplished by phosphorylation of retinol to retinyl phosphate which then functions similarly to dolichol phosphate.

CLINICAL SIGNIFICANCES OF VITAMIN A DEFICIENCY

Vitamin A is stored in the liver and deficiency of the vitamin occurs only after prolonged lack of dietary intake. The earliest symptoms of vitamin A deficiency are night blindness. Additional early symptoms include follicular hyperkeratinosis, increased susceptibility to infection and cancer and anemia equivalent to iron deficient anemia. Prolonged lack of vitamin A leads to deterioration of the eye tissue through progressive keratinization of the cornea, a condition known as xerophthalmia.

The increased risk of cancer in vitamin deficiency is thought to be the result of a depletion in β-carotene. Beta-carotene is a very effective antioxidant and is suspected to reduce the risk of cancers known to be initiated by the production of free radicals.

Of particular interest is the potential benefit of increased β-carotene intake to reduce the risk of lung cancer in smokers. However, caution needs to be taken when increasing the intake of any of the lipid soluble vitamins. Excess accumulation of vitamin A in the liver can lead to toxicity which manifests as bone pain, hepatosplenomegaly, nausea and diarrhea.

VITAMIN D

Vitamin D is a steroid hormone that functions to regulate specific gene expression following interaction with its intracellular receptor. The biologically active form of the hormone is 1,25-dihydroxy vitamin D_3 (1,25-$(OH)_2D_3$, also termed calcitriol). Calcitriol functions primarily to regulate calcium and phosphorous homeostasis.

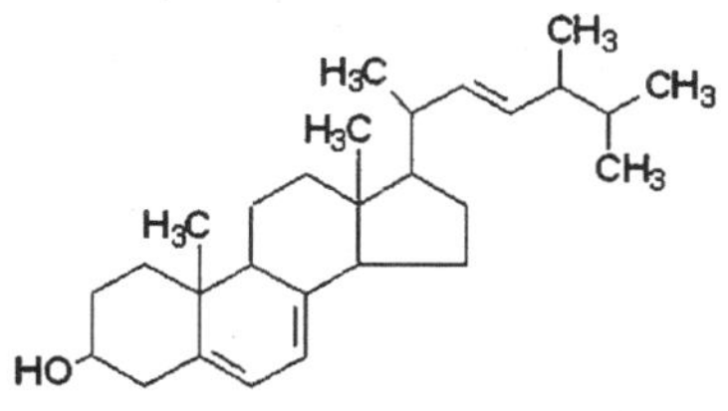

Ergosterol

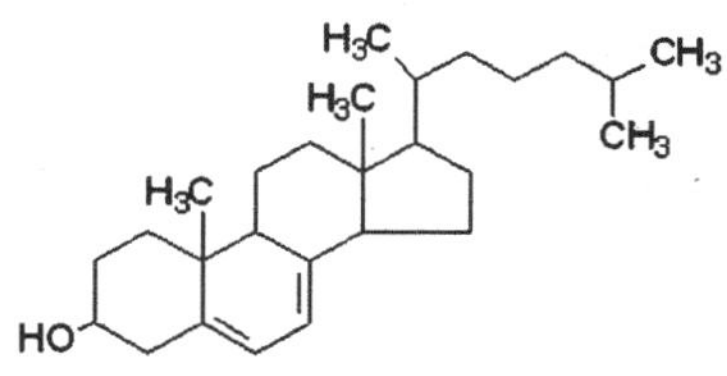

7-Dehydrocholesterol

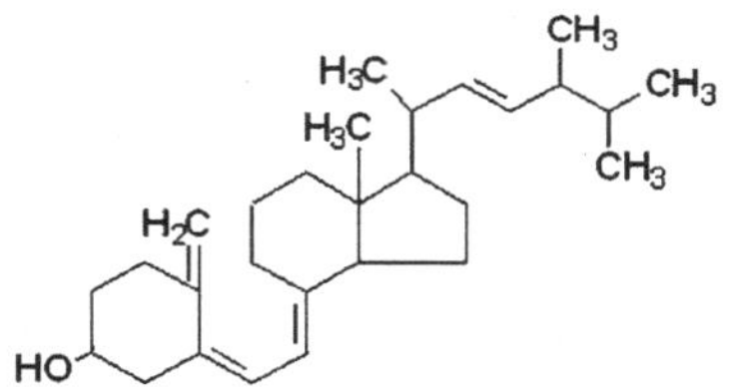

Vitamin D_2

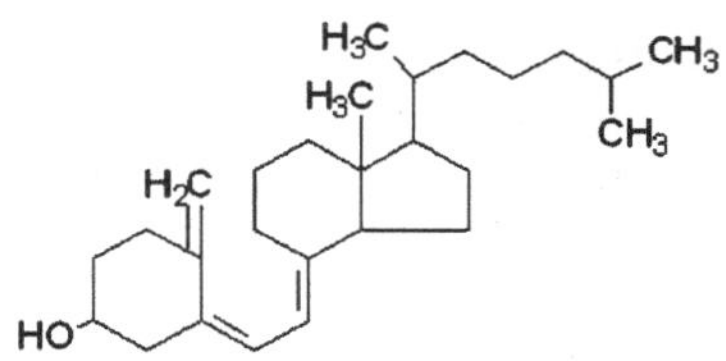

Vitamin D_3

Active calcitriol is derived from ergosterol (produced in plants) and from 7-dehydrocholesterol (produced in the skin). Ergocalciferol (vitamin D_2) is formed by uv irradiation of ergosterol. In the skin 7-dehydrocholesterol is converted to cholecalciferol (vitamin D_3) following *uv* irradiation.

Vitamin D_2 and D_3 are processed to D_2-calcitriol and D_3-calcitriol, respectively, by the same enzymatic pathways in the body. Cholecalciferol (or ergocalciferol) are absorbed from the intestine and transported to the liver bound to a specific vitamin D-binding protein. In the liver cholecalciferol is hydroxylated at the 25 position by a specific D_3-25-hydroxylase generating 25-hydroxy-D_3 [25-$(OH)D_3$] which is the major circulating form of vitamin D.

Conversion of 25-$(OH)D_3$ to its biologically active form, calcitriol, occurs through the activity of a specific D_3-1-hydroxylase present in the proximal convoluted tubules of the kidneys, and in bone and placenta. 25-$(OH)D_3$ can also be hydroxylated at the 24 position by a specific D_3-24-hydroxylase in the kidneys, intestine, placenta and cartilage.

Calcitriol functions in concert with parathyroid hormone (PTH) and calcitonin to regulate serum calcium and phosphorous levels. PTH is released in response to low serum calcium and induces the production of calcitriol. In contrast, reduced levels of PTH stimulate synthesis of the inactive 24,25-$(OH)_2D_3$.

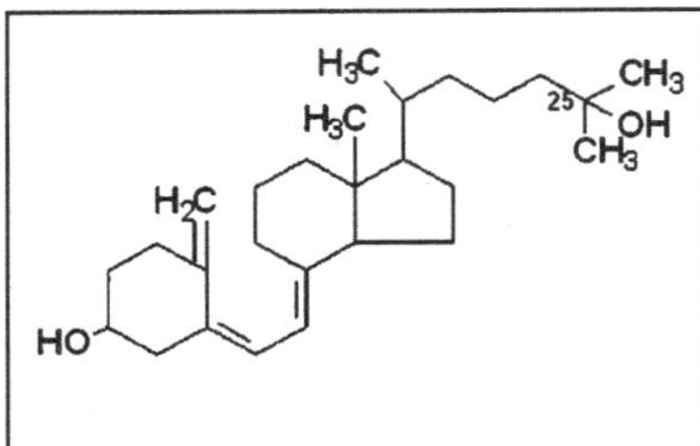

25-hydroxyvitamin D_3

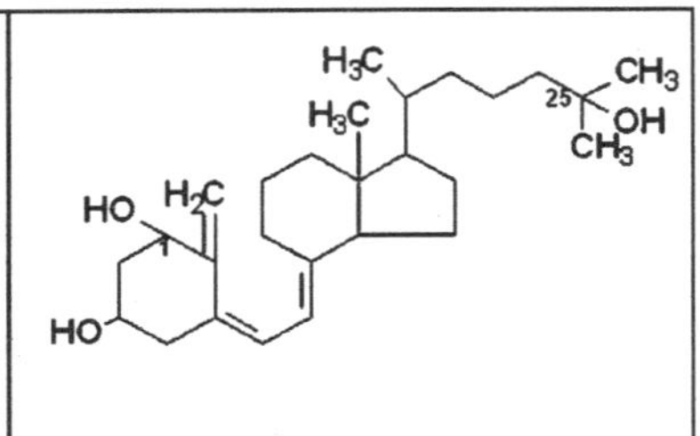

1,25-dihydroxyvitamin D_3

In the intestinal epithelium, calcitriol functions as a steroid hormone in inducing the expression of calbindinD_{28K}, a protein involved in intestinal calcium absorption. The steroid hormone action of vitamin D occurs via the action of calcitriol binding to a specific intracellular receptor that is a member of the nuclear receptor family of hormone receptors called the vitamin D receptor (VDR).

The increased absorption of calcium ions requires concomitant absorption of a negatively charged counter ion to maintain electrical neutrality. The predominant counter ion is Pi. When plasma calcium levels fall the major sites of action of calcitriol and PTH are bone where they stimulate bone resorption and the kidneys where they inhibit calcium excretion by stimulating reabsorption by the distal tubules.

The role of calcitonin in calcium homeostasis is to decrease elevated serum calcium levels by inhibiting bone resorption.

CLINICAL SIGNIFICANCES OF VITAMIN D DEFICIENCY

As a result of the addition of vitamin D to milk, deficiencies in this vitamin are rare in this country. The main symptom of vitamin D deficiency in children is rickets and in adults is osteomalacia.

Rickets is characterized improper mineralization during the development of the bones resulting in soft bones. Osteomalacia is characterized by demineralization of previously formed bone leading to increased softness and susceptibility to fracture.

VITAMIN E

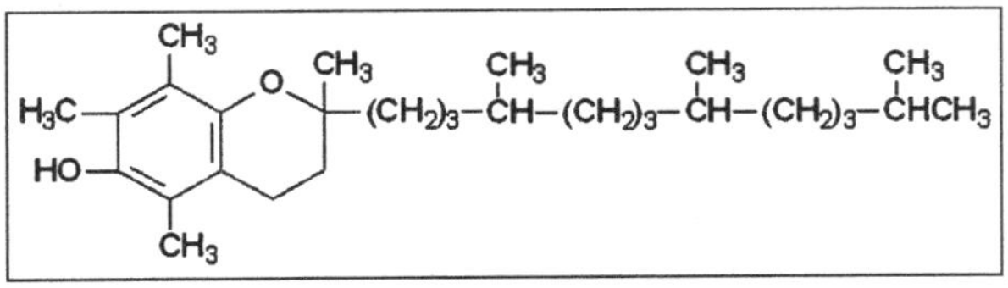

α-Tocopherol

Vitamin E is a mixture of several related compounds known as tocopherols. The α-tocopherol molecule is the most potent of the tocopherols.

Vitamin E is absorbed from the intestines packaged in chylomicrons. It is delivered to the tissues via chylomicron transport and then to the liver through chylomicron remnant uptake. The liver can export vitamin E in VLDLs. Due to its lipophilic nature, vitamin E accumulates in cellular membranes, fat deposits and other circulating lipoproteins. The major site of vitamin E storage is in adipose tissue.

The major function of vitamin E is to act as a natural antioxidant by scavenging free radicals and molecular oxygen. In particular vitamin E is important for preventing peroxidation of polyunsaturated membrane fatty acids. The vitamins E and C are interrelated in their antioxidant capabilities. Active α-tocopherol can be regenerated by interaction with vitamin C following scavenge of a peroxy free radical. Alternatively, α-tocopherol can scavenge two peroxy free radicals and then be conjugated to glucuronate for excretion in the bile.

CLINICAL SIGNIFICANCES OF VITAMIN E DEFICIENCY

No major disease states have been found to be associated with vitamin E deficiency due to adequate levels in the average American diet. The major symptom of vitamin E deficiency in humans is an increase in red blood cell fragility. Since vitamin E is absorbed from the intestines in chylomicrons, any fat malabsorption diseases can lead to deficiencies in vitamin E intake. Neurological disorders have been associated with vitamin E deficiencies associated with fat malabsorptive disorders. Increased intake of vitamin E is recommended in premature infants fed formulas that are low in the vitamin as well as in persons consuming a diet high in polyunsaturated fatty acids. Polyunsaturated fatty acids tend to form free radicals upon exposure to oxygen and this may lead to an increased risk of certain cancers.

VITAMIN K

The K vitamins exist naturally as K_1 (phylloquinone) in green vegetables and K_2 (menaquinone) produced by intestinal bacteria and K_3 is synthetic menadione. When administered, vitamin K_3 is alkylated to one of the vitamin K_2 forms of menaquinone. Vitamin K_2"n" can be 6, 7 or 9 isoprenoid groups Vitamin K_3 The major function of the K vitamins is in the maintenance of normal levels of the blood clotting proteins, factors II, VII, IX, X and protein C and protein S, which are synthesized in the liver as inactive precursor proteins. Conversion from inactive to active clotting factor requires a posttranslational modification of specific glutamate (E) residues.

This modification is a carboxylation and the enzyme responsible requires vitamin K as a cofactor. The resultant modified E residues are ã-carboxyglutamate (*gla*). This process is most clearly understood for factor II, also called preprothrombin. Prothrombin is modified preprothrombin. The *gla* residues are effective calcium ion chelators.

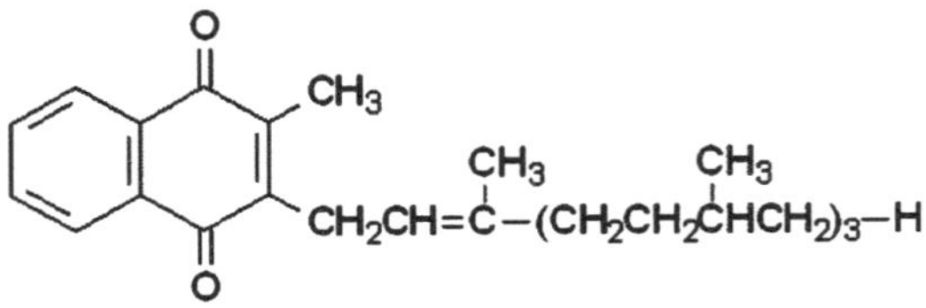

Fig. Vitamin K_1.

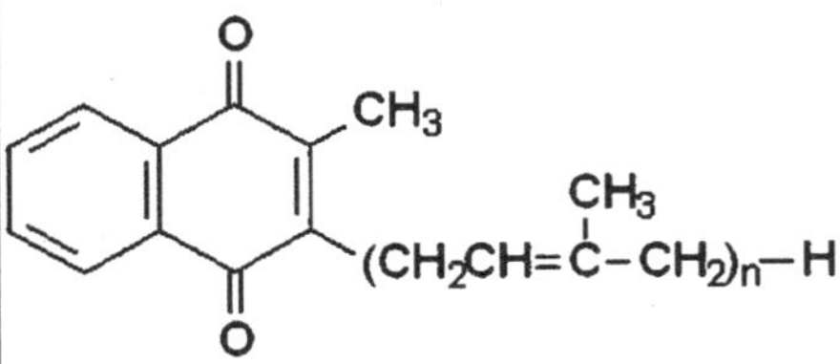

Vitamin K_2

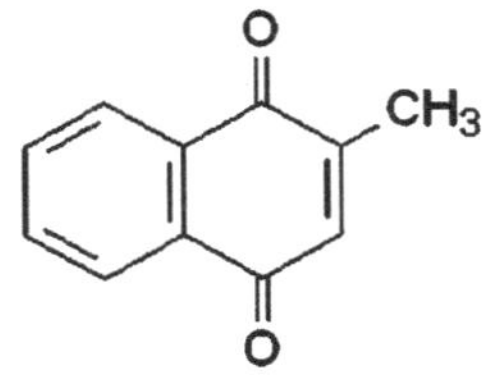

Vitamin K_3

Upon chelation of calcium, prothrombin interacts with phospholipids in membranes and is proteolysed to thrombin through the action of activated factor X (Xa).

During the carboxylation reaction reduced hydroquinone form of vitamin K is converted to a 2,3-epoxide form. The regeneration of the hydroquinone form requires an uncharacterized reductase. This latter reaction is the site of action of the coumarin based anticoagulants such as warfarin.

CLINICAL SIGNIFICANCES OF VITAMIN K DEFICIENCY

Naturally occurring vitamin K is absorbed from the intestines only in the presence of bile salts and other lipids through interaction with chylomicrons. Therefore, fat malabsorptive diseases can result in vitamin K deficiency. The synthetic vitamin K_3 is water soluble and absorbed irrespective of the presence of intestinal lipids and bile. Since the vitamin K_2 form is synthesized by intestinal bacteria, deficiency of the vitamin in adults is rare. However, long term antibiotic treatment can lead to deficiency in adults. The intestine of newborn infants is sterile, therefore, vitamin K deficiency in infants is possible if lacking from the early diet. The primary symptom of a deficiency in infants is a hemorrhagic syndrome.

MINERALS

The functions of the "minerals" are numerous and either quite broad or highly specific. Sodium is a systemic electrolyte and also functions in the regulation of ATP. Magnesium is required for bone mineralization as well as for the processing of ATP. Phosphorous is the most important systemic electrolyte acting as a significant buffer in the blood (in the form of phosphate ion), is required for bone mineralization, and is necessary for energy utilization. Sulfur has a primary function in amino acid metabolism.

Chlorine (as chloride ion) is important to maintain the function of numerous cellular pumps and is used in the production of hydrochloric acid (HCl) in the stomach. Potassium is a systemic electrolyte and is necessary for regulation of ATP. Calcium is required for bone mineralization, cardiac function, muscle contraction, and the synthesis and function of blood cells. Iron, although considered a trace element, has a critical role in the transport of oxygen. Iron is the functional centre of the heme moiety found in each of the protein subunit of hemoglobin. The function of iron is to coordinate the oxygen molecule into heme of hemoglobin so that it can be transported from the lungs to the tissues.

The trace minerals function primarily as cofactors or regulators of enzyme function. Iodine is required for the synthesis of the thyroid hormones. The primary role of selenium is to serve as a modifier of the activity of glutathione peroxidase through its incorporation into the protein in the form of selenocysteine. The role of selenocysteine in protein synthesis is described in the Protein Synthesis page.

FAT-SOLUBLE VITAMINS A, D, E, AND K

Fat-soluble vitamins A, D, E, and K differ from the water-soluble vitamins in their accumulative action. Little evidence has been recorded for hypervitaminosis with the water-soluble vitamins since these compounds are rapidly metabolized and excreted when intake exceeds liver or tissue storage capacity, but hypervitaminosis is a common occurrence in fish and other animals when large quantities of any one of the fat-soluble vitamins are ingested.

Toxicity symptoms involving vitamins A and D are indistinguishable from deficiency symptoms for the same vitamins. On the other hand, symptoms of excess vitamins E and K intake are more discrete. Fish rations may often be enriched with fish oils to increase caloric density of the ration resulting in excessive intake of the fat-soluble vitamins.

VITAMIN A

A fat-soluble factor that promoted growth in rats by Hopkins and by Osborne and Mendel at the beginning of this century. McCollum and Simmonds cured xerophthalmia, a disease of the eye, with this material. The chemical structure of vitamin A and its relationship to -carotene was shown by Von Euler in 1928. Active vitamin A was synthesized in the mid-thirties.

The relationship of the vitamin A alcohols to naturally occurring -carotene Retinene, the aldehyde form of vitamin A, has been isolated from the retina of dark adapted eyes and is involved in vision in dim light. Retinoic acid, which is the oxidized form of vitamin A alcohol, has been shown to have some vitamin A activity. Vitamin A1 is found in saltwater fish, whereas vitamin A is more abundant in freshwater fish. Interconversion of the two forms in living fish tissue has been demonstrated.

Fish oils contain vitamin A as free alcohols or esters. Vitamin A alcohol occurs as a light coloured viscous oil which is heat labile and subject to air oxidation. Beta-carotene occurs as an orange, crystalline compound which is more stable to heat and oxidation. Vitamin A is water insoluble but is soluble in fat and organic solvent.

POSITIVE FUNCTIONS

Vitamin A is essential in maintaining epithelial cells, preventing atrophy and keratinization of epithelial tissues. It is combined with a protein in visual purple and is important in night vision. Vitamin A also prevents xerophthalmia in rats and young children. Vitamin A promotes growth of new cells and aids in maintaining resistance to infection.

THE SYNDROME OF DEFICIENCY OR EXCESS

Hypovitaminosis A is characterized by poor growth, poor vision, keratinization of epithelial tissue, xerophthalmia, night blindness, haemorrhage in the anterior chamber of the eye, haemorrhage of the base of the fins, and abnormal bone formation. Nerve degeneration has been reported in pigs, chickens, rats, rabbits, and ducks, but only occasionally observed in fish after long periods of deficiency.

Hypervitaminosis A fish and in other animals and involved enlargement of liver and spleen, abnormal growth, skin lesions, epithelial keratinization, hyperplasia. of head cartilage, and abnormal bone formation resulting in ankylosis and fusion of vertebrae. Hypervitaminosis A is reflected in very high liver oil vitamin A content and elevated serum alkaline phosphatase. Removal of excess vitamin A from the diet promotes rapid recovery.

SOURCES AND PROTECTION

Cod liver oil is the best known source of vitamin A although black sea bass, swordfish, or ling cod oils contain much more of the vitamin. Whale liver oil contains kitol which has little or no biological activity until heated above 200° C when one molecule of biologically active vitamin A is generated per molecule of whale kitol.

This biologically inactive kitol may be deposited in the whale to avoid hypervitaminosis A during excessive vitamin A intake. The possibility of hypervitaminosis A occurs when tuna, shark, or ling cod viscera are used in preparation of moist fish diets. Synthetic vitamin A preparations, such as vitamin A palmitate, are available and are often used to supplement rations low in fish meal, fish viscera, or carotenes. Some fish species seem able to utilize b-carotene as a vitamin A source, whereas others are unable to split the g-carotene molecule and vitamin A must be added to the diet.

CLINICAL ASSESSMENT

Vitamin A status can best be assessed by the absence of deficiency signs and by assay of liver oil for vitamin A content. Assay for the vitamin in blood or plasma has not been found to be useful.

VITAMIN D

Rickets was induced with test diets by Hopkins, and Mellanby cured the disease in dogs by adding cod liver oil to the ration. Ultraviolet light exposure was shown to have an anti-rachitic effect, and provitamin D activity had been ascribed to ergosterol. Crystalline vitamin D was isolated by Angus and activated 7-dehydrocholesterol was isolated by Windaus.

Vitamin D_2 or ergocalciferol, is one of several biologically active forms of vitamin D. Vitamin D_3 or activated 7-dehydrocholesterol has the chemical formula $C_{27}H_{44}O$ and contains a more simplified, unsaturated 8-carbon side chain. Vitamin D_3, also known as cholecalciferol, is formed in most animal tissue by the rupture of one of the ring bonds of 7-dehydrocholesterol by ultraviolet radiation. Cholecalciferol is a white, crystalline compound soluble in fat and organic solvents and is stable to heat and oxidation in mild alkali or acid solution.

POSITIVE FUNCTIONS

Vitamin D is essential for maintaining calcium and inorganic phosphate homeostasis. It is involved in alkaline phosphatase activity, promotes intestinal absorption of calcium, and influences the action of parathyroid hormone on bone. Fish may sequester calcium from water through the gill membrane; thus the major function of vitamin D for other animals may not be necessary to satisfy calcium requirements for fish.

SYNDROME OF DEFICIENCY OR EXCESS

Rickets and abnormal bone formation for animals has not been observed in fish fed low vitamin D diets. However little work has been done under carefully controlled conditions with young growing fish and only alteration in alkaline phosphatase activity has been reported on different vitamin D diet intake.

Hypervitaminosis D, however, has been reported. Brook trout fed large doses of vitamin D showed impaired growth, lethargy, and dark colouration. High intake of vitamin D mobilizes phosphorus and calcium from the bone and tissues and may result in fragile bones, poor growth, and poor appetite related to the nausea in man afflicted with hyper-vitaminosis D. This area needs to be explored because of the potential for hypervitaminosis D in fish fed diets containing various fish viscera which might contain large amounts of the vitamin D. Tuna liver oil may contain, for example, 100 to 1 000 times as much active vitamin D as cod liver oil.

SOURCES AND PROTECTION

Vitamin D requirements of many animals can be met by ultraviolet irradiation of the skin where cholesterol derivatives are converted to calciferol. Since the vitamin is fat soluble and accumulates in lipid stores, fish liver oil is a rich source of the material.

Content varies tremendously in liver oil, however, with values of about 25 I.U./g present in shark liver oil and over 200 000 I.U./g in albacore tuna liver oil. Cod liver oil contains from 100-500 I.U./g and animal liver contains some vitamin D. One international unit (I.U.) is equal to 0.025 mg of crystalline vitamin D.

CLINICAL ASSESSMENT

Absorption maxima in the ultraviolet region can be used to detect provitamins D in the non-saponifiable fraction of oils. Concentrated preparations of vitamin D can be assayed by the Carr-Price antimony trichloride reaction when assay is necessary to determine biologically active materials in liver oil of fish on different treatments.

The chick assay may not apply to fish liver storage levels because the biologically active form for fish has not been determined. Therefore, clinical assessment for hyper-vitaminosis A must rely on crude methods for determination of vitamin D in liver oil samples by the Carr-Price reaction or by measuring absorption in the UV spectrum.

VITAMIN E

The existence of an antisterility vitamin was postulated by Evans and Bishop. This factor was named'vitamin E' by Sure. The active tocopherol was isolated, characterized, and synthesized by Karrer. Vitamin E is composed of a class of compounds known as tocopherols. One of the most important tocopherols is -tocopherol. Eight naturally occurring tocopherol derivatives have been isolated and all belong to the D series. Synthetic -tocopherol is a racemic DL- -tocoperhol mixture. The derivatives of tocol or of tocotrienol are named, alpha, beta, gamma, delta, epsilon, eta, zeta1 and zeta2-tocopherol.

The pure tocopherols are fat-soluble oils which are capable of esterification to form crystalline compounds. The tocopherols are stable to heat and acids in the absence of oxygen, but are rapidly oxidized in the presence of nascent oxygen, peroxides, or other oxidizing agents. The tocopherols are sensitive to ultraviolet light and are excellent antioxidants in the free form. The esters are more stable and are commonly used as dietary supplements-anticipating hydrolysis in the gut and absorption of the free alcohol to act as an active intra- and intercellular antioxidant. Ethyl derivatives on the aromatic ring are also active. Oxidation products of a-tocopherol can be reduced with hydrosulphite to -tocopherylhydroquinone or, in the presence of ascorbic acid, to -tocopherol.

Table: The Tocopherols

Tocopherol (alpha)	5,7,8-trimethyltocol
Tocopherol (beta)	5,8-dimethyltocol
Tocopherol (gamma)	7,8-dimethyltocol
Tocopherol ($zeta_2$)	5,7-dimethyltocol
Tocopherol (eta)	7-methyltocol
Tocopherol (delta)	8-methyltocol
Tocopherol (epsilon)	5,8-dimethyltocotrienol
Tocopherol ($zeta_1$)	5,7,8-trimethyltocotrienol

POSITIVE FUNCTIONS

The tocopherols act as extra and intracellular antioxidants to maintain homeostasis of labile metabolites in the cell and tissue plasma. As physiological antioxidants, these usually protect oxidizable vitamins and labile unsaturated fatty acids. Vitamin E prevents encephalomalacia in chicks, erythrocyte haemolysis in several animals and steatitis in mink, pigs, and farm animals. The tocopherols also prevent exudative diathesis in chicks, white muscle disease in lambs and calves, and dietary liver necrosis in rats. They act with selenium and with vitamin C for normal reproductive activity and are involved in prevention of nutritional muscular dystrophy in the chick, the yellowtail and carp.

The tocopherols act as free radical traps to stop the chain reaction during peroxide formation and stabilize unsaturated carbon bonds of polyunsaturated fatty acids and other long-chain labile compounds. Vitamin E is involved in the maintenance of normal blood capillary permeability and the integrity of heart muscle. It was first shown to be involved in prevention of sterility and foetal resorption in rats. It may likewise affect embryo membrane permeability and hatch-ability of fish eggs.

SYNDROME OF DEFICIENCY OR EXCESS

One of the first signs is erythrocyte fragility closely followed by anaemia, ascites, xerophthalmia, poor growth, poor food conversion, epicarditis, and ceroid deposits in spleen and liver. Muscle dystrophy and xerophthalmia have been in yellowtail and carp. Non-specific forms of degenerative conditions have been described in several species of fish fed large quantities of polyunsaturated fatty acids with inadequate tocopherol in the ration. Hypervitaminosis E results in poor growth, toxic liver reaction, and death.

SOURCES AND PROTECTION

Vegetable oils are rich sources of tocopherols. Synthetic -tocopherol in esterfied acetate or phosphate form is commonly used as a diet supplement. These esters are much more stable than the free form which is rapidly lost by air oxidation or in the presence of labile compounds like the polyunsaturated fish oils. Addition of antioxidants such as BHT (butylated hydroxyanisole) and BHT (butylated hydroxytoluene) protects fats and other labile compounds in the ration from oxidation, but these antioxidants have no vitamin E activity.

CLINICAL ASSESSMENT

The erythrocyte fragility test indicates the physiological state of fish. Absence of histologically detectable ceroid in liver and spleen from representative samples in the population is a good clue on the presence of adequate amounts of physiological antioxidants in the fish.

A barbituric acid test for oxidation of components has not been applied to fish tissues as a clinical tool for assessment of nutrition state except when liver oils become saturated with very labile polyunsaturated fatty acids such as when feeding squid or saury oils.

These assays are good indicators of the state of oxidation of the finished ration, but the absence of deficiency signs and normal erythrocyte fragility are better clinical tests. Analysis for tocopherol is difficult and time-consuming and only applicable under critical research experiment situation.

VITAMIN K

The name vitamin K (for'koagulation') was proposed by Dam. Dam isolated the vitamin from alfalfa and from fish meal in 1939. It was synthesized later that year. The K vitamins K_2 contain 6,7, or 9 isoprene units in the side chain which varies from 30-45 carbon atoms. Many isosteres of vitamin K have been identified in animal tissues, plant tissues, and micro-organisms. Although fairly stable compounds, they are destroyed upon oxidation and exposure to ultraviolet radition. Menadione is very reactive and is subject in aqueous media to chemical interaction.

POSITIVE FUNCTIONS

Vitamin K is involved in the hepatic synthesis of blood clotting proteins - prothrombin and proconvertin. Substituted forms of vitamin K are strongly bacteriostatic and may serve as an alternate defence mechanism for bacterial infections. Vitamin K is structurally related to flavo compounds which function as electron carriers in the electron transport system. The primary role of vitamin K is to maintain a fast normal blood clotting rate which is so important to fish living in a water environment.

SYNDROME OF DEFICIENCY OR EXCESS

Prothrombin time in salmon fed diets devoid of vitamin K was increased three to five times and, during prolonged deficiency states, anaemia and haemorrhagic areas appeared in the gills, eyes, and vascular tissues. Increased blood clotting time has also been reported for other fish reared on diets with low vitamin K content. Interrelationships with other vitamins have not been documented in fish where the primary deficiency signs are slow blood clotting and haemorrhage, severe anaemia, or death in wounded fish. Haemorrhagic areas often appear in fragile tissues such as the gills.

SOURCES AND PROTECTION

Vitamin K sources are green, leafy vegetables. Alfalfa leaves are among the best sources of vitamin K. Low levels are found in soybeans and animal liver. Synthetic menadione is also available. Although vitamin K in ground alfalfa is fairly stable, synthetic material should be protected from exposure to ultraviolet light and to excessive oxidizing or reducing conditions. The use of rapidly cured dehydrated alfalfa is essential to minimize formation of the physiological antagonist, dicumarol. The diet should be kept dry, prepared with minimum exposure to air oxidation, and fed as soon as practicable after manufacture to minimize vitamin K loss.

ANTIMETABOLITES AND INACTIVATION

Dicumarol was isolated from spoiled sweet clover hay and shown to prevent the normal function of vitamin K in maintenance of normal blood clotting times. Early work was with cattle and hogs but was soon extended to rats, other experimental animals, and man. Dicumarol is an anticoagulant and has been used to prevent thrombosis in animals and man. It is not an antimetabolite competing for vitamin K sites but plays another role in preventing normal blood clotting. Vitamin K counte-racts the dicumarol effect quantitatively. Another anticoa-gulant, warfarin, is a common rat poison which has five to ten times the anticoagulant activity of dicumarol. The effects of warfarin can also be reversed by administration of vitamin K.

CLINICAL ASSESSMENT

Vitamin K status of fish is determined by measuring blood clotting time. Chemical methods for determining menadione content of active metabolite after storage are available but have not been applied to measure vitamin K activity in the liver or tissues of young, growing fish.

9

Structures of Protein

COMPLEX PROTEIN STRUCTURES

Proteins also are found to be covalently conjugated with carbohydrates. These modifications occur following the synthesis (translation) of proteins and are, therefore, termed post-translational modifications. These forms of modification impart specialized functions upon the resultant proteins. Proteins covalently associated with carbohydrates are termed glycoproteins.

Glycoproteins are of two classes, *N*-linked and *O*-linked, referring to the site of covalent attachment of the sugar moieties. *N*-linked sugars are attached to the amide nitrogen of the R-group of asparagine; *O*-linked sugars are attached to the hydroxyl groups of either serine or threonine and occasionally to the hydroxyl group of the modified amino acid, hydroxylysine. There are extremely important glycoproteins found on the surface of erythrocytes. It is the variability in the composition of the carbohydrate portions of many glycoproteins and glycolipids of erythrocytes that determines blood group specificities.

There are at least 100 blood group determinants, most of which are due to carbohydrate differences. The most common blood groups, A, B, and O, are specified by the activity of specific gene products whose activities are to incorporate distinct sugar groups onto RBC membrane glycoshpingolipids as well as secreted glycoproteins.

Structural complexes involving protein associated with lipid via noncovalent interactions are termed lipoproteins. The distinct roles of lipoproteins are described on the linked page. Their major function in the body is to aid in the storage transport of lipid and cholesterol.

CLINICAL SIGNIFICANCES

This discussion is not intended to be a complete review of all disorders that result from defects in protein structure and function.

The substitution of a hydrophobic amino acid (V) for an acidic amino acid (E) in the β-chain of hemoglobin results in sickle cell anemia (HbS). This change of a single amino acid alters the structure of hemoglobin molecules in such a way that the deoxygenated proteins polymerize and precipitate within the erythrocyte, leading to their characteristic sickle shape.

Collagens are the most abundant proteins in the body. Alterations in collagen structure arising from abnormal genes or abnormal processing of collagen proteins results in numerous diseases, including Larsen syndrome, scurvy, osteogenesis imperfecta and Ehlers-Danlos syndrome.

Ehlers-Danlos syndrome is actually the name associated with at least ten distinct disorders that are biochemically and clinically distinct yet all manifest structural weakness in connective tissue as a result of defective collagen structure. Osteogenesis imperfecta also encompasses more than one disorder.

At least four biochemically and clinically distinguishable maladies have been identified as osteogenesis imperfecta, all of which are characterized by multiple fractures and resultant bone deformities. Marfan syndrome manifests itself as a disorder of the connective tissue and was originally believed to be the result of abnormal collagens. However, recent evidence has shown that Marfan syndrome results from mutations in the extracellular protein, fibrillin, which is an integral constituent of the non-collagenous microfibrils of the extracellular matrix.

Several forms of familial hypercholesterolemia are the result of genetic defects in the gene encoding the receptor for low-density lipoprotein (LDL). These defects result in the synthesis of abnormal LDL receptors that are incapable of binding to LDLs, or that bind LDLs but the receptor/LDL complexes are not properly internalized and degraded.

The outcome is an elevation in serum cholesterol levels and increased propensity towards the development of atherosclerosis. A number of proteins can contribute to cellular transformation and carcinogenesis when their basic structure is disrupted by mutations in their genes. These genes are termed proto-oncogenes.

For some of these proteins, all that is required to convert them to the oncogenic form is a single amino acid substitution. The cellular gene, RAS, is observed to sustain single amino acid substitutions at positions 12 or 61 with high frequency in colon carcinomas. Mutations in RAS are most frequently observed genetic alterations in colon cancer

FORCES CONTROLLING PROTEIN STRUCTURE

Hydrogen Bonding

Polypeptides contain numerous proton donors and acceptors both in their backbone and in the R-groups of the amino acids. The environment in which proteins are found also contains the ample H-bond donors and acceptors of the water molecule. H-bonding, therefore, occurs not only within and between polypeptide chains but with the surrounding aqueous medium.

Hydrophobic Forces

Proteins are composed of amino acids that contain either hydrophilic or hydrophobic R-groups. It is the nature of the interaction of the different R-groups with the aqueous environment that plays the major role in shaping protein structure. The spontaneous folded state of globular proteins is a reflection of a balance between the opposing energetics of H-bonding between hydrophilic R-groups and the aqueous environment and the repulsion from the aqueous environment by the hydrophobic R-groups.

The hydrophobicity of certain amino acid R-groups tends to drive them away from the exterior of proteins and into the interior. This driving force restricts the available conformations into which a protein may fold.

Electrostatic Forces

Electrostatic forces are mainly of three types; charge-charge, charge-dipole and dipole-dipole. Typical charge-charge interactions that favour protein folding are those between oppositely charged R-groups such as K or R and D or E. A substantial component of the energy involved in protein folding is charge-dipole interactions.

This refers to the interaction of ionized R-groups of amino acids with the dipole of the water molecule. The slight dipole moment that exist in the polar R-groups of amino acid also influences their interaction with water. It is, therefore, understandable that the majority of the amino acids found on the exterior surfaces of globular proteins contain charged or polar R-groups.

Van der Waals Forces

There are both attractive and repulsive van der Waals forces that control protein folding. Attractive van der Waals forces involve the interactions among induced dipoles that arise from fluctuations in the charge densities that occur between adjacent uncharged non-bonded atoms. Repulsive van der Waals forces involve the interactions that occur when uncharged non-bonded atoms come very close together but do not induce dipoles. The repulsion is the result of the electron-electron repulsion that occurs as two clouds of electrons begin to overlap.

Although van der Waals forces are extremely weak, relative to other forces governing conformation, it is the huge number of such interactions that occur in large protein molecules that make them significant to the folding of proteins.

QUATERNARY STRUCTURE

Many proteins contain 2 or more different polypeptide chains that are held in association by the same non-covalent forces that stabilize the tertiary structures of proteins. Proteins with multiple polypetide chains are oligomeric proteins. The structure formed by monomer-monomer interaction in an oligomeric protein is known as quaternary structure.

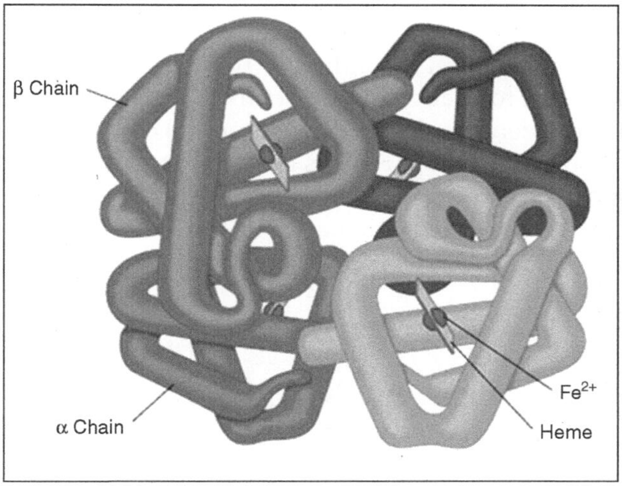

Fig. Structure of Hemoglobin.

Oligomeric proteins can be composed of multiple identical polypeptide chains or multiple distinct polypeptide chains. Proteins with identical subunits are termed homo-oligomers. Proteins containing several distinct polypeptide chains are termed hetero-oligomers.

Hemoglobin, the oxygen carrying protein of the blood, contains two α and two β subunits arranged with a quaternary structure in the form, $\alpha_2\beta_2$. Hemoglobin is, therefore, a hetero-oligomeric protein.

PRIMARY STRUCTURE OF PROTEINS

The primary structure of peptides and proteins refers to the linear number and order of the amino acids present. The convention for the designation of the order of amino acids is that the N-terminal end (*i.e.* the end bearing the residue with the free α-amino group) is to the left (and the number 1 amino acid) and the C-terminal end (*i.e.* the end with the residue containing a free α-carboxyl group) is to the right.

SECONDARY STRUCTURE IN PROTEINS

The ordered array of amino acids in a protein confer regular conformational forms upon that protein. These conformations constitute the secondary structures of a protein. In general proteins fold into two broad classes of structure termed, globular proteins or fibrous proteins. Globular proteins are compactly folded and coiled, whereas, fibrous proteins are more filamentous or elongated. It is the partial double-bond character of the peptide bond that defines the conformations a polypeptide chain may assume. Within a single protein different regions of the polypeptide chain may assume different conformations determined by the primary sequence of the amino acids.

The α-Helix

The α-helix is a common secondary structure encountered in proteins of the globular class. The formation of the α-helix is spontaneous and is stabilized by H-bonding between amide nitrogens and carbonyl carbons of peptide bonds spaced four residues apart. This orientation of H-bonding produces a helical coiling of the peptide backbone such that the R-groups lie on the exterior of the helix and perpendicular to its axis.

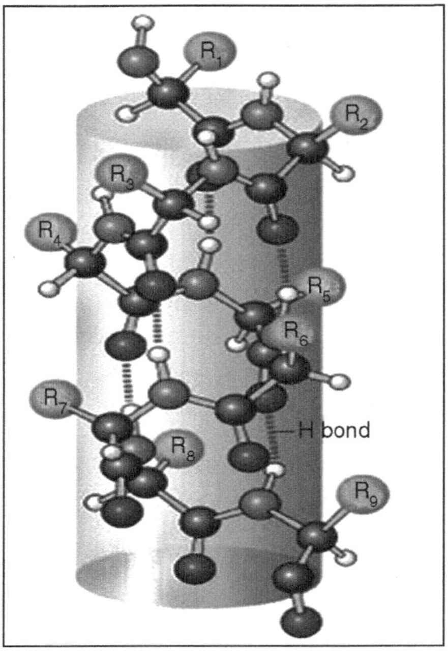

Fig. Typical a-Helix.

Not all amino acids favour the formation of the (α-helix due to steric constraints of the R-groups. Amino acids such as A, D, E, I, L and M favour the formation of α-helices, whereas, G and P favour disruption of the helix.

This is particularly true for P since it is a pyrrolidine based imino acid (HN=) whose structure significantly restricts movement about the peptide bond in which it is present, thereby, interfering with extension of the helix. The disruption of the helix is important as it introduces additional folding of the polypeptide backbone to allow the formation of globular proteins.

β-Sheets

Whereas an α-helix is composed of a single linear array of helically disposed amino acids, β-sheets are composed of 2 or more different regions of stretches of at least 5-10 amino acids. The folding and alignment of stretches of the polypeptide backbone aside one another to form β-sheets is stabilized by H-bonding between amide nitrogens and carbonyl carbons.

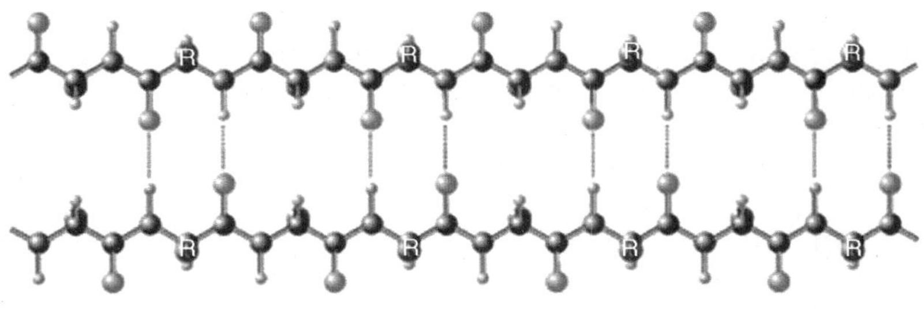

Ball and stick Represention of a β- sheet

Fig. Ball and Stick Representation of a b-Sheet
Ribbon Depiction of b-Sheet.

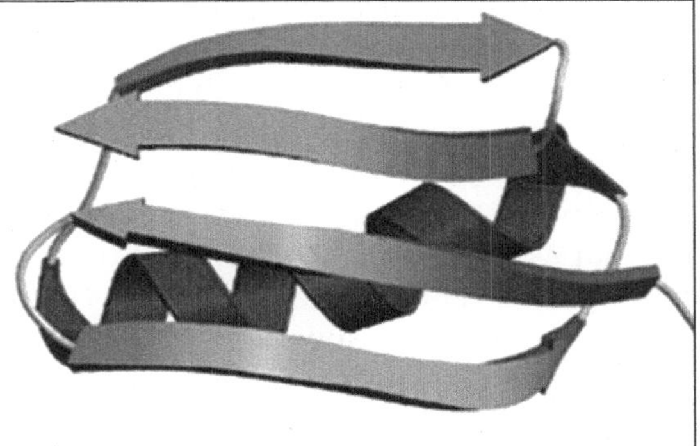

Ribbon Depiction of β-Sheet

However, the H-bonding residues are present in adjacently opposed stretches of the polypetide backbone as opposed to a linearly contiguous region of the backbone in the α-helix. β-sheets are said to be pleated. This is due to positioning of the α-carbons of the peptide bond which alternates above and below the plane of the sheet. β-sheets are either parallel or antiparallel. In parallel sheets adjacent peptide chains proceed in the same direction (*i.e.* the direction of N-terminal to C-terminal ends is the same), whereas, in antiparallel sheets adjacent chains are aligned in opposite directions. β-sheets can be depicted in ball and stick format or as ribbons in certain protein formats.

Super-Secondary Structure

Some proteins contain an ordered organization of secondary structures that form distinct functional domains or structural motifs. Examples include the helix-turn-helix domain of bacterial proteins that regulate transcription and the leucine zipper, helix-loop-helix and zinc finger domains of eukaryotic transcriptional regulators. These domains are termed super-secondary structures.

TERTIARY STRUCTURE OF PROTEINS

Tertiary structure refers to the complete three-dimensional structure of the polypeptide units of a given protein. Included in this description is the spatial relationship of different secondary structures to one another within a polypeptide chain and how these secondary structures themselves fold into the three-dimensional form of the protein. Secondary structures of proteins often constitute distinct domains. Therefore, tertiary structure also describes the relationship of different domains to one another within a protein. The interactions of different domains is governed by several forces: These include hydrogen bonding, hydrophobic interactions, electrostatic interactions and van der Waals forces.

PROTEIN FOLDING AND TERTIARY STRUCTURE

The folding of a single polypeptide chain in three–dimensional space is referred to as its tertiary structure. As discussed, all of the information needed to fold the protein into its native tertiary structure is contained within the primary structure of the peptide chain itself. With this in mind, it was disappointing to the biochemists of the 1950s when the early protein structures did not reveal the governing principles in any particular detail. It soon became apparent that the proteins knew how they were supposed to fold into tertiary shapes, even if the biochemists did not. Vigourous work in many laboratories has slowly brought important principles to light. First, secondary structures—helices and sheets—form whenever possible as a consequence of the formation of large numbers of hydrogen bonds. Second, α–helices and β–sheets often associate and pack close together in the protein. No protein is stable as a single–layer structure, for reasons that become apparent later.

There are a few common methods for such packing to occur. Third, because the peptide segments between secondary structures in the protein tend to be short and direct, the peptide does not execute complicated twists and knots as it moves from one region of a secondary structure to another. A consequence of these three principles is that protein chains are usually folded so that the secondary structures are arranged in one of a few common patterns. For this reason, there are families of proteins that have similar tertiary structure, with little apparent evolutionary or functional relationship among them. Finally, proteins generally fold so as to form the most stable structures possible.

The stability of most proteins arises from:

- The formation of large numbers of intramolecular hydrogen bonds and
- The reduction in the surface area accessible to solvent that occurs upon folding.

FIBROUS PROTEINS

Fibrous proteins contain polypeptide chains organized approximately parallel along a single axis, producing long fibres or large sheets. Such proteins tend to be mechanically strong and resistant to solubilization in water and dilute salt solutions. Fibrous proteins often play a structural role in nature.

α–Keratin

As their name suggests, the structure of the α–keratins is dominated by α–helical segments of polypeptide. The amino acid sequence of α–keratin subunits is composed of central α–helix—rich rod domains about 311 to 314 residues in length, flanked by non-helical N– and C–terminal domains of varying size and composition (Figure). The structure of the central rod domain of a typical α–keratin is shown in Figure. It consists of four helical strands arranged as twisted pairs of two–stranded coiled coils. X–ray diffraction patterns show that these structures resemble α–helices, but with a pitch of 0.51 nm rather than the expected 0.54 nm. This is consistent with a tilt of the helix relative to the long axis of the fibre, as in the two–stranded "rope" in Figure.

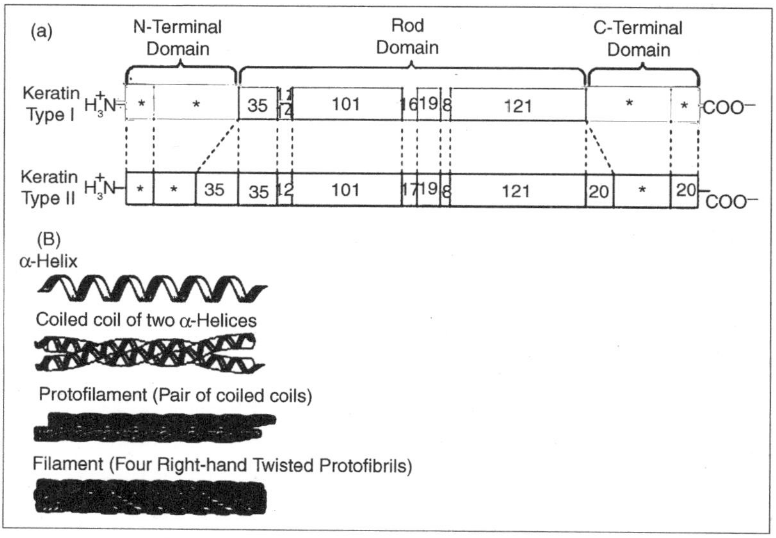

Fig. (a) Both Type I and Type II α–keratin Molecules have Sequences Consisting of Long, Central Rod Domains with Terminal Cap Domains. The Numbers of Amino Acid Residues in each Domain are Indicated. Asterisks Denote Domains of Variable Length. (b) The Rod Domains form Coiled Coils Consisting of Intertwined Right–handed α–helices.

These coiled coils then wind around each other in a left–handed twist.

The primary structure of the central rod segments of α–keratin consists of quasi–repeating seven–residue segments of the form $(a–b–c–d–e–f–g)_n$. These units are not true repeats, but residues *a* and *d* are usually non-polar amino acids. In α–helices, with 3.6 residues per turn, these non-polar residues are arranged in an inclined row or stripe that twists around the helix axis.

These non-polar residues would make the helix highly unstable if they were exposed to solvent, but the association of hydrophobic strips on two coiled coils to form the two–stranded rope effectively buries the hydrophobic residues and forms a highly stable structure. The helices clearly sacrifice some stability in assuming this twisted conformation, but they gain stabilization energy from the packing of side chains between the helices. In other forms of keratin, covalent disulfide bonds form between cysteine residues of adjacent molecules, making the overall structure rigid, inextensible, and insoluble—important properties for structures such as claws, fingernails, hair, and horns in animals. How and where these disulfides form determines the amount of curling in hair and wool fibres. When a hairstylist creates a permanent wave (simply called a "permanent") in a hair salon, disulfides in the hair are first reduced and cleaved, then reorganized and reoxidized to change the degree of curl or wave. In contrast, a "set" that is created by wetting the hair, setting it with curlers, and then drying it represents merely a rearrangement of the hydrogen bonds between helices and between fibres.

Fibroin and β–Keratin: β–Sheet Proteins

The fibroin proteins found in silk fibres represent another type of fibrous protein. These are composed of stacked antiparallel β–sheets, as shown in Figure. The primary structure of fibroin molecules consists of long stretches of alternating glycine and alanine or serine residues. When the sheets stack, the more bulky alanine and serine residues on one side of a sheet interdigitate with similar residues on an adjoining sheet. Glycine hydrogens on the alternating faces interdigitate in a similar manner, but with a smaller intersheet spacing. In the polypeptide sequence of silk proteins, there are large stretches in which every other residue is a glycine. As previously mentioned, the residues of a β–sheet extend alternately above and below the plane of the sheet.

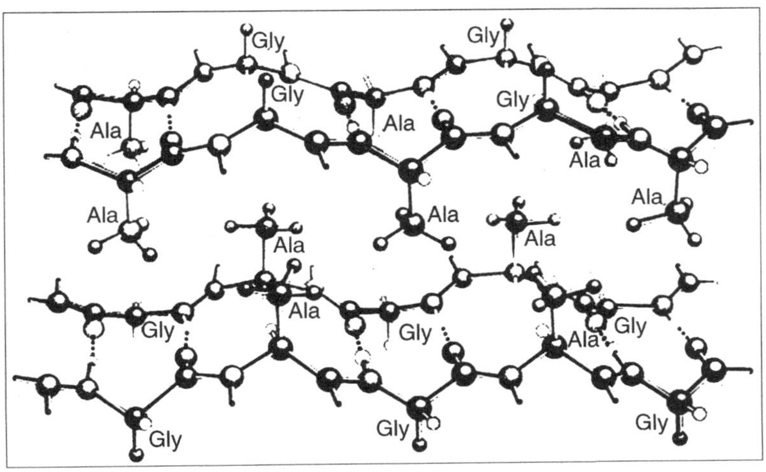

Fig. Silk Fibroin Consists of a Unique Stacked Array of β –sheets

As a result, the glycines all end up on one side of the sheet and the other residues (mainly alanines and serines) compose the opposite surface of the sheet. Pairs of β–sheets can then pack snugly together (glycine surface to glycine surface or alanine—serine surface to alanine—serine surface). The β–keratins found in bird feathers are also made up of stacked β–sheets.

AMINO ACID SEQUENCE IN PROTEIN STRUCTURE

It can be inferred from the first section of this chapter that many different forces work together in a delicate balance to determine the overall three–dimensional structure of a protein. These forces operate both within the protein structure itself and between the protein and the water solvent. How, then, does nature dictate the manner of protein folding to generate the three–dimensional structure that optimizes and balances these many forces? *All of the information necessary for folding the peptide chain into its "native" structure is contained in the amino acid sequence of the peptide.* This principle was first appreciated by C. B. Anfinsen and F. White, whose work in the early 1960s dealt with the chemical denaturation and subsequent renaturation of bovine pancreatic ribonuclease. Ribonuclease was first denatured with urea and mercaptoethanol, a treatment that cleaved the four covalent disulfide (S–S) cross–bridges in the protein.

Subsequent air oxidation permitted random formation of disulfide cross–bridges, most of which were incorrect.

Thus, the air–oxidized material showed little enzymatic activity. However, treatment of these inactive preparations with small amounts of mercaptoethanol allowed a reshuffling of the disulfide bonds and permitted formation of significant amounts of active native enzyme. In such experiments, the only road map for the protein, that is, the only "instructions" it has, are those directed by its primary structure, the linear sequence of its amino acid residues. Just how proteins recognize and interpret the information that is stored in the polypeptide sequence is not well understood yet. It may be assumed that certain loci along the peptide chain act as nucleation points, which initiate folding processes that eventually lead to the correct structures.

Regardless of how this process operates, it must take the protein correctly to the final native structure, without getting trapped in a local energy–minimum state which, although stable, may be different from the native state itself. A long–range goal of many researchers in the protein structure field is the prediction of three–dimensional conformation from the amino acid sequence.

As the details of secondary and tertiary structure are described in this chapter, the complexity and immensity of such a prediction will be more fully appreciated. This area is perhaps the greatest uncharted frontier remaining in molecular biology.

10

Role of Enzymes

A fundamental task of proteins is to act as enzymes—catalysts that increase the rate of virtually all the chemical reactions within cells. Although RNAs are capable of catalyzing some reactions, most biological reactions are catalyzed by proteins. In the absence of enzymatic catalysis, most biochemical reactions are so slow that they would not occur under the mild conditions of temperature and pressure that are compatible with life. Enzymes accelerate the rates of such reactions by well over a million-fold, so reactions that would take years in the absence of catalysis can occur in fractions of seconds if catalyzed by the appropriate enzyme. Cells contain thousands of different enzymes, and their activities determine which of the many possible chemical reactions actually take place within the cell.

FUNCTIONAL ROLE OF COENZYMES

The functional role of coenzymes is to act as transporters of chemical groups from one reactant to another. The chemical groups carried can be as simple as the hydride ion ($H^+ + 2e^-$) carried by NAD or the mole of hydrogen carried by FAD; or they can be even more complex than the amine ($-NH_2$) carried by pyridoxal phosphate.

Since coenzymes are chemically changed as a consequence of enzyme action, it is often useful to consider coenzymes to be a special class of substrates, or second substrates, which are common to many different holoenzymes. In all cases, the coenzymes donate the carried chemical grouping to an acceptor molecule and are thus regenerated to their original form. This regeneration of coenzyme and holoenzyme fulfills the definition of an enzyme as a chemical catalyst, since (unlike the usual substrates, which are used up during the course of a reaction) coenzymes are generally regenerated.

ENZYMES IN THE DIAGNOSIS OF PATHOLOGY

The measurement of the serum levels of numerous enzymes has been shown to be of diagnostic significance. This is because the presence of these enzymes in the serum indicates that tissue or cellular damage has occurred resulting in the release of intracellular components into the blood. Hence, when a physician indicates that he/she is going to assay for liver enzymes, the purpose is to ascertain the potential for liver cell damage. Commonly assayed enzymes are the amino transferases: alanine transaminase, ALT (sometimes still referred to as serum glutamate-pyruvate aminotransferase, SGPT) and aspartate aminotransferase, AST (also referred to as serum glutamate-oxaloacetate aminotransferase, SGOT); lactate dehydrogenase, LDH; creatine kinase, CK (also called creatine phosphokinase, CPK); gamma-glutamyl transpeptidase, GGT. Other enzymes are assayed under a variety of different clinical situations but they will not be covered here.

The typical liver enzymes measured are AST and ALT. ALT is particularly diagnostic of liver involvement as this enzyme is found predominantly in hepatocytes. When assaying for both ALT and AST the ratio of the level of these two enzymes can also be diagnostic. Normally in liver disease or damage that is not of viral origin the ratio of ALT/AST is less than 1. However, with viral hepatitis the ALT/AST ratio will be greater than 1. Measurement of AST is useful not only for liver involvement but also for heart disease or damage. The level of AST elevation in the serum is directly proportional to the number of cells involved as well as on the time following injury that the AST assay was performed. Following injury, levels of AST rise within 8 hours and peak 24–36 hours later.

Within 3–7 days the level of AST should return to pre-injury levels, provided a continuous insult is not present or further injury occurs. Although measurement of AST is not, in and of itself, diagnostic for myocardial infarction, taken together with LDH and CK measurements the level of AST is useful for timing of the infarct. The measurement of LDH is especially diagnostic for myocardial infarction because this enzyme exist in 5 closely related, but slightly different forms (isozymes). The 5 types and their normal distribution and levels in non-disease/injury are listed below.

- *LDH* 1: Found in heart and red-blood cells and is 17% – 27% of the normal serum total.
- *LDH 2:* Found in heart and red-blood cells and is 27% – 37% of the normal serum total.
- *LDH* 3: Found in a variety of organs and is 18% – 25% of the normal serum total.
- *LDH* 4: Found in a variety of organs and is 3% – 8% of the normal serum total.
- *LDH* 5: Found in liver and skeletal muscle and is 0% – 5% of the normal serum total.

Following a myocardial infarct the serum levels of LDH rise within 24-48 hours reaching a peak by 2–3 days and return to normal in 5-10 days. Especially diagnostic is a comparison of the LDH-1/LDH-2 ratio. Normally, this ration is less than 1. A reversal of this ration is referred to as a "flipped LDH". Following an acute myocardial infarct the flipped LDH ratio will appear in 12–24 hours and is definitely present by 48 hours in over 80% of patients. Also important is the fact that persons suffering chest pain due to angina only will not likely have altered LDH levels.

CPK is found primarily in heart and skeletal muscle as well as the brain. Therefore, measurement of serum CPK levels is a good diagnostic for injury to these tissues. The levels of CPK will rise within 6 hours of injury and peak by around 18 hours. If the injury is not persistent the level of CK returns to normal within 2–3 days. Like LDH, there are tissue-specific isozymes of CPK and there designations are described below.

- CPK3 (CPK-MM) is the predominant isozyme in muscle and is 100% of the normal serum total.
- CPK2 (CPK-MB) accounts for about 35% of the CPK activity in cardiac muscle, but less than 5% in skeletal muscle and is 0% of the normal serum total.
- CPK1 (CPK-BB) is the characteristic isozyme in brain and is in significant amounts in smooth muscle and is 0% of the normal serum total.

Since most of the released CPK after a myocardial infarction is CPK-MB, an increased ratio of CPK-MB to total CPK may help in diagnosis of an acute infarction, but an increase of total CPK in itself may not. CPK-MB levels rise 3–6 hours after a myocardial infarct and peak 12–24 hours later if no further damage occurs and returns to normal 12–48 hours after the infarct.

REGULATION OF ENZYME ACTIVITY

While it is clear that enzymes are responsible for the catalysis of almost all biochemical reactions, it is important to also recognize that rarely, if ever, do enzymatic reactions proceed in isolation. The most common scenario is that enzymes catalyze individual steps of multi-step metabolic pathways, as is the case with glycolysis, gluconeogenesis or the synthesis of fatty acids. As a consequence of these lock- step sequences of reactions, any given enzyme is dependent on the activity of preceding reaction steps for its substrate.

In humans, substrate concentration is dependent on food supply and is not usually a physiologically important mechanism for the routine regulation of enzyme activity. Enzyme concentration, by contrast, is continually modulated in response to physiological needs. Three principal mechanisms are known to regulate the concentration of active enzyme in tissues:

- Regulation of gene expression controls the quantity and rate of enzyme synthesis.
- Proteolytic enzyme activity determines the rate of enzyme degradation.
- Covalent modification of preexisting pools of inactive proenzymes produces active enzymes.

Enzyme synthesis and proteolytic degradation are comparatively slow mechanisms for regulating enzyme concentration, with response times of hours, days or even weeks. Proenzyme activation is a more rapid method of increasing enzyme activity but, as a regulatory mechanism, it has the disadvantage of not being a reversible process. Proenzymes are generally synthesized in abundance, stored in secretory granules and covalently activated upon release from their storage sites. Examples of important proenzymes include pepsinogen, trypsinogen and chymotrypsinogen, which give rise to the proteolytic digestive enzymes. Likewise, many of the proteins involved in the cascade of chemical reactions responsible for blood clotting are synthesized as proenzymes. Other important proteins, such as peptide hormones and collagen, are also derived by covalent modification of precursors.

Another mechanism of regulating enzyme activity is to sequester enzymes in compartments where access to their substrates is limited. For example, the proteolysis of cell proteins and glycolipids by enzymes responsible for their degradation is controlled by sequestering these enzymes within the lysosome. In contrast to regulatory mechanisms that alter enzyme concentration, there is an important group of regulatory mechanisms that do not affect enzyme concentration, are reversible and rapid in action, and actually carry out most of the moment to moment physiological regulation of enzyme activity.

These mechanisms include allosteric regulation, regulation by reversible covalent modification and regulation by control proteins such as calmodulin. Reversible covalent modification is a major mechanism for the rapid and transient regulation of enzyme activity. The best examples, again, come from studies on the regulation of glycogen metabolism where phosphorylation of glycogen synthase and glycogen phosphorylase kinase results in the stimulation of glycogen degradation while glycogen synthesis is coordinately inhibited.

Numerous other enzymes of intermediary metabolism are affected by phosphorylation, either positively or negatively. These covalent phosphorylations can be reversed by a separate sub-subclass of enzymes known as phosphatases.

Recent research has indicated that the aberrant phosphorylation of growth factor and hormone receptors, as well as of proteins that regulate cell division, often leads to unregulated cell growth or cancer. The usual sites for phosphate addition to proteins are the serine, threonine and tyrosine R group hydroxyl residues.

ALLOSTERIC ENZYMES

In addition to simple enzymes that interact only with substrates and inhibitors, there is a class of enzymes that bind small, physiologically important molecules and modulate activity in ways other than those described above. These are known as allosteric enzymes; the small regulatory molecules to which they bind are known as effectors. Allosteric effectors bring about catalytic modification by binding to the enzyme at distinct allosteric sites, well removed from the catalytic site, and causing conformational changes that are transmitted through the bulk of the protein to the catalytically active site(s). The hallmark of effectors is that when they bind to enzymes, they alter the catalytic properties of an enzyme's active site. Those that increase catalytic activity are known as positive effectors. Effectors that reduce or inhibit catalytic activity are negative effectors.

Most allosteric enzymes are oligomeric (consisting of multiple subunits); generally they are located at or near branch points in metabolic pathways, where they are influential in directing substrates along one or another of the available metabolic paths. The effectors that modulate the activity of these allosteric enzymes are of two types. Those activating and inhibiting effectors that bind at allosteric sites are called heterotropic effectors. (Thus there exist both positive and negative heterotropic effectors.) These effectors can assume a vast diversity of chemical forms, ranging from simple inorganic molecules to complex nucleotides such as cyclic adenosine monophosphate (cAMP). Their single defining feature is that they are not identical to the substrate. In many cases the substrate itself induces distant allosteric effects when it binds to the catalytic site. Substrates acting as effectors are said to be homotropic effectors. When the substrate is the effector, it can act as such, either by binding to the substrate-binding site, or to an allosteric effector site. When the substrate binds to the catalytic site it transmits an activity-modulating effect to other subunits of the molecule. Often used as the model of a homotropic effector is hemoglobin, although it is not a branch-point enzyme and thus does not fit the definition on all counts.

There are two ways that enzymatic activity can be altered by effectors: the V_{max} can be increased or decreased, or the K_m can be raised or lowered. Enzymes whose K_m is altered by effectors are said to be K-type enzymes and the effector a K-type effector. If V_{max} is altered, the enzyme and effector are said to be V-type. Many allosteric enzymes respond to multiple effectors with V-type and K-type behaviour. Here again, hemoglobin is often used as a model to study allosteric interactions, although it is not strictly an enzyme. In the preceding discussion we assumed that allosteric sites and catalytic sites were homogeneously present on every subunit of an allosteric enzyme. While this is often the case, there is another class of allosteric enzymes that are comprised of separate catalytic and regulatory subunits.

The archetype of this class of enzymes is cAMP-dependent protein kinase (PKA), whose mechanism of activation is illustrated in the Figure below. The enzyme is tetrameric, containing two catalytic subunits and two regulatory subunits, and enzymatically inactive. When intracellular cAMP levels rise, one molecule of cAMP binds to each regulatory subunit, causing the tetramer to dissociate into one regulatory dimer and two catalytic monomers. In the dissociated form, the catalytic subunits are fully active; they catalyze the phosphorylation of a number of other enzymes, such as those involved in regulating glycogen metabolism. The regulatory subunits have no catalytic activity.

Representative pathway for the activation of cAMP-dependent protein kinase (PKA). In this example glucagon binds to its' cell-surface receptor, thereby activating the receptor. Activation of the receptor is coupled to the activation of a receptor-coupled G-protein (GTP-binding and hydrolyzing protein) composed of 3 subunits. Upon activation the α-subunit dissociates and binds to and activates adenylate cyclase. Adenylate cylcase then converts ATP to cyclic-AMP (cAMP). The cAMP thus produced then binds to the regulatory subunits of PKA leading to dissociation of the associated catalytic subunits.

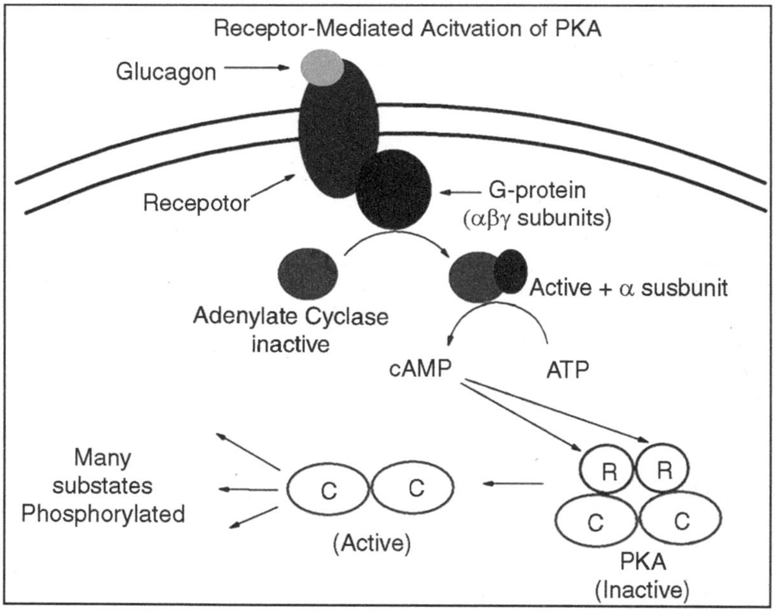

The catalytic subunits are inactive until dissociated from the regulatory subunits. Once released the catalytic subunits of PKA phosphorylate numerous substrate using ATP as the phosphate donor.

ENZYMES AS BIOLOGICAL CATALYSTS

In cells and organisms most reactions are catalyzed by enzymes, which are regenerated during the course of a reaction. These biological catalysts are physiologically important because they speed up the rates of reactions that would otherwise be too slow to support life. Enzymes increase reaction rates, sometimes by as much as one million-fold, but more typically by about one thousand fold. Catalysts speed up the forward and reverse reactions proportionately so that, although the magnitude of the rate constants of the forward and reverse reactions is are increased, the ratio of the rate constants remains the same in the presence or absence of enzyme.

Since the equilibrium constant is equal to a ratio of rate constants, it is apparent that enzymes and other catalysts have no effect on the equilibrium constant of the reactions they catalyze.

Enzymes increase reaction rates by decreasing the amount of energy required to form a complex of reactants that is competent to produce reaction products. This complex is known as the activated state or transition state complex for the reaction. Enzymes and other catalysts accelerate reactions by lowering the energy of the transition state.

The free energy required to form an activated complex is much lower in the catalyzed reaction. The amount of energy required to achieve the transition state is lowered; consequently, at any instant a greater proportion of the molecules in the population can achieve the transition state. The result is that the reaction rate is increased.

MICHAELIS-MENTEN KINETICS

In typical enzyme-catalyzed reactions, reactant and product concentrations are usually hundreds or thousands of times greater than the enzyme concentration. Consequently, each enzyme molecule catalyzes the conversion to product of many reactant molecules. In biochemical reactions, reactants are commonly known as substrates.

The catalytic event that converts substrate to product involves the formation of a transition state, and it occurs most easily at a specific binding site on the enzyme. This site, called the catalytic site of the enzyme, has been evolutionarily structured to provide specific, high-affinity binding of substrate(s) and to provide an environment that favors the catalytic events.

The complex that forms, when substrate(s) and enzyme combine, is called the enzyme substrate (ES) complex. Reaction products arise when the ES complex breaks down releasing free enzyme.

Between the binding of substrate to enzyme, and the reappearance of free enzyme and product, a series of complex events must take place. At a minimum an ES complex must be formed; this complex must pass to the transition state (ES*); and the transition state complex must advance to an enzyme product complex (EP).

The latter is finally competent to dissociate to product and free enzyme. The series of events can be shown thus:

$$E + S \leftrightarrow ES \leftrightarrow ES^* \leftrightarrow EP \leftrightarrow E + P$$

The kinetics of simple reactions like that above were first characterized by biochemists Michaelis and Menten. The concepts underlying their analysis of enzyme kinetics continue to provide the cornerstone for understanding metabolism today, and for the development and clinical use of drugs aimed at selectively altering rate constants and interfering with the progress of disease states.

The Michaelis-Menten equation is a quantitative description of the relationship among the rate of an enzyme- catalyzed reaction [v_1], the concentration of substrate [S] and two constants, V_{max} and K_m (which are set by the particular equation).

The symbols used in the Michaelis-Menten equation refer to the reaction rate [v_1], maximum reaction rate (V_{max}), substrate concentration [S] and the Michaelis-Menten constant (K_m).

$$V_1 = \frac{V_{max}[S]}{\{K_m + [S]\}}$$

The Michaelis-Menten equation can be used to demonstrate that at the substrate concentration that produces exactly half of the maximum reaction rate, i.e. ½ V_{max}, the substrate concentration is numerically equal to K_m.

This fact provides a simple yet powerful bioanalytical tool that has been used to characterize both normal and altered enzymes, such as those that produce the symptoms of genetic diseases.

Rearranging the Michaelis-Menten equation leads to:

$$K_m = [S]\left\{\left[\frac{V_{max}}{V_1}\right] - 1\right\}$$

From this equation it should be apparent that when the substrate concentration is half that required to support the maximum rate of reaction, the observed rate, v_1, will, be equal to V_{max} divided by 2; in other words, v_1 = [$V_{max}/2$]. At this substrate concentration V_{max}/v_1 will be exactly equal to 2, with the result that:

$$[S](1) = K_m$$

The latter is an algebraic statement of the fact that, for enzymes of the Michaelis-Menten type, when the observed reaction rate is half of the maximum possible reaction rate, the substrate concentration is numerically equal to the Michaelis-Menten constant. In this derivation, the units of K_m are those used to specify the concentration of S, usually Molarity.

The Michaelis-Menten equation has the same form as the equation for a rectangular hyperbola; graphical analysis of reaction rate (v) versus substrate concentration [S] produces a hyperbolic rate plot.

The key features of the plot are marked by points A, B and C. At high substrate concentrations the rate represented by point C the rate of the reaction is almost equal to V_{max}, and the difference in rate at nearby concentrations of substrate is almost negligible.

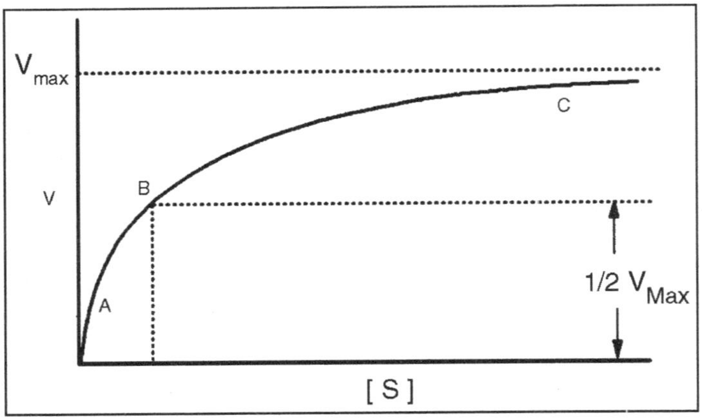

Fig. Plot of Substrate Concentration Versus Reaction Velocity.

If the Michaelis-Menten plot is extrapolated to infinitely high substrate concentrations, the extrapolated rate is equal to V_{max}. When the reaction rate becomes independent of substrate concentration, or nearly so, the rate is said to be zero order. (Note that the reaction is zero order only with respect to this substrate.

If the reaction has two substrates, it may or may not be zero order with respect to the second substrate). The very small differences in reaction velocity at substrate concentrations around point C (near V_{max}) reflect the fact that at these concentrations almost all of the enzyme molecules are bound to substrate and the rate is virtually independent of substrate, hence zero order. At lower substrate concentrations, such as at points A and B, the lower reaction velocities indicate that at any moment only a portion of the enzyme molecules are bound to the substrate.

In fact, at the substrate concentration denoted by point B, exactly half the enzyme molecules are in an ES complex at any instant and the rate is exactly one half of V_{max}.

At substrate concentrations near point A the rate appears to be directly proportional to substrate concentration, and the reaction rate is said to be first order.

INHIBITION OF ENZYME CATALYZED REACTIONS

To avoid dealing with curvilinear plots of enzyme catalyzed reactions, biochemists Lineweaver and Burk introduced an analysis of enzyme kinetics based on the following rearrangement of the Michaelis-Menten equation:

$$\frac{1}{V} = \left[\frac{Km(1)}{V\max[S]} + \frac{1}{V\max}\right]$$

Plots of 1/v versus 1/[S] yield straight lines having a slope of K_m/V_{max} and an intercept on the ordinate at $1/V_{max}$.

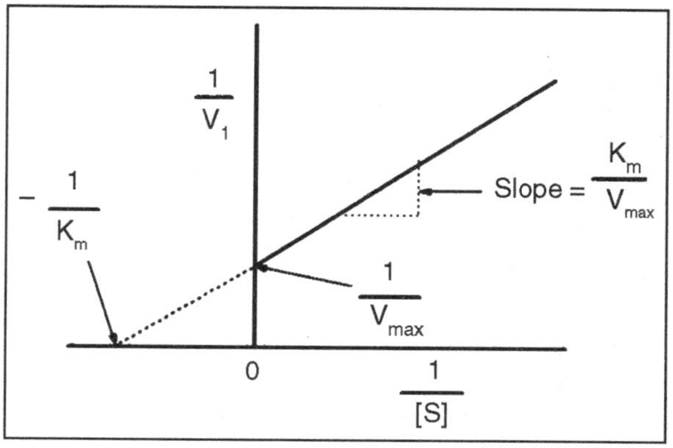

Fig. A Lineweaver-Burk Plot.

An alternative linear transformation of the Michaelis-Menten equation is the Eadie-Hofstee transformation:

$$\frac{v}{[S]} = -v\left[\frac{1}{K_m}\right] + \left[\frac{V_{max}}{K_m}\right]$$

and when v/[S] is plotted on the y-axis versus v on the x-axis, the result is a linear plot with a slope of $-1/K_m$ and the value V_{max}/K_m as the intercept on the y-axis and V_{max} as the intercept on the x-axis.

Both the Lineweaver-Burk and Eadie-Hofstee transformation of the Michaelis-Menton equation are useful in the analysis of enzyme inhibition. Since most clinical drug therapy is based on inhibiting the activity of enzymes, analysis of enzyme reactions using the tools described above has been fundamental to the modern design of pharmaceuticals.

Well- known examples of such therapy include the use of methotrexate in cancer chemotherapy to semi-selectively inhibit DNA synthesis of malignant cells, the use of aspirin to inhibit the synthesis of prostaglandins which are at least partly responsible for the aches and pains of arthritis, and the use of sulfa drugs to inhibit the folic acid synthesis that is essential for the metabolism and growth of disease-causing bacteria.

In addition, many poisons, such as cyanide, carbon monoxide and polychlorinated biphenols (PCBs) produce their life-threatening effects by means of enzyme inhibition.

Enzyme inhibitors fall into two broad classes: those causing irreversible inactivation of enzymes and those whose inhibitory effects can be reversed. Inhibitors of the first class usually cause an inactivating, covalent modification of enzyme structure. Cyanide is a classic example of an irreversible enzyme inhibitor: by covalently binding mitochondrial cytochrome oxidase, it inhibits all the reactions associated with electron transport. The kinetic effect of irreversible inhibitors is to decrease the concentration of active enzyme, thus decreasing the maximum possible concentration of ES complex. Since the limiting enzyme reaction rate is often k_2[ES], it is clear that under these circumstances the reduction of enzyme concentration will lead to decreased reaction rates.

Note that when enzymes in cells are only partially inhibited by irreversible inhibitors, the remaining unmodified enzyme molecules are not distinguishable from those in untreated cells; in particular, they have the same turnover number and the same K_m.

Turnover number, related to V_{max}, is defined as the maximum number of moles of substrate that can be converted to product per mole of catalytic site per second. Irreversible inhibitors are usually considered to be poisons and are generally unsuitable for therapeutic purposes.

Reversible inhibitors can be divided into two main categories; competitive inhibitors and noncompetitive inhibitors, with a third category, uncompetitive inhibitors, rarely encountered.

Inhibitor Type	Binding Site on Enzyme	Kinetic effect
Competitive Inhibitor	Specifically at the catalytic site, where it competes with substrate for binding in a dynamic equilibrium- like process. Inhibition is reversible by substrate.	V_{max} is unchanged; K_m, as defined by [S] required for ½ maximal activity, is increased.
Noncompetitive Inhibitor	Binds E or ES complex other than at the catalytic site. Substrate binding unaltered, but ESI complex cannot form products. Inhibition cannot be reversed by substrate.	K_m appears unaltered; V_{max} is decreased proportionately to inhibitor concentration.
Uncompetitive Inhibitor	Binds only to ES complexes at locations other than the catalytic site. Substrate binding modifies enzyme structure, making inhibitor- binding site available. Inhibition cannot be reversed by substrate.	Apparent V_{max} decreased; K_m, as defined by [S] required for ½ maximal activity, is decreased.

The hallmark of all the reversible inhibitors is that when the inhibitor concentration drops, enzyme activity is regenerated. Usually these inhibitors bind to enzymes by non-covalent forces and the inhibitor maintains a reversible equilibrium with the enzyme. The equilibrium constant for the dissociation of enzyme inhibitor complexes is known as K_i:

$$K_i = \frac{[E][I]}{[EI]}$$

The importance of K_I is that in all enzyme reactions where substrate, inhibitor and enzyme interact, the normal K_m and or V_{max} for substrate enzyme interaction appear to be altered. These changes are a consequence of the influence of K_i on the overall rate equation for the reaction. The effects of K_i are best observed in Lineweaver-Burk plots.

Probably the best known reversible inhibitors are competitive inhibitors, which always bind at the catalytic or active site of the enzyme. Most drugs that alter enzyme activity are of this type.

Competitive inhibitors are especially attractive as clinical modulators of enzyme activity because they offer two routes for the reversal of enzyme inhibition, while other reversible inhibitors offer only one.

First, as with all kinds of reversible inhibitors, a decreasing concentration of the inhibitor reverses the equilibrium regenerating active free enzyme. Second, since substrate and competitive inhibitors both bind at the same site they compete with one another for binding

Raising the concentration of substrate (S), while holding the concentration of inhibitor constant, provides the second route for reversal of competitive inhibition. The greater the proportion of substrate, the greater the proportion of enzyme present in competent ES complexes. As noted earlier, high concentrations of substrate can displace virtually all competitive inhibitor bound to active sites. Thus, it is apparent that V_{max} should be unchanged by competitive inhibitors. This characteristic of competitive inhibitors is reflected in the identical vertical-axis intercepts of Lineweaver-Burk plots, with and without inhibitor. Since attaining V_{max} requires appreciably higher substrate concentrations in the presence of competitive inhibitor, K_m (the substrate concentration at half maximal velocity) is also higher, as demonstrated by the differing negative intercepts on the horizontal axis in panel B.

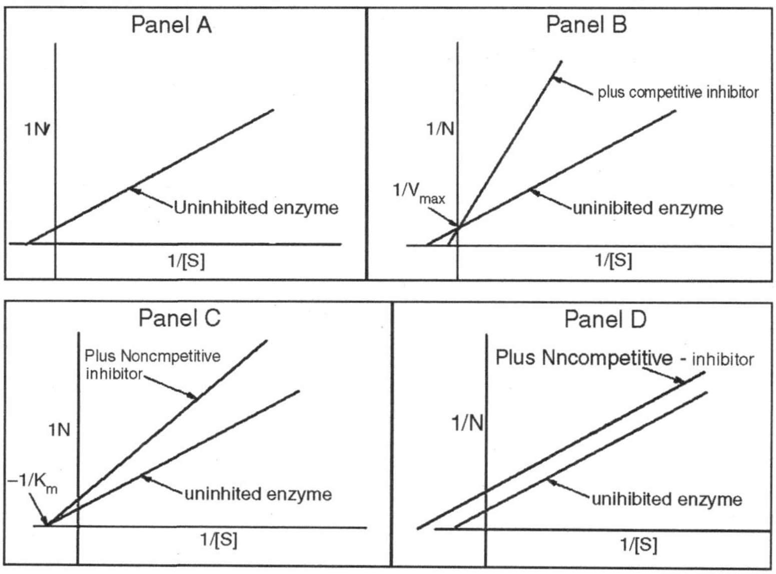

Fig. Lineweaver-Burk Plots of Inhibited Enzymes.

Analogously, panel C illustrates that noncompetitive inhibitors appear to have no effect on the intercept at the x-axis implying that noncompetitive inhibitors have no effect on the K_m of the enzymes they inhibit.

Since noncompetitive inhibitors do not interfere in the equilibration of enzyme, substrate and ES complexes, the K_m's of Michaelis-Menten type enzymes are not expected to be affected by noncompetitive inhibitors, as demonstrated by x-axis intercepts in panel C. However, because complexes that contain inhibitor (ESI) are incapable of progressing to reaction products, the effect of a noncompetitive inhibitor is to reduce the concentration of ES complexes that can advance to product.

Since $V_{max} = k_2[E_{total}]$, and the concentration of competent E_{total} is diminished by the amount of ESI formed, noncompetitive inhibitors are expected to decrease V_{max}, as illustrated by the y-axis intercepts in panel C.

A corresponding analysis of uncompetitive inhibition leads to the expectation that these inhibitors should change the apparent values of K_m as well as V_{max}. Changing both constants leads to double reciprocal plots, in which intercepts on the x and y axes are proportionately changed; this leads to the production of parallel lines in inhibited and uninhibited reactions.

ENZYME RELATIVE TO SUBSTRATE TYPE

Although enzymes are highly specific for the kind of reaction they catalyze, the same is not always true of substrates they attack. For example, while succinic dehydrogenase (SDH) always catalyzes an oxidation-reduction reaction and its substrate is invariably succinic acid, alcohol dehydrogenase (ADH) always catalyzes oxidation-reduction reactions but attacks a number of different alcohols, ranging from methanol to butanol.

Generally, enzymes having broad substrate specificity are most active against one particular substrate. In the case of ADH, ethanol is the preferred substrate.

Enzymes also are generally specific for a particular steric configuration (optical isomer) of a substrate. Enzymes that attack D sugars will not attack the corresponding L isomer. Enzymes that act on L amino acids will not employ the corresponding D optical isomer as a substrate.

The enzymes known as racemases provide a striking exception to these generalities; in fact, the role of racemases is to convert D isomers to L isomers and vice versa. Thus racemases attack both D and L forms of their substrate.

As enzymes have a more or less broad range of substrate specificity, it follows that a given substrate may be acted on by a number of different enzymes, each of which uses the same substrate(s) and produces the same product(s). The individual members of a set of enzymes sharing such characteristics are known as isozymes.

These are the products of genes that vary only slightly; often, various isozymes of a group are expressed in different tissues of the body. The best studied set of isozymes is the lactate dehydrogenase (LDH) system. LDH is a tetrameric enzyme composed of all possible arrangements of two different protein subunits; the subunits are known as H (for heart) and M (for skeletal muscle).

These subunits combine in various combinations leading to 5 distinct isozymes. The all H isozyme is characteristic of that from heart tissue, and the all M isozyme is typically found in skeletal muscle and liver.

These isozymes all catalyze the same chemical reaction, but they exhibit differing degrees of efficiency. The detection of specific LDH isozymes in the blood is highly diagnostic of tissue damage such as occurs during caelow).

ENZYME-SUBSTRATE INTERACTIONS

The favored model of enzyme substrate interaction is known as the induced fit model. This model proposes that the initial interaction between enzyme and substrate is relatively weak, but that these weak interactions rapidly induce conformational changes in the enzyme that strengthen binding and bring catalytic sites close to substrate bonds to be altered.

After binding takes place, one or more mechanisms of catalysis generates transition- state complexes and reaction products. The possible mechanisms of catalysis are four in number:

- Catalysis by Bond Strain: In this form of catalysis, the induced structural rearrangements that take place with the binding of substrate and enzyme ultimately produce strained substrate bonds, which more easily attain the transition state. The new conformation often forces substrate atoms and bulky catalytic groups, such as aspartate and glutamate, into conformations that strain existing substrate bonds.
- Catalysis by Proximity and Orientation: Enzyme-substrate interactions orient reactive groups and bring them into proximity with one another. In addition to inducing strain, groups such as aspartate are frequently chemically reactive as well, and their proximity and orientation toward the substrate thus favors their participation in catalysis.
- Catalysis Involving Proton Donors (Acids) and Acceptors (Bases): Other mechanisms also contribute significantly to the completion of catalytic events initiated by a strain mechanism, for example, the use of glutamate as a general acid catalyst (proton donor).
- Covalent Catalysis: In catalysis that takes place by covalent mechanisms, the substrate is oriented to active sites on the enzymes in such a way that a covalent intermediate forms between the enzyme or coenzyme and the substrate. One of the best-known examples of this mechanism is that involving proteolysis by serine proteases, which include both digestive enzymes (trypsin,

chymotrypsin, and elastase) and several enzymes of the blood clotting cascade. These proteases contain an active site serine whose R group hydroxyl forms a covalent bond with a carbonyl carbon of a peptide bond, thereby causing hydrolysis of the peptide bond.

CHEMICAL REACTIONS AND RATES

According to the conventions of biochemistry, the rate of a chemical reaction is described by the number of molecules of reactant(s) that are converted into product(s) in a specified time period.

Reaction rate is always dependent on the concentration of the chemicals involved in the process and on rate constants that are characteristic of the reaction. For example, the reaction in which A is converted to B is written as follows:

$$A \dashrightarrow B$$

The rate of this reaction is expressed algebraically as either a decrease in the concentration of reactant A:

$$-[A] = k[B]$$

or an increase in the concentration of product B:

$$[B] = k[A]$$

In the second equation (of the 3 above) the negative sign signifies a decrease in concentration of A as the reaction progresses, brackets define concentration in molarity and the *k* is known as a rate constant. Rate constants are simply proportionality constants that provide a quantitative connection between chemical concentrations and reaction rates.

Each chemical reaction has characteristic values for its rate constants; these in turn directly relate to the equilibrium constant for that reaction. Thus, reaction can be rewritten as an equilibrium expression in order to show the relationship between reaction rates, rate constants and the equilibrium constant for this simple case. The rate constant for the forward reaction is defined as k_{+1} and the reverse as k_{-1}.

At equilibrium the rate (v) of the forward reaction ($A \rightarrow B$) is, by definition, equal to that of the reverse or back reaction ($B \rightarrow A$), a relationship which is algebraically symbolized as:

$$v_{forward} = v_{reverse}$$

where, for the forward reaction:

$$v_{forward} = k_{+1}[A]$$

and for the reverse reaction:

$$v_{reverse} = k_{-1}[B]$$

In the above equations, k_{+1} and k_{-1} represent rate constants for the forward and reverse reactions, respectively. The negative subscript refers only to a reverse reaction, not to an actual negative value for the constant. To put the relationships of the two equations into words, we state that the rate of the forward reaction [$v_{forward}$] is equal to the product of the forward rate constant k_{+1} and the molar concentration of A. The rate of the reverse reaction is equal to the product of the reverse rate constant k_{-1} and the molar concentration of B.

At equilibrium, the rate of the forward reaction is equal to the rate of the reverse reaction leading to the equilibrium constant of the reaction and is expressed by:

$$Keq = \frac{K_{+1}}{K_1} = \frac{[B]}{[A]}$$

This equation demonstrates that the equilibrium constant for a chemical reaction is not only equal to the equilibrium ratio of product and reactant concentrations, but is also equal to the ratio of the characteristic rate constants of the reaction.

CHEMICAL REACTION ORDER

Reaction order refers to the number of molecules involved in forming a reaction complex that is competent to proceed to product(s). Empirically, order is easily determined by summing the exponents of each concentration term in the rate equation for a reaction.

A reaction characterized by the conversion of one molecule of A to one molecule of B with no influence from any other reactant or solvent is a first-order reaction. The exponent on the substrate concentration in the rate equation for this type of reaction is 1.

A reaction with two substrates forming two products would a second-order reaction. However, the reactants in second– and higher– order reactions need not be different chemical species.

An example of a second order reaction is the formation of ATP through the condensation of ADP with orthophosphate:

$$ADP + H_2PO_4^- \leftrightarrow ATP + H_2O$$

For this reaction the forward reaction rate would be written as:

$$v_{forward} = k_1[ADP][H_2PO_4^-]$$

11

Animal Models in Drug Discovery

Twenty years ago, the initial screening of new compounds for pharmacologic activity was conducted using whole animals or tissues and organs isolated from animals. Today most of the initial screening for new drugs is done in vitro using the techniques of biochemistry and molecular biology. Only after a new drug candidate has been identified and studied in vitro are studies in animals initiated. The purpose of the animal studies is to verify the pharmacologic activity of the new drug, identify any unexpected pharmacologic activity, and develop an initial data base on the action of the drug in vivo.

An ideal approach would be to have an in vivo screening programme designed to allow for the detection of unique profiles of activity or combinations of activities. Therefore, in addition to a set of initial screening models, relevant secondary tests would be conducted to generate additional information on specificity, mechanism of action, and possible side effects. A complete in vivo pharmacology screening programme would include models for the detection of the major therapeutic classes of compounds shown in Table.

Anti-inflammatory

Two commonly used models to detect anti-inflammatory activity are carrageenan-induced paw edema and adjuvant- induced polyarthritis in rats. The former represents an acute and the latter a chronic inflammatory process. Non-steroidal anti-inflammatory agents inhibit the formation of carrageenan-induced paw edema.

However, to detect activity in the developing and established phases of chronic inflammation, the polyarthritis model is well accepted. In both tests the measured endpoint is volume of the hind paw.

This is done by immersing the hind paw in the well of a mercury displacement plethysmograph. In the carrageenan test, 0.1 ml of a 1% suspension of carrageenan is injected into the plantar surface of one hind paw of a rat. After 1 h the test compound is administered, and after a further 3 h the volumes of both the carrageenan-treated and nontreated hind paws are measured. The net difference between the two measurements is an indication of the amount of edema remaining after test compound treatment.

The smaller the net difference, the greater the anti-inflammatory activity of the test compound. In the polyarthritis model, 0.1 ml of a sterile 0.75% suspension of finely ground Mycobacterium tuberculosis or Mycobacterium butyricum (Freund's adjuvant) in paraffin oil is injected into the hind paw as done for carrageenan.

The polyarthritis disease state takes about 20 days to develop. Therefore, to test a compound for activity in the developing stages of chronic inflammation, the test compound is administered during the first 20 days after the single injection of Freund's adjuvant.

To test a compound for activity against fully developed chronic inflammation, the test compound is administered for 4–10 days, starting about 20 days after the Freund's adjuvant injection. In both the carrageenan and polyarthritis models, drug effects are expressed as the percentage inhibition of paw edema in drug-treated animals compared to non-drug-treated controls.

In the croton oil topical inflammation test in mice, 20 *m*l of a 3% solution of croton oil is applied topically to the anterior and posterior surfaces of one ear and 20 *m*l of the test drug is applied to the same ear 30 min later. Inhibition of ear swelling by more than 50% relative to vehicle-treated controls measured 2 h after croton oil application indicates acute topical anti-inflammatory activity.

Immunomodulation

Inhibition of oxazolone-induced delayed-type hypersensitivity is used as an in vivo test for inhibition of the cellular immune response. The shaved abdominal surfaces of groups of mice are sensitized by topical application of 0.1 ml of 5% oxazolone.

The test drug is given intraperitoneally an hour later and then daily for five consecutive doses. After an additional 4 days, the animals are challenged by application of 0.5 ml oxazolone to one ear. Ear thickness is measured 24 h later. A decrease in ear thickness relative to untreated controls is an indication of immunosuppresion.

Inhibition of sheep red blood cell lysis by serum from sensitized mice is an indication of suppression of the humoral immune response. Groups of mice are sensitized by the intravenous injection of 0.2 ml of a 2% sheep red blood cell suspension.

The test drug or vehicle is administered intraperitoneally to groups of mice for 3 consecutive days beginning 2 h after sensitization to sheep red blood cells. Nine days after sensitization, blood samples are collected, complementinactivated serum prepared, and serial dilutions made. Sheep red blood cells and complement are added to aliquots of the serum dilutions. The serum titer is then expressed as the reciprocal of the dilution causing complete hemolysis. A lower than normal titer may indicate immunosuppression, and a higher than normal titer may indicate immunostimulation.

Antitussive

Dogs are surgically prepared by implantation of a small iron slug in the lumen of the trachea. After a 2-week recovery period, the cough response of the dogs is determined in response to electromagnetic stimulation of the implanted tracheal slug. The stimulus is imposed for 3- to 5-s periods, and the total number of coughs elicited during and 30 s after stimulation is recorded. After establishing each dog's control cough response, test substances can be administered at weekly intervals and compared with the effects observed after administration of a standard antitussive agent.

Endocrine

A simple assay for antidiabetic activity is the determination of changes that occur in blood glucose levels in rats after oral administration of a test compound followed by a subcutaneous injection of glucose. Rats are fasted overnight, administered the test compound, usually by gavage, and then immediately given a subcutaneous injection of 125 mg glucose.

Two hours after test compound administration, the rats are anesthetized and blood drawn for plasma glucose determination. Decreased plasma glucose relative to controls not treated with test compound is an indication of antidiabetic activity.

Antimicrobial

Many animal models use infections artificially introduced into the animal either systemically or into specific organs or tissues. The behaviour of infections and the efficacy of antibiotics in eliminating those infections differ, depending on the organ or tissue involved.

The type of local pathology, the penetration and pharmacokinetics of antibiotics, the local host defence system, and the clearance rates for the bacterial inocula differ among the various organs and tissues of an animal. Therefore, to get a reliable prediction of how effective an antibiotic will be clinically it is mandatory to test new antibiotics in vivo. There are a great variety of such models. For descriptions of many of these, the reader is referred to a three-volume series.

Anticancer

New drugs suspected of having anticancer activity are tested in standard animal models in which tumors are transplanted from one generation of animals, generally mice, to the next and are thus perpetuated in vivo. In vitro tumor cell culture systems are commonly used to determine if a new drug has cytotoxic or growthinhibiting activity. However, the in vivo test is considered to be a better predictor of activity in humans. The L1210 lymphoid leukemia model is commonly used. This tumor is carried as an ascites tumor in BDF1 or CDF1 mice. Ascites fluid from carrier mice containing 100,000 cells is implanted in the test mice. Twenty-four hours later, treatment with the test drug is started. After 6 or 7 days of treatment, the numbers of survivors are determined and the animals weighed. The results are evaluated by comparing drug-treated groups to untreated ones. Drugs resulting in a greater proportion of surviving mice compared with nondrug- treated controls may have anticancer activity. Many other similar models are available.

PROPER CARE OF LABORATORY ANIMALS

Proper care of laboratory animals used in research is a basic requirement to assure the validity and reproducibility of the results obtained. Animals used in drug research are subject to the strictest standards of care beginning with the animal supplier. For the most commonly used laboratory animals, these standards of care often apply for the entire life of the animal. Guidelines for proper care of research animals are provided by the Department of Health and Human Services. The American Association for the Accreditation of Laboratory Animal Care provides the service of certifying laboratories complying with those guidelines. All reputable industrial drug development houses strive to achieve and maintain certification by the association.

SOURCES OF LABORATORY ANIMALS

Laboratory animals for drug research in industry are virtually always obtained from reputable animal suppliers who observe strict standards of care of the animals they provide. Animals that are not properly cared for often result in lost time—and therefore lost money—in drug development.

Governmental Guidelines

The Animal Welfare Act enacted in 1966 and amended in 1970 and 1976 contains provisions to ensure that animals intended for use in research receive humane care and treatment. Implementing rules and regulations are published in the Code of Federal Regulations (CER), Title 9 (Animals and Animal Products), Subchapter A-Animal Welfare, Parts 1, 2, and 3.

The National Institutes of Health (NIH) requires that grantees and contractors using live, vertebrate animals in projects supported by NIH follow the guidelines prescribed in the guide. The Public Health Service further requires that grant-seeking institutions either be accredited by the American Association for Accreditation of Laboratory Animal Care or have an institutional committee that reviews its animal facilities and practices for compliance with the guide.

American Association for Accreditation of Laboratory Animal Care

AAALAC is a non-profit corporation directed by representatives of 24 scientific and professional organizations that are members of the corporation. It was organized in 1965 to conduct a voluntary programme for the accreditation of laboratory animal care facilities and programs. AAALAC encourages optimal care for laboratory animals by providing a mechanism for peer evaluation of animal care programs by the scientific community.

Humane treatment of laboratory animals, protection of personnel from hazards associated with the use of animals, and control of variables that could affect animal research adversely are among the principal objectives of the accreditation programme. Animal care facilities of applicant institutions are visited and a thoroughly evaluated by two experts in laboratory animal science, who submit a detailed report to the Council on Accreditation. Following the standard listed in the Guide for the Care and Use of Laboratory Animals, the Council determines whether AAALAC accreditation should be granted.

Accredited facilities submit annual reports on the status of their animal facilities to AAALAC, and site visits to accredited facilities are conducted at least every 3 years. These annual reports and site visits determine whether accreditation will be continued. Full accreditation by AAALAC is accepted by the NIH as assurance that the animal facilities are evaluated in accordance with their policy on laboratory animals.

PRINCIPLES OF ANIMALS IN DRUG SAFETY TESTING

Before a new drug can be given to people, it must be tested in animals to determine its side effects and at what dosage those side effects will appear. This testing must be done in vivo because the effects of the processes of absorption, distribution, metabolism, excretion, the interactions among these processes, and the interactions among the various organs and neuroendocrine systems within the whole animal cannot be duplicated in vitro. To characterize the nature of the side effects to be expected from a new drug, it is usually necessary to give much higher dosages than would be given clinically and sometimes to give the drug over prolonged periods of time. This is because the dosages of drugs needed to elicit pharmacological or toxicological effects are often higher in laboratory animals than in humans, and because the side effects of drugs suitable for clinical use are usually provoked only by exaggerated dosages.

In some cases the drug may accumulate within the body or within particular organs or tissues and thus give rise to toxic manifestations. Some side effects appear only after long periods of repeated administration. In addition to the requirement to test new drugs in animals, it is also necessary to test in more than one species of animal, because extrapolation of the results of testing in animals to humans is less than perfect. Certain species may be more sensitive or predictive than others. To ensure that the safety of a new drug has not been overpredicted and that any potential side effects have not been overlooked, it is routine practice to test new drugs for safety in at least two species. Inmost cases, these will be a rodent and a non-rodent species. The variety of animal tests and the length of the studies required for a new drug depend on the nature of the drug (pharmacological or chemical class), the intended clinical use of the drug (for example, length of the usual course of treatment), and, to some extent, on the requirements of the countries in which the new drug will be registered for marketing. These factors will be discussed further in the following sections.

ACUTE TOXICOLOGY

In acute toxicology testing, animals are given single doses of a drug. The most common study design is to give groups of rats or mice (such as 5 per sex) single treatments over a wide range of dosages and then observe them over a period of 7–14 days for survival and for physical or behavioral signs of toxicity. In most cases a LD50 (the dosage at which 50% of the animals die) will be determined; however, this is not always feasible or necessary. The acute toxicity test may be conducted for several reasons. This is the first toxicity test that will be done for a new drug, and it provides the basis for choosing dosages that will be used in subsequent toxicity studies involving repeat dosing.

This test provides information regarding the effects of single large doses of a new drug. This information may be used to predict the effects of overdosing (either accidental or purposeful) in humans. The acute toxicity test is a convenient safety check on new batches of drug. Even though extensive chemical analyses are conducted on new batches of any drug, a quick test in live animals to assure that there have been no dramatic changes in acute toxicity provides a desirable level of comfort before making the new drug batch available for human use.

The acute toxicity test is also a convenient method to determine any interaction between two drugs that would cause unexpected toxicity in clinical use. Thus, if a new drug will be commonly used clinically in combination with other drugs or in patients who commonly receive certain medications, it might be prudent to test the new drug in combination with these other drugs in an acute toxicity test in order to predict such an interaction before using the new drug clinically. Apositive finding in such a combination study would likely be followed up by longer-term, repeat-dosing studies.

Acute toxicity studies may be conducted by administering the drug by any of several routes (oral, intravenous, intramuscular, intraperitoneal, subcutaneous, or dermal). The route chosen usually is that intended to be used clinically. Rats and mice are generally used for acute toxicity work; however, rabbits are commonly used when the route of administration is dermal. If a drug is to be given parenterally, acute vein and muscle irritation studies are commonly conducted in rabbits. These studies help assure that a new parenteral formulation will not cause unnecessary pain or tissue damage in people. They are also useful in developing new formulations for parenteral use with minimal potential for such effects.

In a vein irritation study, groups of drug-treated and vehicle-treated rabbits are given injections via the lateral ear veins. The reason rabbits are used for this test is because the lateral ear veins are large enough, long enough, and easily visible, so as to facilitate making the injections and identifying the injected portion for subsequent sampling for histological examination. The animals are observed during the injection and at intervals (such as 2–4 h) afterward for indications of discomfort during the injection and of swelling, erythema, or tenderness of the injection site. Twenty-four hours after injection, the animals are killed, and the injection site veins are examined histologically for evidence of tissue damage or inflammation.

In a typical muscle irritation study, groups of male rabbits are given intramuscular injections of 1.0 ml of drug formulation or vehicle with a 23-gauge needle inserted vertically to a depth of about 15–20mm into the sacrospinalis muscle. The injection site is marked by drawing a circle of ink around the site. The animals are observed during the injection and daily thereafter for 7 days for discomfort during injection and for signs of swelling, erythema, or tenderness at the injection site. On 1, 3, and 7 days after injections, blood samples are taken for measurement of creatine phosphokinase activity. Elevation of this enzyme in blood is an indication of muscle damage. Also, 1 and 7 days after injection, rabbits from each group are sacrificed and the muscle tissue surrounding the injection site taken for histological examination for evidence of tissue damage or inflammation.

Subchronic and Chronic Toxicology

Repeat-dosing toxicity studies are conducted to determine what side effects will arise from repeated administration of a drug at lower dosages than those used in acute toxicity studies and to determine safe dosages to be used in the initial human clinical trials. These studies range in duration from 1 to 2 weeks to 1 to 2 years. The length of studies required for a drug depends mainly upon the duration of treatment and the intended clinical dosing regimen. (This is discussed further under Governmental Guidelines). These studies are conducted in stages so that the results of one study can be used to design the subsequent study of longer duration.

The first are usually 2 weeks in length followed by 1-month, 3-month, 6-month, and then 1-year studies. Parallel subchronic and chronic studies are almost always conducted in two species, usually the rat and dog, because there is a large historical data base for these species and they are easy to work with, relatively economical to house, and readily available from commercial vendors. However, special circumstances may dictate that other species be used. For instance, if a drug causes excessive vomiting in dogs, which are known to be particularly sensitive to such effects, then some other non-rodent species such as the monkey may have to be used.

Also, if the absorption or metabolic handling of a drug in the rat or dog is found to be markedly different from that in humans such that one of these species would not be a reasonable predictor of toxicity in humans, then another species would be used. The usual protocol for subchronic and chronic studies includes groups of animals containing equal numbers of both sexes receiving at least three dosage levels of drug plus vehicle or other control groups. These animals are observed daily for clinical signs of toxicity.

Their body weights and food consumption are measured frequently. These three parameters— clinical signs, body weight, and food consumption— can be very sensitive indicators of toxicity. Complete hematology and serum chemistry profiles are determined at least at the end of the administration period and in some cases at intervals during the period of administration. Thorough physical examinations by a veterinarian or a trained technician are conducted at regular intervals. Periodic electrocardiograms are commonly recorded in studies with dogs. At the end of the period of drug administration, all the animals are subjected to a complete necropsy under the supervision of a veterinary pathologist.

The tissues are then subjected to complete microscopic examination by a veterinary pathologist to detect morphologic alterations in the tissues that may have resulted from drug administration.

In some cases it may be desirable to allow some animals from the drug-treated groups to live for a period of time after the end of drug administration to determine if any drug-related changes will disappear upon withdrawal of the drug. As mentioned, the results of the subchronic and chronic studies are used to help determine the dosage to be used in the initial human clinical trials.

To do this, the lowest dosage causing no toxicity (the notoxic- effect dosage) is determined for each study, and a safety factor is applied depending on the species. For instance, for rats the no-toxic-effect dosage is divided by 10 and for dogs by 6 to arrive at an estimate for the initial human dosage. These factors are derived from the observations that laboratory animals can usually tolerate higher dosages of drugs and other chemicals without exhibiting toxicity than can humans and that the differences in tolerance vary with differences in basal metabolic rate, which in turn varies with body surface area to weight ratio. This ratio varies by a factor of approximately 10 in rats and 6 in dogs relative to humans. The descriptions of toxic effects elicited in the subchronic and chronic studies allow clinicians conducting clinical trials to know which side effects to anticipate so as to protect the patient volunteers.

Reproductive Toxicology

Reproductive toxicology covers the entire process of reproduction from mating to pregnancy, birth, weaning and, sometimes, the reproductive function of subsequent generations. The species most commonly used in these studies are rats and rabbits, although in special circumstances, other species such as mice, dogs, or monkeys may be used. A series of studies is usually conducted so as to cover all the phases reproduction. Within each study, groups of animals corresponding to untreated controls and two or three drug-treated groups are used. One type of typical study, usually conducted in rats, provides information on fertility and general reproductive performance, particularly on gonadal function, estrous cycles, mating behaviour, conception rates, and early stages of gestation.

This study is usually referred to as a fertility study. Mature male rats are treated with the drug for 28 days prior to mating. This period is considered to be sufficient to detect effects on male reproductive organs when used in conjunction with thorough histological evaluation of testes. This fertility study in male animals is conducted prior to large-scale (Phase III) clinical trials in men. Female rats are treated for 14 days and then mated to treated males. Daily vaginal inspections are made and the finding of sperm or a copulation plug is considered as Day 0 of pregnancy.

The females are sacrificed on Day 13, and the number of live and dead embryos, the number of implantation sites, and the number of copora lutea on the ovaries determined. Some of the dams may be allowed to deliver normally, and the newborn pups are counted, examined, sexed, and weighed. The pups are again weighed and counted on Days 4 and 21 (weaning). From this test, preimplantation death, derived from the number of implantation sites in the uterine horns compared to the numbers of corpora lutea on the respective ovaries, postimplantation death, derived from the number of resorption sites compared to the number of implantation sites in the uterus, and the survival and growth of the offspring are determined.

This study is generally conducted prior to any clinical trials involving women of childbearing potential. However, in the USA and Europe, women of childbearing potential confirmed to be non-pregnant and using effective birth control may be enrolled in Phase I and Phase II (usually small-scale) studies prior to completion of the female fertility study. The second typical study, commonly called the teratology study, is usually conducted in two species, rats and rabbits. Rabbits are commonly used as a second species in Segment 2 studies because after the thalidomide tragedy in the early 1960s, rabbits were found to be more susceptible than were other commonly used laboratory species to the teratogenic effects of thalidomide, one of the few proven human teratogens.

In the Segment 2 test, inseminated females are treated only during the organogenesis period, which for rats and mice are Days 6–15 and for rabbits are Days 6–18 of gestation. One day prior to birth, the dams are sacrificed and the fetuses delivered by cesarean section. The fetuses are weighed, sexed, and examined for gross abnormalities. The fetuses from each treatment are then randomly divided into two groups. One group is preserved in fixative for subsequent soft tissue examination. In the other group, the tissues are cleared and the skeletons stained for skeletal examination.

The main endpoints in this study are numbers of live and dead fetuses, sex ratio of the fetuses, fetus weights, grossly observable abnormalities, and soft tissue and skeletal abnormalities. This test procedure was developed based on the well-established principle that the embryo is most susceptible to induction of birth defects during the organogenesis period. Previous exposure of the dam, as in the Segment 1 test, can lead to infertility or elevation of detoxification mechanisms, and thus to underestimation of the potential of a compound to cause birth defects.

A separate phase of testing was developed to specifically assess the occurrence of birth defects. This study is usually completed prior to any human studies involving women of childbearing potential (proven non-pregnant and using effective birth control) in Europe and Japan and prior to large-scale Phase III studies involving women in the USA. In a third type of study, called the perinatal/postnatal test, pregnant females are treated during the last quarter of pregnancy and through lactation until weaning.

The weanlings are necropsied to detect any internal anomalies. The purpose of this study is to determine the effects of the drug on late fetal development, labour, delivery, lactation, neonatal viability, and growth. It also assesses potential effects on neonates due to drug being excreted in the mother's milk. The endpoints for this study are the numbers of live and dead pups, the sex ratio of the pups, the numbers of surviving pups at 4 and 21 days after birth, and the growth and development of the pups.

Postnatal testing of pups commonly includes developmental indices such as days on which the following events occur: balanopreputial separation in males, vaginal opening in females, eye opening, pinna unfolding, and incisor eruption. Tests for reflex development and learning and memory are also commonly included. The reproductive competence of the offspring may also be tested. The potential of a compound to induce changes in these endpoints can be masked by infertility in the fertility and general reproductive performance study, which would lead to insufficient numbers of offspring to detect a change. This study must be completed prior to applying for marketing approval.

Mutagenicity

New drugs are routinely screened for their potential to cause gene mutations and/or chromosome aberrations by in vitro tests. An in vivo test is also required by most regulatory agencies. Further mutagenicity tests in whole animals are usually conducted only if the in vitro tests prove inconclusive or questionably positive. A new drug showing a strong positive response in an in vitro mutagenicity test probably would not be developed. In vivo tests serve to determine whether a drug has potential to cause genetic alterations under conditions closer to those that one would obtain under clinical use. The significance of these studies to the safety of a drug is twofold.

First, a drug that causes genetic damage could produce such damage in the sperm or ovaries of the patient and thus potentially cause genetic abnormalities in the children of that patient.

Second, it is widely accepted that any agent capable of causing genetic damage is also likely to be a carcinogen. The most common in vivo tests for genetic damage are chromosome damage tests and the dominant lethal test in mice or rats.

Chromosome damage can be assessed in mice and rats by bone marrow cytogenetic analysis and by bone marrow micronucleus assay. In both methods, bone marrow is examined after animals have been treated with either a single high dose or after several (2–4) daily doses of thedrug.

After the single dose, bone marrow is collected 6, 24, and 48 h after dosing for cytogenetic analysis or 24, 48, and 72 h after dosing for micronucleus determination. After the repeat dosing schedule, bone marrow is collected 6 h or 24–30 h after the last dose for cytogenetic or micronucleus assay, respectively.

For cytogenetic analysis, the animals are treated with colchicine 2–3 h prior to euthanasia to block dividing bone marrow cells in metaphase, femoral bone marrow is collected and fixed, and metaphase spreads are prepared on microscope slides. After Giemsa staining, the numbers of chromosomes and their normal or abnormal appearance in metaphase spreads are determined. In the micronucleus test, femoral bone marrow smears are stained with Giemsa or a fluorescent dye such as acridine orange, and the numbers of polychromatic erythrocytes containing micronuclei are determined.

Normal polychromatic erythrocytes contain no nuclei. In the process of red blood cell maturation, the nucleus is expelled just prior to the polychromatic erythrocyte stage. Micronuclei represent chromosomes or chromosome fragments that are left behind in the cell after expulsion of the nucleus. The dominant lethal assay is used to detect mutagenic damage to male germ cells.

The principle behind this assay is that theoretically there are a large number of chromosomes that, when damaged, could lead to lethality during early gestation if included in a fertilized egg. Hence, the target size relative to the genome is large for dominant lethality, and therefore the test should be a sensitive indicator of genetic damage. In this test, male animals (usually rats or mice) are treated with the drug daily for 5 days. They are then mated with separate groups of females each week for 8 weeks.

The 8-week interval covers the entire spermatogenic cycle. The females are killed at midpregnancy, about 14 days after the midweek of mating. At necropsy the uteri are examined and the numbers of corpora lutea, living implantations. and dead implantations are counted. The endpoints for this assay are:

- The numbers of preimplantation losses, determined by subtracting the numbers of implants from the numbers of corpora lutei,
- The numbers of dead implants.

Agents that cause increased numbers of preimplantation losses or dead implants compared to vehicle or historical control levels are considered positive in the dominant lethal assay. Because of the serial mating scheme, it is possible to know which stage of spermatogenesis is affected if a positive result is obtained. Several transgenic animal models are available for mutagenicity testing in vivo. The most widely used are the Big BlueTM and MutaTM Mouse models. These animals have portions of the E. coli galactosidase operon inserted into their genome.

This bacterial DNA is used as the target for mutagenicity testing when the animals are treated with test chemicals. After chemical treatment, the bacterial DNA is removed from the tissues and packaged into bacterial viral (phage) particles. The phages are used to infect bacteria growing on agar plates. The bacteria multiply to form colonies that can be counted. Because of a colour reaction that can be catalyzed by the galactosidase protein, a colour reaction formed in the bacterial colonies can be used to assess the presence or absence of mutations formed in the target DNA from virtually any tissue within the treated mice.

Carcinogenicity

Carcinogenicity bioassays are long-term studies in mice and rats conducted according to standard guidelines established by the National Cancer institute and the International Conference on Harmonization. In these studies the animals are treated repeatedly for periods of 18–24 months for mice and of 24 months for rats. Treatment is by a route consistent with the intended clinical route. The study design usually includes vehicle control and two or three drug-treated groups. The highest dosage is selected as the maximum dosage that will be tolerated by the animals.

This dosage may cause minimal signs of toxicity but should cause no more than a 10% decrement in body weight gain compared with control groups and should not cause toxicity, other than that related to a neoplastic response, that would be predicted to shorten the animal's natural life span. This high dosage is necessary to give a maximum test of the potential of a new drug to cause cancer.

The animals are examined frequently for palpable masses. At the end of the treatment period, the animals are necropsied and histologic examinations of thetissues are conducted with particular attention given tothe detection and identification of tumors.

Recently, transgenic strains of mice that allow detection of a carcinogenic response within 6 months of treatment rather than 2 years have become available. Examples of these are the $P53^{\pm}$ (P53 knockout), TG.AC and the rasH2 mouse models. These animals contain deletions of genes involved in suppressing the formation of tumors. The use of these animal models is justified based upon the fact that these same genes are known to be involved in many human tumors. Thus, the mechanisms leading to enhanced tumorigenic response in these animals is known to be relevant to human tumorigenesis.

Primary Irritation Testing

If a new drug is intended to be applied to the skin or eyes, one of the first tests to be conducted would be to determine if the drug, or the formulation containing the drug, will cause irritation of the skin or eyes. Even if a drug is intended only for dermal application, eye irritation testing may also be required because of the possibility of inadvertent exposure to the eyes.

For instance, an antibiotic intended for topical treatment of acne would be applied on the face, and the potential for eye contact would be considerable. These tests are usually conducted in rabbits because of their widely accepted sensitivity to dermal and ocular irritation and because of the scientific literature describing such tests in rabbits. Other species, such as guinea pigs or mice, may be used for dermal irritation studies under special circumstances.

A typical dermal irritation assay is conducted as follows. Six male albino rabbits are be clipped free of hair on the back. One area of skin is left intact, whereas another is abraded in a tic-tac-toe pattern with the point of a hypodermic needle so as to incise the superficial epidermis layer without causing bleeding. The test material, 0.5 ml of liquid or 0.5 g of solid or semisolid is applied to each site under a 1 _ 1 in. gauze pad.

The entire trunk of the animal is wrapped with an impervious material and held in place with tape for 24 h. The patches are then removed and excessive material wiped off. The skin reactions are scored at 24 and 72 h after the initial application according to a scheme such as that listed in Table. The mean values of the six rabbits for erythema and eschar formation at 24 and 72 h for both intact and abraded skin (four values) are added. The mean values of the six rabbits for edema at 24 and 72 h (four values) are also added. The total of eight values is divided by 4 to give the primary irritation index. Values of 5 or greater are considered indicative of a positive irritant.

A typical ocular irritation test is conducted as follows. Six albino rabbits of the same sex are given thorough ophthalmological examinations within 24 h before use to ensure absence of preexisting ocular damage.

The animals would be firmly but gently restrained. The test formulation is be placed in one eye of each animal by gently pulling the lower lid away from the eyeball (conjunctival cul-de-sac) to form a cup into which the test substance is dropped.

Table. Dermal Irritation Scoring System.

Skin reaction	Value[a]
Erythema and eschar formation	
No erythema	0
Very slight erythema (barely perceptible)	1
Well-defined erythema	2
Moderate to severe erythema	3
Severe erythema (beet redness) to slight eschar formation (injuries in depth)	4
Edema formation	
No edema	0
Very slight edema (barely perceptible)	1
Slight edema (edges of area well defined by definite raising)	2
Moderate edema (raised approximately 1mm)	3
Severe edema (raised more than 1mm and extending beyond the area of exposure)	4

[a]The value recorded for each reading is the average value of the six or more animals subjected to the test.

The lids are then gently held together for one second and the animal released. The other eye, remaining untreated, serves as a control. For testing liquids, 0.1 ml is used. For solid, paste, or particulate substances, the amount used must have a volume of 0.1 ml weighing not more than 100mg. The eyes of each rabbit would be examined 24, 48, and 72 h after treatment. After the 24-h examination, the treated eyes may be rinsed with tap water or saline. At this time the eyes may also be examined with the aid of fluorescein stain. One drop of fluorescein sodium ophthalmic solution USP is dropped directly on the cornea.

After flushing out the excess fluorescein with tap water or saline, injured areas of the cornea appear yellow and are best seen under ultraviolet illumination.

At each examination period, the eyes of each rabbit are scored for ocular reaction according to a scheme such as that listed in Table. The test may be considered positive if three or more animals exhibit positive reactions at any observation period. Equivocal tests are repeated. For humane reasons, formulations known to be corrosive or to be severely irritating in the dermal irritation test are assumed to be ocular irritants and not tested.

Antigenicity Testing

New drugs may undergo testing for antigenicity depending upon the chemical structure (suspected of being a sensitizer because of similarity to known sensitizers) or because of the intended use of the drug. Formulations intended for topical use are usually tested for delayed contact sensitivity. However, new drugs intended for systemic use are sometimes tested for their potential to induce anaphylaxis. These tests are commonly conducted in guinea pigs, but rats or mice are sometimes used.

Delayed contact hypersensitivity can be evaluated by several standard protocols in guinea pigs. A typical procedure is to sensitize groups of 10 guinea pigs each by injecting 0.1 ml (0.05 ml for the first injection) of a 0.1% suspension or solution of test material intradermally in the upper dorsal area three times per week for a total of 10 injections.

Table. Ocular Irritation Scoring System.

Ocular reaction	Score
Cornea	
No ulceration or opacity	0
Scattered or diffuse areas of opacity (other than slight dulling of normal luster, details of iris clearly visible)	1[a]
Easily discernible translucent areas, details of iris slightly obscured	2
Nacreous areas, no details of iris, size of pupil barely discernible	3
Opaque cornea, iris not discernible through the opacity	4
Iris	
Normal	0
Markedly deepened rugae, congestion, swelling, moderate circumcorneal hyperemia, or injection, any of these or any combination thereof, iris still reacting to light (sluggish reaction is positive)	1[a]
No reaction to light, hemorrhage,	2

gross destruction (any or all of these)	
Conjuctivae	
Blood vessels normal	0
Some blood vessels definitely	1
hyperemic (injected)	
Diffuse, crimson colour, individual	2[a]
essels not easily discernible	
Diffuse beefy red	3
No swelling	0
Any swelling above normal	1
(includes nictitating membranes)	2[a]
Obvious swelling with partial	
eversion of lids	
Swelling with lids about half closed	3
Swelling with lids more than half closed	4

[a]Readings at these numerical values or greater indicate positive response.

Additional groups are similarly treated with a positive control compound such as dinitrochlorobenzene and a negative control (saline).

About 2 weeks after the tenth injection, the animals are challenged by intradermal injection of 0.05 ml of the respective test compound, positive control, or saline in the dorsal lumbar area. The animals are shaved at least 18 prior to each injection. At 24 and 48 h after each sensitizing and challenge injection, the area of skin around the injection site is graded for erythema and edema according to the scheme shown in Table 4.

The erythema and edema grades are recorded separately. A dermal irritation score for each sensitizing injection and the challenge injection for each animal are calculated by adding the 24-h and 48-h scores for both erythema and edema (four scores) and dividing by 4.

The dermal irritation scores obtained from the initial sensitizing injection and the challenge injection for each animal are used to determine whether sensitization had been produced. When the challenge score exceeds the initial irritation score for a given animal by 1 or more, the animal is considered sensitized.

The potential of a compound to induce anaphylaxis can be evaluated in both active systemic anaphylaxis (ASA) and passive cutaneous anaphylaxis (PCA) tests in guinea pigs or rats. In both tests, the test compounds and positive and negative control substances are injected subcutaneously or intraperitoneally three times per weektogroups of 6–10 animals for a total of

6 injections. In the ASA test, the animals are challenged 2 weeks after the final sensitization injection intravenously with the respective test compound or positive or negative control and observed for signs of anaphylaxis. These would be graded as:

Negative: No anaphylactic symptoms

Mild: Restlessness, rubbing of the nose or ears, coughing, espiratory acceleration, or piloerection

Moderate: Urination, defecation. dyspnea, or ataxic gait

Severe: Convulsion followed by recovery

Fatal: Death by a fatal anaphylactic reaction

Table.Dermal Hypersensitivity Scoring Scheme

Skin reaction	Score
Erythema and eschar formation	
No erythema	0
Very slight erythema (barely perceptible)	1
Well-defined erythema	2
Moderate to severe erythema	3
Severe erythema (beet redness) to slight eschar formation (injuries in depth)	4
Edema	
No edema	0
Very slight edema (barely perceptible)	1
Slight edema (edges of area well defined by definite raising)	2
Moderate edema (raised approximately 1mm)	3
Severe edema (raised more than 1mm and extending beyond area of exposure)	4

In the PCA test, serum is collected from the sensitized animals 2 weeks after the final sensitization injection. This serum is diluted with saline to a dilution of 1/8 or more and injected intradermally on the backs of previously untreated guinea pigs.

Four hours later, these guinea pigs are given intravenous injections of 1.5–2.0 ml of saline containing 0.5% Evans Blue plus the respective test compound or positive or negative control. Thirty minutes after the intravenous injection, the animals are examined for the presence of blue areas of skin surrounding the earlier intradermal injections.

A blue spot having a diameter of 5–10mm is considered a positive reaction.

GOVERNMENTAL REQUIREMENTS

Governmental regulatory agencies such as the USFDA have established guidelines describing the kind of safety tests that should be conducted in animals in order to have a new drug approved for use in clinical trials and in order to get approval of a new drug application (NDA) for marketing.

The rationale and circumstances for conducting reproductive, mutagenicity, carcinogenicity, irritation, and sensitization studies have already been mentioned. Tables summarize the requirements for acute, subacute, and chronic toxicity studies for pharmaceutical products intended for use in humans.

These descriptions are limited to the requirements of the United States, Japan, and Europe because these areas represent the largest pharmaceutical markets in the world today. These requirements have been developed at the International Conference on Harmonization to provide uniformity among the three regions.

Phases I, II, and III refer to the different phases of human clinical trials. Phase I refers to the initial trials, limited to one or a few doses to determine absorption, pharmacokinetics, and an initial estimate of safety. Phase II refers to larger-scale studies to establish safety and to get an initial estimate of clinical efficacy. Phase III refers to the final, large-scale, multicenter trials aimed at establishing efficacy.

SUSTAINABLE SOURCES OF PLANT AND ANIMAL DRUGS

As species face loss of habitat or overharvesting, there have been new issues to deal with in sourcing crude drugs. These include changes to the herb from farming practices, substitution of species or other plants altogether, adulteration and cross-pollination issues. For instance, ginseng which is field farmed may have significant problems with fungus, making contamination with fungicides an issue.

This may be remedied with woods grown programme, but they are insufficient to produce enough ginseng to meet demand. The wildcrafted echinacea, black cohosh and American ginseng often rely upon old growth root, often in excess of 50 years of age and it is not clear that younger stock will have the same pharmaceutical effect.

Black cohosh may be adulterated with the related Chinese actea species, which is not the same. Ginseng may be replaced by ginseniodes from Jiaogulan which has been stated to have a different effect than the full panax root.

The problem may be exacerbated by the growth of pills and capsules as the preferred method of ingesting medication as they are cheaper and more available than traditional, individually tailored prescriptions of raw medicinals but the contents are harder to track. Seahorses are a case in point: Seahorses once had to be of a certain size and quality before they were accepted by practitioners and consumers.

But declining availability of the preferred large, pale and smooth seahorses has been offset by the shift towards prepackaged medicines, which make it possible for TCM merchants to sell previously unused juvenile, spiny and dark-coloured animals. Today almost a third of the seahorses sold in China are prepackaged.

The farming of plant or animal species, used for medicinal purposes has caused difficulties.

Rob Parry Jones and Amanda Vincent write:

- One solution is to farm medicinal animals and plants. Chinese officials have promoted this as a way of guaranteeing supplies as well as protecting endangered species. And there have been some successes—notably with plant species, such as American ginseng—which is used as a general tonic and for chronic coughs. Red deer, too, have for centuries been farmed for their antlers, which are used to treat impotence and general fatigue. But growing your own is not a universal panacea. Some plants grow so slowly that cultivation in not economically viable. Animals such as musk deer may be difficult to farm, and so generate little profit. Seahorses are difficult to feed and plagued by disease in captivity. Other species cannot be cultivated at all. Even when it works, farming usually fails to match the scale of demand. Overall, cultivated TCM plants in China supply less than 20 per cent of the required 1.6 million tonnes per annum. Similarly, China's demand for animal products such as musk and pangolin scales far exceeds supply from captive-bred sources.
- Farming alone can never resolve conservation concerns, as government authorities and those who use Chinese medicine realise. For a start, consumers often prefer ingredients taken from the wild, believing them to be more potent. This is reflected in the price, with wild oriental ginseng fetching up to 32 times as much as cultivated plants. Then there are welfare concerns. Bear farming in China is particularly controversial. Around 7600 captive bears have their bile "milked" through tubes inserted into their gall bladders. The World Society for the Protection of Animals states that bear farming is surrounded by "appalling levels of cruelty and neglect". Chinese officials state that 10 000 wild bears would need to be killed each year to produce as much bile, making bear farming the more desirable option. The World society for the Protection of Animals, however, states that "it is commonly believed in China that the bile from a wild bear is the most potent, and so farming bears for their bile cannot replace the demand for the product extracted from wild animals".
- One alternative to farming involves replacing medical ingredients from threatened species with manufactured chemical compounds. In general, this sort of substitution is difficult to achieve because the active ingredient is often not known. In addition, most TCM users believe

that TCM compounds may act synergistically so several ingredients may interact to give the required effect. Thus TCM users often people prefer the wild source. Tauro ursodeoxycholic acid, the active ingredient of bear bile, can be synthesised and is used by some Western doctors to treat gallstones, but many TCM consumers reject it as being inferior to the natural substance from wild animals.

The Scope of Pharmacognosy

Pharmacognosy is one of the five major divisions of the pharmaceutical curriculum, represents the oldest branch of the profession of pharmacy. The ancients gathered herbs, animals, plants, and minerals and concocted them into ill-flavoured pungent mixtures. Innumerable remedies were known to the early practitioners of pharmacy and medicine, a fact indicated by the writings of Theophrastus, Pliny, Dioscorides, and their contemporaries. Thousands of plant and animal products used for the treatment of ills were described by Dioscorides in his book "Dc Materia Medica." Of these a surprisingly large -number are still of importance in modern therapeutic practice: Aspidium, Cinnamon, Ergot, Ilyoscyamus, and Opium were used in much the same manner as today.

Fig. Rauwolfia Serpentina Bentham.

Materia Medica Meaning Medicinal Materials

From this humble beginning medicine and pharmacy gradually emerged along separate paths: the physician diagnosing the ailment and prescribing the remedy and the apothecary or pharmacist specializing in the collection, preparation, and compounding of the substance.

Thus, the term materia medica (meaning medicinal materials) was synonymous with the substances and products derived from natural sources and employed by the physicians of that era. It was not until 1815 that the term Pharmacognosy was introduced by Seydler.

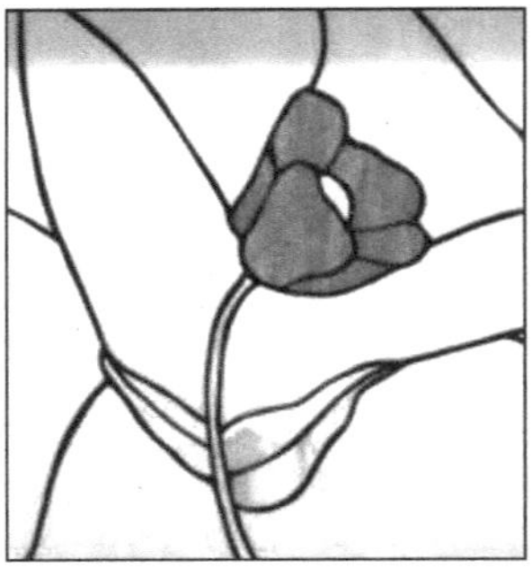

Fig. Opium Poppy Stained Glass Pharmacognosy.

This name is formed from two Greek words, pharmakon, drug, and gnosis, knowledge, and literally means the "entire knowledge of drugs."

The most comprehensive idea of the scope of pharmacognosy was presented by Fluckiger, who stated that it "is the simultaneous application of various scientific disciplines with the object of acquiring knowledge of drugs from every point of view."

Pharmacognosy Embraces a Knowledge of History

In a broad sense, pharmacognosy embraces a knowledge of the history, distribution, cultivation, collection, selection, preparation, commerce, identification, evaluation, preservation, and use of drugs and economic substances affecting the health of man and other animals.

Such economic substances extend beyond the category of crude drugs and their derivatives and include a variety of commercial products: allergens, allergenic extracts, antibiotics, immunizing biologicals, flavouring agents, condiments, beverages, insecticides, rodenticides, and herbicides. In a restricted sense, the definition of pharmacognosy implies a particular knowledge of methods of identification and evaluation of drugs.

As a part of the pharmaceutical curriculum pharmaco-gnosy plays an important role as liaison agent between pharmacology and pharmaceutical chemistry on one hand and between pharmacy and pharmacy administration on the other. Pharmacology, like pharmacognosy, is an outgrowth of the old-time materia medica; it is easy to see the correlation of these two sciences concerning drugs of biological origin.

The pharmacology student has a more complete understanding of the action of belladonna leaf if he is aware of the physical nature of the drug and its chemical constituents. Some of the labouratory procedures in pharmaceutical chemistry have miniature counterparts in the extractions and microchemical tests essential for the identification and evaluation of certain drugs.

Methods of collection, curing, drying, and assaying affect the price of drugs and, insofar as economics are concerned, pharmacognosy is intimately associated with phases of pharmacy administration dealing with prescription pricing. The relationship of pharmacognosy to operative pharmacy and dispensing pharmacy is obvious when one considers the number of plant and animal products handled by the pharmacist in this synthetic age.

Because of his knowledge of drug constituents and their physical and chemical properties, the pharmacist is able to predict incompatibilities in actual compounding. In dispensing non-compounded prescription specialties, he is conversant with a vast store of information permitting him to confer intelligently with his fellow colleagues in the medical, dental, and nursing professions.

12

Classification of Nutraceuticals

Nutraceuticals is a broad umbrella term used to describe any product derived from food sources that provides extra health benefits in addition to the basic nutritional value found in foods. Products typically claim to prevent chronic diseases, improve health, delay the aging process, and increase life expectancy. There is minimal regulation over which products are allowed to display the nutraceutical term on their labels. Because of this, the term is often used to market products with varying uses and effectiveness. The definition of nutraceuticals and related products often depend on the source.

Members of the medical community desire that the nutraceutical term be more clearly established in order to distinguish between the wide varieties of products out there. There are multiple different types of products that fall under the category of nutraceuticals.

Dietary supplements

A dietary supplement is a product that contains nutrients derived from food products that are concentrated in liquid or capsule form. The Dietary Supplement Health and Education Act (DSHEA) of 1994 defined generally what constitutes a dietary supplement. "A dietary supplement is a product taken by mouth that contains a "dietary ingredient" intended to supplement the diet.

The "dietary ingredients" in these products may include: vitamins, minerals, herbs or other botanicals, amino acids, and substances such as enzymes, organ tissues, glandulars, and metabolites. Dietary supplements can also be extracts or concentrates, and may be found in many forms such as tablets, capsules, softgels, gelcaps, liquids, or powders."

Dietary supplements do not have to be approved by the U.S. Food and Drug Administration (FDA) before marketing. Although supplements claim to provide health benefits, products usually include a label that says: "These statements have not been evaluated by the Food and Drug Administration. This product is not intended to diagnose, treat, cure, or prevent any disease."

Functional Foods

Functional foods are designed to allow consumers to eat enriched foods close to their natural state, rather than by taking dietary supplements manufactured in liquid or capsule form. Functional foods have been either enriched or fortified, a process called nutrification. This practice restores the nutrient content in a food back to similar levels from before the food was processed. Sometimes, additional complementary nutrients are added, such as vitamin D to milk.

Health Canada defines functional foods as "ordinary food that has components or ingredients added to give it a specific medical or physiological benefit, other than a purely nutritional effect." In Japan, all functional foods must meet three established requirements.

Foods should be:

1. Present in their naturally-occurring form, rather than a capsule, tablet, or powder;
2. Consumed in the diet as often as daily;
3. Should regulate a biological process in hopes of preventing or controlling disease.

Medical Foods

Medical foods aren't available as an over-the-counter product to consumers. The FDA considers medical foods to be "formulated to be consumed or administered enterally under the supervision of a physician, and which is intended for the specific dietary management of a disease or condition for which distinctive nutritional requirements, on the basis of recognized scientific principles, are established by medical evaluation."

Medical foods can be ingested through the mouth or through tube feeding. These foods are often designed to meet certain nutritional requirements for people diagnosed with specific illnesses. Medical foods are closely monitored by medical supervision.

Farmaceuticals

According to a report written for the United States Congress entitled "Agriculture: A Glossary of Terms, Programme, and Laws", "(Farmaceuticals) is a melding of the words farm and pharmaceuticals.

It refers to medically valuable compounds produced from modified agricultural crops or animals (usually through biotechnology).

Proponents believe that using crops and possibly even animals as pharmaceutical factories could be much more cost effective than conventional methods (i.e., in enclosed manufacturing facilities) and also provide agricultural producers with higher earnings.

Fig. Studies Show that Broccoli may Help in the Prevention of Cancer.

"At issue in the United States has been whether the current system for regulating biotechnology is adequate for ensuring the safety (to humans, animals and crops, and the environment) of newly emerging applications, such as farmaceuticals... The term farmaceuticals is more frequently associated, in agricultural circles, with medical applications of genetically engineered crops or animals."

The following is an incomplete list of foods with reported medicinal value:

- Antioxidants: resveratrol from red grape products; flavonoids inside citrus, tea, wine, and dark chocolate foods; anthocyanins found in berries
- Reducing hypercholesterolemia: soluble dietary fibre products, such as psyllium seed husk
- Cancer prevention: broccoli (sulforaphane)
- Improved arterial health: soy or clover (isoflavonoids)
- Lowered risk of cardiovascular disease:alpha-linolenic acid from flax or Chia seeds

In addition, many botanical and herbal extracts such as ginseng, garlic oil, etc. have been developed as nutraceuticals. Nutraceuticals are often used in nutrient premixes or nutrient systems in the food and pharmaceutical industries.

EFFECTIVENESS AND SAFETY

Regulation

Unlike pharmaceutical drugs, nutraceutical products are widely available and minimally monitored. Companies are not obligated to back claims about the function and effectiveness of their product, but many companies attempt to provide scientific backing of their products to increase credibility.

International Sources

In the global market, there are significant product quality issues

Nutraceuticals from the international market may claim to use organic or exotic ingredients, yet the lack of regulation may compromise the safety and effectiveness of products. Companies looking to create a wide profit margin may create unregulated products overseas with low-quality or ineffective ingredients.

Bioavailability

Bioavailability, which can be thought of as the the "absorption rate" of a supplement product, is one of the main challenges in finding effective nutraceutical products. The bioavailability of nutrients is higher in food eaten in its natural state. Even among unprocessed foods, not all foods are broken down and digested as effectively. Nutraceuticals with poor absorption rates results in nutrients being disposed from the body without providing any nutritional or medicinal benefit.

Impact of Placebo Effect

Similar to pharmaceuticals, part of the effectiveness of nutraceuticals may be attributed to the placebo effect. Consumers using nutraceuticals may inaccurately credit their use of nutraceuticals for healing illness, when the body is often able to recover on its own.

PROCESSED FOODS

Since the Industrial Revolution some two hundred years ago, the food processing industry has invented many technologies that both help keep foods fresh longer and alter the fresh state of food as they appear in nature. Cooling is the primary technology used to maintain freshness, whereas many more technologies have been invented to allow foods to last longer without becoming spoiled.

These latter technologies include pasteurisation, autoclavation, drying, salting, and separation of various components, and all appear to alter the original nutritional contents of food. Pasteurisation and autoclavation (heating techniques) have no doubt improved the safety of many common foods, preventing epidemics of bacterial infection. But some of the (new) food processing technologies undoubtedly have downfalls as well.

Modern separation techniques such as milling, centrifugation, and pressing have enabled concentration of particular components of food, yielding flour, oils, juices and so on, and even separate fatty acids, amino acids, vitamins, and minerals. Inevitably, such large scale concentration changes the nutritional content of food, saving certain nutrients while removing others.

Heating techniques may also reduce food's content of many heat-labile nutrients such as certain vitamins and phytochemicals, and possibly other yet to be discovered substances. Because of reduced nutritional value, processed foods are often 'enriched' or 'fortified' with some of the most critical nutrients (usually certain vitamins) that were lost during processing.

Nonetheless, processed foods tend to have an inferior nutritional profile compared to whole, fresh foods, regarding content of both sugar and high GI starches, potassium/sodium, vitamins, fibre, and of intact, unoxidized (essential) fatty acids. In addition, processed foods often contain potentially harmful substances such as oxidized fats and trans fatty acids. A dramatic example of the effect of food processing on a population's health is the history of epidemics of beri-beri in people subsisting on polished rice.

Removing the outer layer of rice by polishing it removes with it the essential vitamin thiamine, causing beri-beri. Another example is the development of scurvy among infants in the late 1800s in the United States. It turned out that the vast majority of sufferers were being fed milk that had been heat-treated (as suggested by Pasteur) to control bacterial disease.

Pasteurisation was effective against bacteria, but it destroyed the vitamin C. As mentioned, lifestyle- and obesity-related diseases are becoming increasingly prevalent all around the world. There is little doubt that the increasingly widespread application of some modern food processing technologies has contributed to this development.

The food processing industry is a major part of modern economy, and as such it is influential in political decisions (e.g. nutritional recommendations, agricultural subsidising). In any known profit-driven economy, health considerations are hardly a priority; effective production of cheap foods with a long shelf-life is more the trend. In general, whole, fresh foods have a relatively short shelf-life and are less profitable to produce and sell than are more processed foods.

Thus the consumer is left with the choice between more expensive but nutritionally superior whole, fresh foods, and cheap, usually nutritionally inferior processed foods. Because processed foods are often cheaper, more convenient (in both purchasing, storage, and preparation), and more available, the consumption of nutritionally inferior foods has been increasing throughout the world along with many nutrition-related health complications.

History

Humans have evolved as omnivorous hunter-gatherers over the past 250,000 years. The diet of early modern humans varied significantly depending on location and climate. The diet in the tropics tended to be based more heavily on plant foods, while the diet at higher latitudes tended more towards animal products. Analysis of postcranial and cranial remains of humans and animals from the Neolithic, along with detailed bone modification studies have shown that cannibalism was also prevalent among prehistoric humans.

Agriculture developed about 10,000 years ago in multiple locations throughout the world, providing grains such as wheat, rice, potatoes, and maize, with staples such as bread, pasta, and tortillas. Farming also provided milk and dairy products, and sharply increased the availability of meats and the diversity of vegetables.

The importance of food purity was recognized when bulk storage led to infestation and contamination risks. Cooking developed as an often ritualistic activity, due to efficiency and reliability concerns requiring adherence to strict recipes and procedures, and in response to demands for food purity and consistency.

From Antiquity to 1900

The first recorded nutritional experiment is found in the Bible's Book of Daniel. Daniel and his friends were captured by the king of Babylon during an invasion of Israel. Selected as court servants, they were to share in the king's fine foods and wine. But they objected, preferring vegetables (pulses) and water in accordance with their Jewish dietary restrictions. The king's chief steward reluctantly agreed to a trial. Daniel and his friends received their diet for 10 days and were then compared to the king's men. Appearing healthier, they were allowed to continue with their diet.

Around 475 BC, Anaxagoras stated that food is absorbed by the human body and therefore contained "homeomerics" (generative components), suggesting the existence of nutrients. Around 400 BC, Hippocrates said, "Let food be your medicine and medicine be your food."

In the 1500s, scientist and artist Leonardo da Vinci compared metabolism to a burning candle. In 1747, Dr. James Lind, a physician in the British navy, performed the first scientific nutrition experiment, discovering that lime juice saved sailors who had been at sea for years from scurvy, a deadly and painful bleeding disorder.

The discovery was ignored for forty years, after which British sailors became known as "limeys." The essential vitamin C within lime juice would not be identified by scientists until the 1930s. Around 1770, Antoine Lavoisier, the "Father of Nutrition and Chemistry" discovered the details of metabolism, demonstrating that the oxidation of food is the source of body heat.

In 1790, George Fordyce recognized calcium as necessary for fowl survival. In the early 1800s, the elements carbon, nitrogen, hydrogen and oxygen were recognized as the primary components of food, and methods to measure their proportions were developed.

In 1816, François Magendie discovered that dogs fed only carbohydrates and fat lost their body protein and died in a few weeks, but dogs also fed protein survived, identifying protein as an essential dietary component. In 1840, Justus Liebig discovered the chemical makeup of carbohydrates (sugars), fats (fatty acids) and proteins (amino acids.)

In the 1860s, Claude Bernard discovered that body fat can be synthesized from carbohydrate and protein, showing that the energy in blood glucose can be stored as fat or as glycogen.In the early 1880s, Kanehiro Takaki observed that Japanese sailors (whose diets consisted almost entirely of white rice) developed beriberi (or endemic neuritis, a disease causing heart problems and paralysis) but British sailors and Japanese naval officers did not.

Adding various types of vegetables and meats to the diets of Japanese sailors prevented the disease. In 1896, Baumann observed iodine in thyroid glands. In 1897, Christiaan Eijkman worked with natives of Java, who also suffered from beriberi. Eijkman observed that chickens fed the native diet of white rice developed the symptoms of beriberi, but remained healthy when fed unprocessed brown rice with the outer bran intact.

Eijkman cured the natives by feeding them brown rice, discovering that food can cure disease. Over two decades later, nutritionists learned that the outer rice bran contains vitamin B1, also known as thiamine.

From 1900 to the Present

In the early 1900s, Carl Von Voit and Max Rubner independently measured caloric energy expenditure in different species of animals, applying principles of physics in nutrition. In 1906, Wilcock and Hopkins showed that the amino acid tryptophan was necessary for the survival of rats. He fed them a special mixture of food containing all the nutrients he believed were essential for survival, but the rats died.

A second group of rats to which he also fed fed an amount of milk containing vitamins. Gowland Hopkins recognized "accessory food factors" other than calories, protein and minerals, as organic materials essential to health but which the body cannot synthesize. In 1907, Stephen M. Babcock and Edwin B. Hart conducted the single-grain experiment. This experiment runs through 1911.

In 1912, Casimir Funk coined the term vitamin, a vital factor in the diet, from the words "vital" and "amine," because these unknown substances preventing scurvy, beriberi, and pellagra, were thought then to be derived from ammonia. The vitamins were studied in the first half of the twentieth century.

In 1913, Elmer McCollum discovered the first vitamins, fat soluble vitamin A, and water soluble vitamin B (in 1915; now known to be a complex of several water-soluble vitamins) and names vitamin C as the then-unknown substance preventing scurvy. Lafayette Mendel and Thomas Osborne also perform pioneering work on vitamin A and B.

In 1919, Sir Edward Mellanby incorrectly identified rickets as a vitamin A deficiency, because he could cure it in dogs with cod liver oil. In 1922, McCollum destroyed the vitamin A in cod liver oil but finds it still cures rickets, naming vitamin D Also in 1922, H.M.

Evans and L.S. Bishop discover vitamin E as essential for rat pregnancy, originally calling it "food factor X" until 1925.In 1925, Hart discovered that trace amounts of copper are necessary for iron absorption. In 1927, Adolf Otto Reinhold Windaus synthesized vitamin D, for which he won the Nobel Prize in Chemistry in 1928. In 1928, Albert Szent-Györgyi isolated ascorbic acid, and in 1932 proves that it is vitamin C by preventing scurvy.

In 1935 he synthesizes it, and in 1937 he wins a Nobel Prize for his efforts. Szent-Györgyi concurrently elucidates much of the citric acid cycle. In the 1930s, William Cumming Rose identified essential amino acids, necessary protein components which the body cannot synthesize.

In 1935, Underwood and Marston independently discover the necessity of cobalt. In 1936, Eugene Floyd Dubois showed that work and school performance are related to caloric intake. In 1938, Erhard Fernholz discovered the chemical structure of vitamin E. It was synthesised by Paul Karrer.

In 1940, rationing in the United Kingdom during and after World War II took place according to nutritional principles drawn up by Elsie Widdowson and others. In 1941, the first Recommended Dietary Allowances (RDAs) were established by the National Research Council.In 1992, The U.S. Department of Agriculture introduced the Food Guide Pyramid. In 2002, a Natural Justice study showed a relation between nutrition and violent behaviour. In 2005, a study found that obesity may be caused by adenovirus in addition to bad nutrition.

NUTRITIOUS MEAL IN MENTAL AGILITY

Research indicates that improving the awareness of nutritious meal choices and establishing long-term habits of healthy eating has a positive effect on a cognitive and spatial memory capacity, potentially increasing a student's potential to process and retain academic information.

Some organizations have begun working with teachers, policymakers, and managed foodservice contractors to mandate improved nutritional content and increased nutritional resources in school cafeterias from primary to university level institutions.

Health and nutrition have been proven to have close links with overall educational success. Currently less than 10% of American college students report that they eat the recommended five servings of fruit and vegetables daily.

Better nutrition has been shown to have an impact on both cognitive and spatial memory performance; a study showed those with higher blood sugar levels performed better on certain memory tests.

In another study, those who consumed yogurt performed better on thinking tasks when compared to those who consumed caffeine free diet soda or confections. Nutritional deficiencies have been shown to have a negative effect on learning behaviour in mice as far back as 1951. "Better learning performance is associated with diet induced effects on learning and memory ability". The "nutrition-learning nexus" demon-strates the correlation between diet and learning and has application in a higher education setting. "We find that better nourished children perform significantly better in school, partly because they enter school earlier and thus have more time to learn but mostly because of greater learning productivity per year of schooling." 91% of college students feel that they are in good health while only 7% eat their recommended daily allowance of fruits and vegetables.

Nutritional education is an effective and workable model in a higher education setting. ore "engaged" learning models that encompass nutrition is an idea that is picking up steam at all levels of the learning cycle. There is limited research available that directly links a student's Grade Point Average (G.P.A.) to their overall nutritional health. Additional substantive data is needed to prove that overall intellectual health is closely linked to a person's diet, rather than just another correlation fallacy.

Mental Disorders

Nutritional supplement treatment may be appropriate for major depression, bipolar disorder, schizophrenia, and obsessive compulsive disorder, the four most common mental disorders in developed countries.

Supplements that have been studied most for mood elevation and stabilization include eicosapentaenoic acid and docosahexaenoic acid (each of which are an omega-3 fatty acid contained in fish oil, but not in flaxseed oil), vitamin B12, folic acid, and inositol.

Cancer

Cancer is now common in developing countries. According a study by the International Agency for Research on Cancer, "In the developing world, cancers of the liver, stomach and esophagus were more common, often linked to consumption of carcinogenic preserved foods, such as smoked or salted food, and parasitic infections that attack organs." Lung cancer rates are rising rapidly in poorer nations because of increased use of tobacco. Developed countries "tended to have cancers linked to affluence or a 'Western lifestyle' — cancers of the colon, rectum, breast and prostate — that can be caused by obesity, lack of exercise, diet and age."

Metabolic Syndrome

Several lines of evidence indicate lifestyle-induced hyperinsulinemia and reduced insulin function (i.e. insulin resistance) as a decisive factor in many disease states. For example, hyperinsulinemia and insulin resistance are strongly linked to chronic inflammation, which in turn is strongly linked to a variety of adverse developments such as arterial microinjuries and clot formation (i.e. heart disease) and exaggerated cell division (i.e. cancer). Hyperinsulinemia and insulin resistance (the so-called metabolic syndrome) are characterized by a combination of abdominal obesity, elevated blood sugar, elevated blood pressure, elevated blood triglycerides, and reduced HDL cholesterol.

The negative impact of hyperinsulinemia on prostaglandin PGE1/PGE2 balance may be significant. The state of obesity clearly contributes to insulin resistance, which in turn can cause type 2 diabetes. Virtually all obese and most type 2 diabetic individuals have marked insulin resistance. Although the association between overweight and insulin resistance is clear, the exact (likely multifarious) causes of insulin resistance remain less clear.

Importantly, it has been demonstrated that appropriate exercise, more regular food intake and reducing glycemic load all can reverse insulin resistance in overweight individuals (and thereby lower blood sugar levels in those who have type 2 diabetes). Obesity can unfavourably alter hormonal and metabolic status via resistance to the hormone leptin, and a vicious cycle may occur in which insulin/leptin resistance and obesity aggravate one another. The vicious cycle is putatively fuelled by continuously high insulin/leptin stimulation and fat storage, as a result of high intake of strongly insulin/leptin stimulating foods and energy. Both insulin and leptin normally function as satiety signals to the hypothalamus in the brain; however, insulin/leptin resistance may reduce this signal and therefore allow continued overfeeding despite large body fat stores.

In addition, reduced leptin signalling to the brain may reduce leptin's normal effect to maintain an appropriately high metabolic rate. There is a debate about how and to what extent different dietary factors— such as intake of processed carbohydrates, total protein, fat, and carbohydrate intake, intake of saturated and trans fatty acids, and low intake of vitamins/minerals—contribute to the development of insulin and leptin resistance. In any case, analogous to the way modern man-made pollution may potentially overwhelm the environment's ability to maintain homeostasis, the recent explosive introduction of high glycemic index and processed foods into the human diet may potentially overwhelm the body's ability to maintain homeostasis and health (as evidenced by the metabolic syndrome epidemic).

Hyponatremia

Excess water intake, without replenishment of sodium and potassium salts, leads to hyponatremia, which can further lead to water intoxication at more dangerous levels. A well-publicized case occurred in 2007, when Jennifer Strange died while participating in a water-drinking contest. More usually, the condition occurs in long-distance endurance events (such as marathon or triathlon competition and training) and causes gradual mental dulling, headache, drowsiness, weakness, and confusion; extreme cases may result in coma, convulsions, and death. The primary damage comes from swelling of the brain, caused by increased osmosis as blood salinity decreases. Effective fluid replacement techniques include Water aid stations during running/

cycling races, trainers providing water during team games such as Soccer and devices such as Camel Baks which can provide water for a person without making it too hard to drink the water.

FUNCTIONAL BEVERAGE OF DRINK PRODUCTS

Functional beverages can be defined as a drink products that satisfies thirst, is non-alcoholic, is ready to drink and includes in its formulation non-traditional ingredients, such as herbs, vitamins, minerals, amino acids or additional fruit/vegetable raw ingredients, depending on the purpose it is designed for, so as to provide specific health benefits that beyond general nutrition. Sports and performance drinks, energy drinks, ready to drink (RTD) teas, enhanced fruit drinks, soy beverages and enhanced water, among others, are some of the product segments rolled out as functional beverages in the market space.

Functional beverages have become popular due to its appeal to consumers who are seeking specific health benefits in their foods and beverages with it 'healthiness-on-the-go' idea. Both convenience and health have been identified as important factors when consumers make decisions about purchasing foods and beverages. Functional drinks are promoted with benefits such as heart health, improved immunity and digestion, joint health, satiety, and energy-boosting.

INDUSTRY

Functional beverage sector is a subsector of the functional food and non-alcoholic beverage industry and it is the fastest growing sector. This rapid growth is partially due to the combination of maturity of the carbonated soft drink sector and heavy investments by major food and beverage companies. In 2006, the functional beverage per capita consumption have risen to 66.4 gallons, while the carbonated soft drink sector has a decline in their per capita consumption to 50.4 gallons (192.5 gallons was the average per capita conusmption in 2006).

Another reason for this growth is the consumer oriented market scheme in which innovative ideas come from consumers to satisfy consumers. In US, the market share of functional beverages accounted for 48.9% of the 118.3 billion worth non-alcoholic industry in 2008. Industry entrance is mainly driven by the higher growth annual rate of 15% to 20%. This added value is significantly larger than the 2-4% per annum growth rates of general foods and beverages market. Industry players in functional beverage industry are generally categorized into four types. One, is the traditional non-alcoholic beverage companies including PepsiColnc, Fuze (beverage) and Coca-Cola Co.

Another type consists of major food companies such as Nestlé, Altria Group,Kraft FoodsInc, General MillsInc, Campbell SoupCo. The third group is smaller scaled private companies, like the NOVA Group, Austria and some of which specialize in one subcategory or segment of the functional beverage market like Traditional Medicinals and POM Wonderful. The last group is the growers' cooperative such as Ocean Spray and Sunsweet Growers.

Current Product/ trends

Products of functional beverage industry encompass wide range of varieties targeting different health related concerns. One of the product trends is hybrid drinks with functional and sensory benefits such as thirst quenching ability with daily dosage of vitamins or other nutrients. Another one is probiotics, exemplified by Activia, for gut health and boosting natural defences or immune system. Memory and mental sharpness is also coming into focus with the Braniac carambola punch drink in the Function Drinks line. Children's functional drinks also receive much attention in recent years with leading brand Neslie's Boost (beverage). The current trend of single serve is fueled by consumer's preference in convenience.

According to Campbell's director of single serve beverage, "People know they will be seen when they are drinking single-serving beverages, so the package is critical." Weight loss, health and beauty drinks account considerable market share, such as Nova The Essential Drink. Lastly, energy boosting functional beverage products, such as Red Bull and 5-Hour Energy, are rated number one in growth in the functional beverage market.

Market

In this economic downturn, many products belonging the specialty food category are expected to have reduced sales because paying more for those specialty items are not an option for many consumers. These consumers also tend to shop in discount grocery stores and tend to make less grocery trips. Yet, the functional beverage market is predicted to have an increase in growth. This is because people generally reduce restaurant expenses first before they start cutting down grocery expenses. It has also been determined that people become more proactive in illness prevention and control. In the past, people were very reactive in their health because health problems are treated after they arrive.

Nowadays, people will take an initiative and look for a food product that will prevent an illness from occurring. Thus, all of the above features prevent the functional beverage market from experiencing lower sales. Competition in the functional beverage industry are primarily done in four different ways. Companies firstly focus on their own products by clearly distinguishing their health claims from other similar product and specify naturally healthy ingredients. Then, extensions of existing brand lines need to be considered.

The addition of Kids Essentials to the Boost (beverage) line by Nestle is a good representation of this strategy. Thirdly, bigger companies compete for market share by Mergers and acquisitions of smaller companies that may own particular market sector. An example of this strategy is the purchasing of Glaceau from Energy Brands by Coca Cola. Lastly, to increase the competitiveness of the company, one may explore new functional brands by identifying new markets and demands.

Market segments of the functional beverage industry are divided mainly by four parts. Those include hydration, energy/rejuvenation, health and wellness and weight management. Each segment has its own target market and consumers. Overlapping of target consumers does occur, not because of undefined market needs, but due to the consumer acceptance towards functional beverages. The dollar sales share on each of these sub-segments are: 28% (Hydration), 8.4% (Energy/Rejuvenation), 62.2% (Health and Wellness), 1.2%(Weight Management).

Hydration

Currently there is an increasing amount of advertising in the promotion of the importance of hydration. In 2008, Nestle launched the sale of Glowelle, exclusively in Neiman Marcus. Antioxidant vitamins and fruit extracts are among the ingredient list that claims to "hydrate the inner and outer layer of the skin" and protects from free radicals. Gatorade's Thirst Quencher offers a wide assortment of flavours that each contain an "excellent source" (25% DV) of B vitamins (B3, B5, B6) to help meet the demands of energy metabolism; an "excellent source" (20% DV) of the antioxidant vitamin E to help protect the active body; an "excellent source" (20% DV) of vitamin C to help active people as they attack their day.

In tandem with these adjustments, low-calorie G2 was also reformulated. Now it delivers significant nutrition enhancements by providing an "excellent source" (25% DV) of B vitamins (B3, B5, B6) and a "good source" (10% DV) of vitamins C and E, all of which help meet the nutrient needs of active individuals. Gatorade is well-known for its hydrating qualities of athletes. Another example of a functional beverage targeting the hydration sector is Hydro One Premium Beverage. It contains vitamins and minerals as well as herbal extracts that are specifically and scientifically formulated to promote healthy hydration, maintain healthy levels of blood glucose, and also provides the nutrients that are beneficial for the circulatory system.

Hydro One Premium Beverages delivers a form of sugar which is the most advantageous for the body and aids in the nutritional needs of people while hydrating them in a healthy way. Two to four bottles are recommended each day.

Energy/Rejuvenation

There is an increase in the presence of energy beverages by Red Bull, Adrenaline Rush, 180 and many others. These highly-caffeinated, high-energy drinks have exploded on the beverage market in the United States, as well as globally and generated a whole new generation of copycat caloric or – in many cases – sweetened beverages. According to the different brands of energy drinks, ingredients that serve as stimulants include: taurine, glucoronolactone, caffeine, and B vitamins, guarana, ginseng, ginkgo biloba, L-carnitine, sugars, antioxidants, yerba mate, creatine, milk thistle. Although these ingredients have been approved by the FDA, health experts still recommend consumers to read the label to make wiser choice of beverage since these ingredients may not be beneficial for health.

Health and Wellness

Functional beverage companies are more aware of the 'health conscious' individuals and have introduced functional beverages with less sugar and therefore less calories. For example, Vitaminwater 10 contains only 10 calories per serving (25 calories for a 351mL bottle, 7.5 grams of sugar and 250% of daily allowance of Vitamin C) On the other hand, it has the same 25% of the daily allowance of Vitamins B3, B5, B6 & B12 as the original.

Vitaminwater 10 has an all natural sweetener extracted from the stevia plant, which is a benefit in lowering calorie content as well as naming the product as natural.

Weight Management

With increased worries about obesity and its implications on health, combined with demand for convenience goods, consumers are naturally looking towards easy weight loss methods that they can easily integrate into their lifestyles. As such, functional beverages are striving to achieve that through addition of ingredients that promote weight loss. Coca Cola and Nestlé have partnered to produce Enviga, one of the new 'calorie-burning beverages'. Skinny Water (owned by Skinny Nutritional Corp.) and Nutrisoda's Slenderized (owned by PepsiCo) have both used included polyphenol called epigallocatechin gallate (EGCG) in their ingredient for their "metabolize boosting effect". With the increase in demand, ingredient manufaturer are now offering a 95% EGCG content product, a tea extract called "Blue California".

Consumer Demographic

The consumer group is characterized as well educated, females, ages 35-55 years of age belonging to upper middle, middle, and lower middle classes (ABC1 social class groupings). This is due to their perceptions towards positive health beliefs, as well as a high disposable income. Within the energy and stimulant drink sector, young adults aged 18 to 34 are considered to be the main target market, which can be seen by the high consumption rates. However, there could be potential downsides when companies rely too heavily on trends due to constant attitudinal differences of functional beverages across categories. Therefore, there are significant challenges presented to companies that want to pursue opportunities pertaining to functional beverages.

Target market consumers can be further segmented into four groupings: a functionality-driven segment; a price-sensitive segment; a natural-driven segment; and a grouping of carrier attribute-driven segments.

Functionality-Driven Segment: offers the greatest potential for functional beverages as it is least willing to make trade-offs between functionality and other intrinsic and extrinsic product design attributes, in order to obtain or achieve the desired benefits.

Price-Sensitive Segment: most willing to make trade-offs between functionality, carrier specific attributes and price. Natural-Driven Segment: does not perceive value from added functional ingredients. This segment represents those consumers that consider functional products less natural than conventional products, and instead, value wholesome foods and beverages.

Carrier Attribute-driven Segments: consumers that desire products associated with the maintenance of health and well-being, and therefore represents potential target markets for functional products. However, they are less willing or are unwilling to compromise on carrier-specific related attributes such as flavour, texture or convenience for reputed functional benefits. Companies must also identify consumer's rising needs as well as respond quickly to consumer's enquiries to functional beverages. This will help companies to recognize underdeveloped market niche and be able to create new opportunities based on other competitor's mistake· Therefore, benefit segmentation (allowing the consumer to benefit from seeking or deriving from specific functional beverages) will identify and target viable consumers.

Health Concerns

Health experts are concerned about the ready consumption of the rising functional beverage. As much as these beverages essentially still serve to hydrate the individual, it may not lessen or even address major health issues today such as obesity, heart disease, and cancer that are associated with the consumption of sweetened beverages, of which, most of these beverages are. This is because most of these drinks contain significant amounts of sugars and hence calories, which would add to discretionary and total caloric intake. As such, these ingredients pose health risks because of what they contain (sugar and caffeine) or what they replace in the diet (vitamin and mineral rich foods).

Another aspect for concern is that some functional beverages contain ingredients that have not been sufficiently studied for health benefits, safety, and dosage or have higher levels of a certain ingredient, like caffeine, heart disease, and cancer that are associated with the consumption of sweetened beverages, of which, most of these beverages are. Thus, consumption of these beverages might lead to undesired outcomes due to unforeseen interactions.

Sugar Content

A 20oz bottle of Glacéau's Vitaminwater has been reported to contain approximately 33g of sugar, which is similar to the sugar content of a can of coke. As such, such products may not be as a healthy alternative to other commonly consumed beverages. In addition, the sugar content of such beverages promote dental caries amongst frequent consumers.

Energy Drinks

In some particular functional drinks, particularly energy drinks, the caffeine content can be high, ranging from 0mg to 141.1mg/serving, of which an average 8oz cup of coffee contains about 133mg of caffeine. There have been reports to Health Canada of adverse reactions involving energy drinks.

13

Nutrient Utilization

Under healthy, normal circumstances, consumed food passes through the gastrointestinal (GI) tract for the processing (mechanical and chemical) and absorption of nutrients. Once absorbed, these nutrients are transformed by tissues and organs into forms that are needed for normal respiratory and metabolic purposes.

There are circumstances, however, when this normal and usual process can be corrupted by disease or by the consumption of substances that cause a malabsorption of nutrients, either through a direct irritation of the GI tract or through a chemical sequestering of nutrients that makes them unavailable for absorption.

Other substances may corrupt the normal processing of nutrients by tissues, often in the liver, so that the potential metabolic benefits of the consumed foods are not realized. It is also possible that the timing of consumption of one food may conflict with the absorption of nutrients in another food. We will discusses the major factors that influence how well we consume, absorb, and use energy substrates, with a secondary focus on strategies that can help athletes get the most out of what they eat.

CARBOHYDRATE UTILIZATION DURING EXERCISE

Low carbohydrate levels result in exercise fatigue. Since carbohydrate stores (or, *glycogen stores*) are limited (~350 kilocalories of glycogen in the liver; ~1,400 kilocalories of glycogen in muscles), athletes should consider how to initiate exercise with glycogen stores full and should establish a routine that keeps glycogen stores from running low. Even if muscle glycogen stores are adequate, low liver glycogen stores will result in hypoglycemia and mental fatigue, and mental fatigue leads to muscle fatigue.

The higher the exercise intensity, the greater the reliance athletes have on carbohydrate as an energy substrate. However, even low-intensity (i.e., mainly aerobic) exercise that derives most of its fuel from fat still requires some level of carbohydrate for the complete combustion of fat and to maintain blood glucose. Therefore, all modes of physical activity have some degree of carbohydrate dependence.

Several factors influence the proportionate contribution of carbohydrate to total fuel requirements during exercise.

Factors that increase the reliance on carbohydrate include the following:

- High-intensity activity
- Long-duration activity
- Exercise in hot and cold temperature extremes
- Exercise at high altitude
- Age (higher in young boys than in men)

Factors that decrease the relative energy expenditure from carbohydrate include the following:

- Endurance training
- Good conditioning
- Temperature adaptation
- Gender

There is a common misconception that low-intensity activity (up to ~65 percent.VO_2max) is the most efficient means of fat loss. In fact, many popular exercise programs have been organized around the idea that a greater amount of fat is most efficiently "Burned" with low-intensity aerobic exercise. However, the proportion of fat burned should not be confused with the volume of fat burned. While you're sitting there reading this sentence, you are very likely deriving the vast majority of your energy requirements from fat. However, the total volume of fat being burned is extremely low. (If this were not so, sitting in front of a TV would be a terrific means of initiating a fat-loss program.) When exercise intensity is increased, the proportion of energy derived from fat is decreased, and the proportion of energy derived from carbohydrate is increased, but some level of fat is always being burned. The total caloric requirement per unit of time is much greater in high-intensity activity than in low-intensity activity, and the volume of fat burned is greater in high-intensity activity (despite a lower proportion of fat meeting total energy requirements.) Therefore, athletes should work as intensely as possible within a given time frame to increase fat loss and optimize body composition.

Carbohydrate is a critical fuel for athletes because we can more efficiently create energy per unit of oxygen from carbohydrate than from any other fuel. One liter of oxygen can yield approximately 5 calories from carbohydrate but only 4.7 calories from fat. In addition, aerobic glycolysis can produce ATP for muscular work in a larger quantity and at a faster rate than the oxidation of fat can produce it. The increased energy efficiency of carbohydrate helps explain the muscular fatigue that quickly occurs during high-intensity activities when muscle glycogen is nearly depleted. We simply cannot supply sufficient ATP to working muscles to maintain the workload.

CENTRAL FATIGUE THEORIES

The conversion rate of adenosine diphosphate (ADP) to ATP is a critical step in the supply of energy to working muscles. Inadequate carbohydrate availability lowers the rate of ADP to ATP conversion, making it impossible for muscles to continue exercise at a high-intensity level. In addition, the failure to convert ADP to ATP causes a buildup of ADP that also contributes to muscle fatigue.

A number of other factors involving the central nervous system are also involved in muscle fatigue. As a group, these factors are referred to as *Central Fatigue Theory*. All of these theories involve mechanisms that cause more than the usual amount of the amino acid *tryptophan* to pass the blood-brain barrier, which stimulates an increase in the amount of serotonin (5-HT) that is produced. 5-HT is a neurotransmitter that causes people to feel relaxed and, if enough is produced, to feel sleepy. For the athlete, this could translate into muscular fatigue. Put simply, mental fatigue leads to muscle fatigue.

Theory 1: Low blood sugar and low muscle glycogen stores stimulate the muscle break down for gluconeogenesis. This results in an increased catabolism of branched-chain amino acids (BCAAs), which causes a reduction in circulating blood BCAAs. BCAAs and tryptophan compete for the same receptor carriers that enable their passage through the blood-brain barrier. When BCAAs are high, the tryptophan passage to the brain is controlled. However, when the BCAA blood level is low (as happens when they are catabolized for energy), tryptophan can sequester more of the receptor carrier, and more tryptophan enters the brain. Tryptophan stimulates the formation of 5-HT. To prevent this from happening, blood and muscle glucose levels must be maintained to avoid gluconeogenesis.

Theory2: Consumption of foods high in tryptophan (such as turkey) can increase the volume of tryptophan passing the blood-brain barrier, causing an increase in 5HT production. The increase in 5-HT leads to premature fatigue.

Theory 3: Fats compete for the same protein carrier in the blood as tryptophan. High fat intakes preferentially compete for this protein carrier, leaving a higher proportion of free tryptophan that can cross the blood-brain barrier. This causes an increase in 5-HT production, which may lead to premature fatigue.

While it may be logical that the intake of both BCAAs and carbohydrate reduces 5-HT and, therefore, inhibits both mental and physical fatigue, studies have not been conclusive because of the difficulty in distinguishing between brain and muscle effects. In addition, there may be interference from compounds such as caffeine, the ingestion of which has been shown to delay fatigue by temporarily stimulating the central nervous system.

CARBOHYDRATE REQUIREMENTS

The Institute of Medicine recommends 130 grams (520 kilocalories) of carbohydrate per day, which is the average minimal usage of glucose by the brain. The desirable range of carbohydrate intake is 45 to 65 percent of total caloric intake (also referred to as the Acceptable Macronutrient Distribution Range, or AMDR), and the Daily Value (DV) for carbohydrate on food labels is based on a recommended intake of 60 percent of total caloric consumption. These recommendations also generally advise that no more than 25 percent of carbohydrate intake be derived from sugars (mono- and disaccharides).

Dietary fibre consumption (from indigestible and partially digestible polysaccharides) should be at the level of 38 grams per day for adult men and 25 grams per day for adult women. Adequate fibre consumption aids in the maintenance of normal blood sugar, reduces heart disease risk, and lowers constipation risk. The difference between genders in recommended fibre consumption is based on the lower total food mass and calories typically consumed by women.

It has been suggested that our only true requirement for carbohydrate is for vitamin C, a six-carbon, glucose-like substance that most animals can derive through an enzyme conversion of glucose. There is some historical evidence that our human ancestors consumed very little carbohydrate and survived. However, when considering athletes and the mountain of research demonstrating that carbohydrate is the limiting substrate in athletic performance, it becomes clear that human survival and human performance are entirely different issues. Athletes need carbohydrate.

Athlete requirements for carbohydrate are based on several factors. Athletes must consume enough carbohydrate:

- To provide energy to satisfy the majority of caloric needs;
- Optimize glycogen stores;
- Allow for muscle recovery after physical activity;
- Provide a well-tolerated source of energy during practice and competition; and
- Provide a quick and easy source of energy between meals to maintain blood sugar.

The traditional guideline for determining caloric intake has been to consider the amount of carbohydrate to be consumed as a proportion of total caloric intake. The recommendation for the general population is that carbohydrate should supply 50 to 55 percent of total calories, and the Dietary Reference Intake (DRI) is 130 grams per day (520 calories per day) for male and female adults.

However, the amount typically recommended for athletes is between 55 and 65 percent of total calories, assuming an adequate total caloric intake. Another, and clearly better, way of determining carbohydrate requirement is by taking into consideration the amount of carbohydrate to be consumed (in grams) per kilogram of body mass. The carbohydrate intake recommendations for endurance-trained athletes range from between 7 and 8 grams of carbohydrate per kilogram of body weight per day. Studies of carbohydrate consumption of different athlete groups have found differences in carbohydrate intakes.

These data suggest that an athlete should consume between 5 and 10 grams of carbohydrate per kilogram of body mass, or between 20 and 40 kilocalories of carbohydrate, per kilogram of body mass. A hypothetical 155-pound (70 kilogram) athlete would take in between 1,400 and 2,800 calories from carbohydrate, which represents a far greater carbohydrate consumption than the DRI of 520 calories. Assuming this represents approximately 60 percent of total calories from carbohydrate, this athlete would consume a total of 2,300 to 4,700 kilocalories per day.

Following the same logic, a 300-pound (136 kilogram) lineman on a football team would require 2,700 to 5,400 kilocalories just from carbohydrate each day, an amount that would be difficult to consume because carbohydrate has a relatively low energy density (i.e., only 4 kilocalories per gram). The generally recommended carbohydrate intake is based on the intensity and duration of exercise, with a higher requirement for greater duration and greater intensity. Most carbohydrates are derived from cereals, legumes, fruits, and vegetables. There is no discernable amount of carbohydrate in meats, and only a small amount of carbohydrate in milk and cheese. A number of dairy products have sugars added (yogurt, ice cream) to make them more widely acceptable.

The glycemic index is a measure of how quickly consumed carbohydrates manifest themselves as blood glucose. Foods are compared with the ingestion of glucose, which enters the blood quickly because it requires no digestion and is readily absorbed. Glucose has a glycemic index score of 100, which is the basis of comparison for other foods. Foods are compared on an isocaloric basis (i.e., all providing the same number of calories), which is logical but is also the cause of some of the confusion associated with the glycemic index. For instance, carrots have a high glycemic index (>85) but the amount of carrots typically consumed is so low that the total calories of glucose from carrots entering the blood would be small.

Some foods have a surprisingly high glycemic index, while others are surprisingly low. Corn Flakes cereal, for instance, has a much higher glycemic index (84) than table sugar (65). When you assess the carbohydrate makeup of these two foods, however, it makes sense. The cereal grain in Corn Flakes is mainly composed of the disaccharide maltose, which is made of two glucose molecules. Table sugar, on the other hand, is made of sucrose (glucose plus fructose). The fructose must be converted by the liver to glucose, and this extra conversion slows the speed with which it manifests itself as blood glucose.

Because the volume of glucose and the speed at which it enters the blood can influence the amount of insulin produced, it is generally desirable for people (including athletes) to consume carbohydrates that have a medium to low glycemic index. However, there are times, such as during and immediately after exercise, when high-glycemic foods might be better for athletes.

Generally speaking, carbohydrate foods higher in fibre have a lower glycemic index, so they are good choices for athletes. However, dietary fibre may be a source of gas and distention, making them poor choices for consumption just before or during competition. Soluble fibre foods may create less of a problem, but athletes should experiment to determine which foods are most easily tolerated. Athletes often find that starchy carbohydrates with low fibre concentrations, such as pasta, are the most easily tolerated and deliver the high volume of carbohydrates athletes require.

Table. Foods High in Soluble and Nonsoluble Fibre.

Good Sources of Soluble Fibre	Good Sources of Nonsoluble Fibre
Bananas	Barley
Barley	Beets
Beans and legumes	Brussels sprouts
Carrots	Cabbage
Citrus fruits	Cauliflower
Oat bran	Fruits and vegetables with skin
Oatmeal	Rice (except for white rice)
Peas	Tumips
Rice bran	Wheat bran
Strawberries	Wheat cereal
Sweet potatoes	Whole wheat breads

CARBOHYDRATES AND PHYSICAL ACTIVITY

Because physical activity dramatically increases the rate of energy expenditure, athletes must strategize in order to best supply the needed energy to achieve success. It is critical for athletes to obtain sufficient total energy intake to support total energy requirements, including those for normal tissue maintenance, growth (in children and adolescents), tissue repair, and the energy requirements of the activity itself. It is impossible to discuss the ideal distribution of energy substrates without first conceptualizing how to best meet total energy needs.

Although this may seem a logical and simple act, virtually all surveys have found that athletes fail to consume sufficient energy to fully satisfy their needs. This is much like planning to take a Ferrari on a 100-mile trip, and you appropriately put high-octane gas in the fuel tank-but only enough of it to go 80 miles. The Ferrari just won't get there, and poorly fueled athletes will also have difficulties doing what's required to be optimally competitive. Once a strategy has been established for obtaining sufficient energy, then athletes can reasonably consider how to parse the energy into the optimal distribution of energy substrates: carbohydrate, protein, and fat. It is generally accepted that athletes should consume sufficient carbohydrate to meet the majority of their exercise-related energy needs, plus enough carbohydrate to restore muscle glycogen stores between exercise sessions. Ideally, athletes should consume complex carbohydrates when possible but should consume simple carbohydrates during and immediately after exercise. Other energy substrates (protein and fat) should also be consumed to fulfill total nutrient requirements, but carbohydrate should remain the predominant energy source. It is difficult for athletes to take in sufficient energy and carbohydrate unless there is a well- established plan to do so. Athletes would do well to remember that training alone, without a sound nutrition plan to support the training, will be self-limiting.

FAT (LIPIDS)

Despite some recent literature wrongly espousing the benefits of high fat intakes (i.e., intakes of 30 percent or more of total calories from fat), fat is a highly concentrated fuel that does nothing to improve athletic performance, body composition, or weight when taken in excess. The adult AMDR for total fat intake is 20 to 35 percent of total calories, and there is no scientific information suggesting that more than 25 percent of total calories from fat is generally better for athletes.

However, for athletes who may have difficulty sustaining weight because of a massive energy expenditure (such as cross-country skiers) or who must sustain high weights (such as linemen on football teams), higher fat intakes (up to the AMDR limit of 35 percent) may be necessary. Few Americans consume less than 35 percent of total calories from fat, so consumption of less fat is not easy, and unless steps are taken to provide sufficient energy from other substrates (mainly from more complex carbohydrates) to replace the eliminated fat, athletes may place themselves in an energy-deficit state that is, in itself, a detriment to performance. Therefore, while a reduction in fat intake is generally useful, a conscious effort should be made to provide enough total energy when fat intake is reduced.

Since fat is more than twice as concentrated in calories than either protein or carbohydrate (9 calories per gram versus 4 calories per gram), more than twice as much food must be consumed to make up the difference in reduced fat. Cholesterol, oils, butter, and margarine are all fats, or lipids, but each has slightly different characteristics. The one common attribute shared by lipids is that they are soluble in organic solvents but not soluble in water. (Anyone who has tried to mix Italian dressing knows this to be true.

The oil in the dressing eventually rises to the top, no matter how hard the bottle is shaken.) The term *fat* is usually applied to lipids that are solid at room temperature, and the term *oil* is applied to lipids that are liquid at room temperature. The most commonly consumed form of lipid is triglyceride, which consists of three fatty acids and one glycerol molecule (thus the name tri glyceride). Despite the numerous forms of lipids, we can obtain them all from the food supply, and we are also capable of making many types of lipids by combining carbon units from other substances. Nearly every cell in the body has the capacity to make cholesterol, which is why a person can have a high blood cholesterol level even when on a low-cholesterol diet. We can also manufacture phospholipids, triglycerides, and oils. In fact, it is this ability to effectively manufacture different types of lipids that limits the necessity to consume large amounts of lipids.

FAT FUNCTIONS

A certain amount of fat, between 20 and 35 percent of total consumed calories, is necessary to ensure a sufficient energy and nutrient intake. The fat-soluble vitamins-vitamins A,D,E, and K-must be delivered in a fat package. The essential fatty acids, which are needed for specific body functions but that we are incapable of synthesizing, must also be consumed. Some dietary fat is also needed to give us a feeling of satiety during the meal, creating the important physiological signal that it is time to stop eating. Dietary fats have a longer gastric emptying time than do carbohydrates, which contributes to the feeling of satiety. Of course, fat also makes foods taste good.

LIPID STRUCTURE

Lipids have different levels of saturation, a term that refers to the number of double bonds in the carbon chain. Fatty acids with no double bonds are saturated, those with one double bond are monounsaturated, and those with more than one double bond are polyunsaturated. Single bonds are stronger and less chemically reactive than double bonds, so the greater the number of double bonds, the greater the opportunity for the fatty acid to react with its chemical environment.

It is this differential reactive capacity that makes the number of double bonds an important factor in human nutrition. Saturated fatty acids are most prevalent in fats of animal origin, palm kernel oil, and coconut oil. Monounsaturated fats are highest in olive oil and canola oil but are also present in fats of animal origin. Polyunsaturated fats are highest in vegetable oils (with the exception of olive oil, which is more than 75 percent monounsaturated). In the context of a fat intake that does not exceed 35 percent of total calories, mono- and polyunsaturated fats should make up the majority of the fats consumed. Saturated fats are associated with higher cholesterol levels, so they should be minimized when possible. This is most easily achieved by reducing the consumption of animal fats, chocolate candies (often high in saturated tropical oils), fried foods, and high-fat dairy products.

FACTORS INFLUENCING ENERGY METABOLISM

Nutrients that are consumed, digested, and absorbed must still be delivered to the correct tissues for metabolic purposes. A number of factors may inhibit the normal metabolism of nutrients, including nutrient-nutrient interactions, drug-nutrient interactions, and excess alcohol consumption. Of these, regular alcohol consumption is likely to present the greatest difficulties for athletes.

Alcohol

Although alcohol is a nutrient that provides 7 calories per gram, it must also be considered an antinutrient because of the way it inhibits the normal metabolism of vitamins and, therefore, the main energy substrates (carbohydrate, protein, and fat) if consumed in excess. Most important, alcohol should not be mistaken for an essential nutrient, which it is not. It is a toxic substance that humans have a limited capacity to detoxify, and in the process, alcohol leaves other toxic substances in its wake.

A regular high intake of alcohol increases disease risk, including cancers of the liver, mouth, throat, and esophagus (these latter three cancers are even more likely if combined with smoking) as well as cirrhosis of the liver (a condition where increasing portions of the liver become fibrotic and are no longer able to function). In addition, alcohol can be an irritant to all segments of the GI tract and, as a result, can cause the malabsorption of nutrients. To make matters worse, alcohol increases the urinary excretion of calcium and magnesium. Magnesium is a cofactor in enzymes transferring phosphate groups, so it is a needed ingredient in energy metabolism. Regular alcohol intake lowers the resorption of magnesium (increasing urinary losses) and also increases magnesium excretion in sweat. The result is an increase in muscle cramps, weakness, and cardiac arrhythmias. Athletes can ill afford creating the magnesium deficiency associated with regular alcohol consumption.

Because liver function is impaired, alcohol may also interfere with the normal metabolism and storage of nutrients. Taken together, alcohol increases nutrient requirements so as to repair the damage it creates and to counteract the mal absorption it produces. Chronic alcohol abuse may weaken the heart; alter brain and nerve function; increase blood lipids (particularly triglycerides); result in a fatty, cirrhotic liver that malfunctions; and cause pancreatitis (which can have a major impact on blood glucose control and digestive processes.) Besides the increased risk of athletic injury induced by alcohol consumption, there is also evidence that more than one drink per day can have a clear impact on performance by negatively affecting reaction time, coordination, and energy metabolism.

Alcohol dehydrogenase, a liver enzyme, is made by the liver to dehydrogenate the active form of vitamin A (retinol, an alcohol). However, with alcohol (ethanol) consumption, the limited production of alcohol dehydrogenase is shunted to dehydrogenate ethanol, thereby leaving the potentially toxic form of vitamin A in the alcohol form. The result is an adverse interaction between alcohol and vitamin A that can result in liver toxicity and increased cancer risk. The risk of alcohol-related health effects is real in athletes. Although elite athletes consume alcohol half as often as age-equivalent nonathletes, many athletes (particularly those in team sports) still consume far more alcohol than would support good health, good nutrition, and optimal athletic performance. However, non-elite, adolescent athlete groups are more likely to engage in problem drinking than their nonathlete counterparts. Male athletes appear more likely than female athletes to consume alcohol daily.

Nutrients

Nutrient intake and availability has a clear impact on energy metabolism. Even a single low nutrient from this list will corrupt the normal metabolic pathways for energy utilization and therefore affect athletic performance. Given the potential hazards of inadequate intake, maldigestion, malabsorption, and altered metabolic processes from drug or alcohol ingestion, it is a credit to the human system that most people are able to supply nutrients to the tissues that need them at the time they're required. However, any chronic insult from poor intake, heavy alcohol abuse, or untreated illness will eventually have a negative impact on performance, a fact that athletes and their coaches should constantly bear in mind.

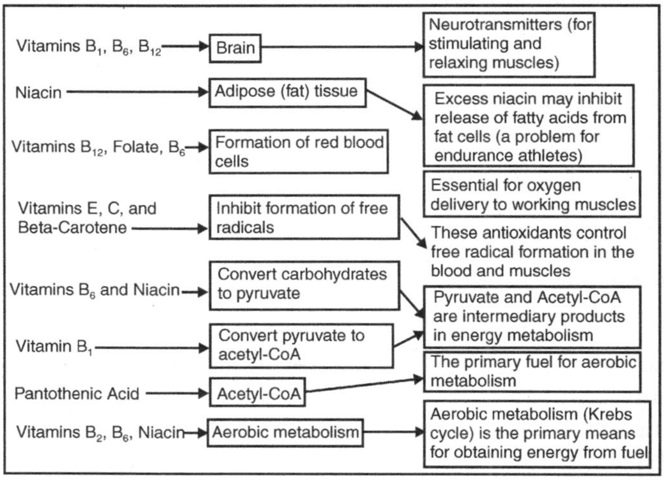

Physical activity increases the need for fuel and for the metabolic processes that are involved in its utilization. Anything that either limits the provision of adequate calories to support the cellular requirement or alters the cells' capacity to properly metabolize the provided fuel will have a negative impact on performance. Some factors are within athletes' control, including adequate food consumption, careful consumption of drugs and supplements, and avoidance of regular alcohol consumption; other factors are not within athletes' control, including disease states that may alter food intake or food absorption. This book is filled with information on how to best provide the fuels needed for successfully pursuing athletic endeavors. For those conditions that are not within the direct control of the athlete (e.g., celiac and Crohn's disease and other GI tract disorders), an increasing number of medical solutions are available; athletes should not hesitate to seek medical advice that can ameliorate the impact these conditions have on health and performance. Ultimately, if the fuel doesn't make it to a cell that has the capacity to burn it, athletic endeavors cannot be successfully pursued.

FACTORS INFLUENCING DIGESTION AND ABSORPTION OF NUTRIENTS

Any number of factors may result in maldigestion or malabsorption, either of which will diminish nutrient delivery to needy cells. It is important to consider that a problem early in the GI tract (i.e., the mouth or esophagus) is likely to create a cascade of problems later in the GI tract. For instance, the pain associated with esophagitis (an irritation in the esophagus from frequent vomiting, frequent alcohol consumption, or gastric reflux) may inhibit normal drinking patterns to the point of changing normal bowel habits, causing colonic irritation and, of course, dehydration. The most common problem areas include the following:

- *Mouth*: Irritations, dental caries, mouth sores
- *Esophagus*: Irritation of the lower esophageal sphincter from gastric reflux or alcohol
- *Stomach*: Gastritis from alcohol, stress, or overtraining
- *Small intestine*: Malabsorption from celiac disease or Crohn's disease or other inflammation (often from alcohol)
- *Large intestine*: Crohn's disease irritation, with possible bowel obstruction associated with a fluid and electrolyte imbalance

Diseases of the Mouth

Any condition that affects the mouth can have an impact on food intake. Dental caries, cold sores, sensitive gums, and swollen tongues all have the potential of limiting food intake and, therefore, restricting nutrient and fuel exposure to needy tissues. Regular visits to a dentist will resolve the majority of these problems. However, a B-vitamin deficiency (particularly deficiencies of vitamins B2 and B6) may also lead to mouth and tongue problems that inhibit food intakes.

Celiac Disease

Celiac disease is an intestinal intolerance of the protein gluten (found in wheat, barley, rye, and oats). It is associated with dermatitis herpetiformis, which occurs in 70 to 80 percent of people with gluten-induced GI tract damage. Celiac disease has a strong genetic association, so relatives of someone who has celiac disease may also develop the condition. Uncontrolled celiac disease results in intestinal damage, with associated malabsorption, and causes diarrhea, steatorrhea (fat malabsorption), iron-deficiency anemia, other vitamin deficiencies, and eventually weight loss.

It may be asymptomatic for years before these symptoms occur, or some signs, such as milk intolerance or iron deficiency, may be present without any of the other signs. Recent studies suggest that celiac disease is underdiagnosed around the world, with incidence rates much higher than previously thought. In people diagnosed with iron-deficiency anemia, for instance, 10 to 15 percent have celiac disease. Individuals with malabsorption, frequent diarrhea, iron deficiency, or any of the other symptoms of GI dysfunction associated with celiac disease should consider the possibility that they do not tolerate gluten. It is quite possible that at least a proportion of those who report feeling so much better on the Atkins diet (a high-protein, low-carbohydrate diet that restricts bread intake) improve because of the elimination of gluten rather than any of the other purported benefits of the Atkins regimen. For those wishing a firm diagnosis of celiac disease, clinically accepted tests are available, often involving a small bowel biopsy, to confirm the presence of the disease.

Patients with celiac disease who remove gluten-containing products from the diet are initially distressed at having this limitation in food consumption, but the improvement in bowel function with gluten restriction often allows them to comfortably consume other foods that may previously have caused GI distress. Virtually all celiac patients feel so much better with the restriction of gluten-containing products that they are self-motivated to continue with a gluten-free diet. In addition, numerous gluten-free bread products on the market, typically made of rice or corn, are excellent substitutes for wheat and rye breads. There are numerous online resources on gluten-free foods and recipes for readers who wish additional information.

Crohn's Disease

Crohn's disease is a regional inflammation of the ileum but may affect the entire small or large intestine. It is associated with abdominal pain and frequent diarrhea, with bowel obstruction a serious problem for the Crohn's patient. This form of irritable bowel syndrome (IBD) causes a thickening of the intestinal wall that reduces the internal transit diameter of the affected portion of the intestines.

This reduced interior intestinal diameter is responsible for the bowel obstruction. Crohn's disease equally affects men and women and appears to run in families, although about only 20 percent of Crohn's patients have a relative with the disease. The disease has no known cause, but it is theorized that an immune system reaction against a bacteria or virus causes the inflammation. It appears clear that, unlike some intestinal disorders, Crohn's disease is not related to stress. For an athlete, Crohn's can have a debilitating effect on the capability to absorb sufficient nutrients, and the associated diarrhea affects fluid and electrolyte balance.

The ileum (the main intestinal area affected by Crohn's disease) is the site of vitamin B12 absorption. A failed absorption of vitamin B12 will eventually lead to megaloblastic, hypochromic anemia, which negatively affects oxygen-carrying capacity. Athletes who have been diagnosed with Crohn's disease are most commonly treated with an anti-inflammatory drug (commonly containing mesalamine; sulfasalazine is the most common of these drugs), along with treatments that aim to correct the nutrition deficiencies and relieve the pain and diarrhea.

Treatment with fluids and electrolytes is common for patients who are suffering from frequent diarrhea. Some drugs also reduce the immune response so as to reduce the cause of the inflammation. An irritated GI tract may require that no solid foods be consumed to allow for a reduction in the inflammation and to reduce the chance for a bowel obstruction. Liquid full-nourishment meals are often consumed during these periods. No foods appear to universally increase the GI inflammation of Crohn's disease, but doctors often ask patients to limit the intake of foods that are not well tolerated in large portions of the population or foods that are known irritants (e.g., milk, alcohol, spicy foods).

In the presence of a vitamin B12 deficiency, consumption of oral supplements or foods high in vitamin B12 (foods of animal origin) does not resolve the deficiency because, regardless of the amount consumed, the absorption of vitamin B12 is sufficiently corrupted that it will not enter the blood. Periodic injections of vitamin B12, which bypass the GI tract, are typically needed to correct the vitamin B12 deficiency.

Drugs

Certain drugs may also have an impact on the digestion and absorption of nutrients. Antibiotics destroy the intestinal microflora (bacteria) that assist in digestive and absorptive processes and that are even involved in creating certain nutrients, such as vitamin B12. The commonly prescribed antibiotic neomycin, for instance, causes a malabsorption of fat, protein, sodium, potassium, and calcium.

Because of the competitive absorption of divalent minerals (calcium, iron, magnesium, and zinc), a high intake of calcium-containing antacids could, for instance, take up most of the absorption site and interfere with the absorption of the other minerals. Nonsteroidal anti-inflammatory drugs (NSAIDs) that are commonly taken by athletes to resolve the bumps, bruises, aches, and pains of athletic endeavors may create a GI irritation that results in blood loss and an iron-deficiency anemia. These are but a few examples of how drugs can result in an altered fuel intake and utilization. Athletes should be fully aware that virtually every drug taken, including over-the-counter drugs, is likely to have a digestive, absorptive, or metabolic impact that could be a detriment to performance. Athletes should therefore consult with an appropriate health care professional rather that pursue self-diagnosis and self-prescription.

FACTORS INFLUENCING FOOD CONSUMPTION

Several factors may influence food consumption and thereby the total nutrient in take of any individual. These factors range from appetite loss, which may result from conditions as varied as transitional depression to consumption of certain drugs; micronutrient deficiency or toxicity, which may alter appetite through the impact it has on taste sensitivity; dieting, which purposefully lowers total food consumption with the hope of achieving a desirable weight loss; and overtraining, which is a major problem for many athletes.

Appetite Loss

A loss of appetite (anorexia), regardless of the cause, interferes with food consumption and may result in malnutrition if it progresses for a sufficient duration of time. Appetite can be affected by both nutrition and non-nutrition factors.

A death in the family, for instance, often leads to some reduction in appetite, while a zinc deficiency is also associated with appetite loss. Inadequate consumption of carbohydrates, as is possible on high-protein, high-fat, low-carbohydrate diets, may result in some degree of ketosis, which is associated with nausea and a loss of appetite. Consumption of certain drugs is also known to result in an altered taste perception or an appetite loss, both of which may result in lower energy consumption.

Even something as seemingly unimportant as a small toothache can reduce food intake or, at the very least, reduce the variety of foods consumed so as to create a below-optimal nutrient and energy intake.

NUTRITION TACTICS

Endurance events such as road cycling, long-distance swimming, marathon, triathlon, and 10K run all require a high level of endurance and place a relatively low premium on anaerobic power. These events force competitors to perform at the margin of their maximal aerobic capabilities over long distances. As training, nutrition, and selection of athletes in endurance sports improve, records continue to fall. This suggests that doing the right things can and will result in moving the known envelope of speed in endurance events. The winner of the marathon at the Atlanta Olympic Games in 1996 won with an average running pace of slightly under a 5-minute mile, a feat that is not commonly achieved. Despite this incredible speed, the athlete had to maintain this pace at a level that allowed a sufficient oxygen uptake to sustain, primarily, aerobic muscular metabolism. That is, the majority of all muscular work took place with fuel being burned in the presence of oxygen. This is an efficient means of obtaining energy, allowing the athlete to sustain muscular work for long periods of time.

Aerobic training does some wonderful things to the athlete's ability to use oxygen. The intermediary (type IIa) fibers, which tend to behave more like fast-twitch (power) fibers than slow-twitch (endurance) fibers, dramatically increase in mitochondrial content and the in enzymes involved in oxidative metabolism. The training impact on oxygen usage is well known. In studies looking at blood lactate than are untrainded subjects doing the same intensity of work. The conversion of the behaviour of the intermediary fibers results in an improvement in the athlete's aerobic endurance. The increased ability to use oxygen results in an improvement in the ability to burn fat as a primary fuel, reducing the reliance on carbohydrates.

Since even the leanest athletes have a great deal of energy stored as fat, this increased ability to burn fat dramatically improves endurance. However, since carbohydrate is needed for the complete combustion of fat, carbohydrate is still the limiting energy source for endurance work because athletes have relatively low carbohydrate stores. This is clearly demonstrated by findings that athletes consuming a high-fat diet have a maximal endurance time of 57 minutes; on a normal mixed diet their endurance rises to 114 minutes; and on a high-carbohydrate diet, their maximal endurance rises to 167 minutes.

Aerobic Metabolic Pathways

Aerobic metabolic pathways are the means we have for obtaining energy from fuels (carbohydrate, protein, and fat) in the presence of oxygen. The controlled release of energy during aerobic metabolism allows for a large amount of the energy in glucose to be stored as energy in ATP. The chemical reaction for the full oxidation of glucose produces energy, carbon dioxide, and water:

Glucose + 6 O_2 + 38 ADP + 39 Phosphate = >6 CO_2 + 6 H_2O + 38 ATP

[Glucose = $C_6H_{12}O_6$]

In anaerobic metabolism, pyruvate is converted to lactate. However, in the presence of sufficient oxygen, pyruvate can be oxidized for energy in the mitochondria (often referred to as the energy factories of the cells). Glucose, a six-carbon molecule, is converted to two molecules of pyruvate, a three-carbon molecule. When pyruvate enters the mitochondria, it undergoes further conversion to a two-carbon molecule to form acetyl-coenzyme A (commonly abbreviated to acetyl-CoA). Acetyl-CoA can also be created from the beta-oxidation of fatty acids that reside in the mitochondria. During beta-oxidation, carbon is cleaved from the long carbon chains of fatty acids in two-carbon units.

These two-carbon molecules form acetyl-CoA. The newly created acetyl-CoA from pyruvate or beta-oxidation of fats can be oxidized to carbon dioxide (CO_2) in the tricarboxylic acid cycle (TCA). The critical aspect of the TCA cycle is producing hydrogen atoms for transport to the electron transport chain. It is in the electron transport chain that oxidative phosphorylation occurs to create ATP from ADP. With sufficient hydrogen to feed into the electron transport chain and enough oxygen for oxidative phosphorylation, the electron transport chain can continuously produce energy in the form of ATP.

If there is excess production of acetyl-CoA (i.e., inadequate oxidative enzymes to process the acetyl-CoA for energy or inadequate oxygen delivery), the excess can be converted to either fat for storage or to the amino acid alanine. Alanine can be converted by the liver to glucose or can be made part of larger protein structures. Anaerobic metabolic processes have the capacity to provide ATP energy immediately but only for a short duration, while aerobic metabolic processes begin providing ATP energy more slowly but for long durations, provided there is sufficient substrate and oxygen available to the cells. We have large stores of energy that we can call upon to create ATP energy for muscular work.

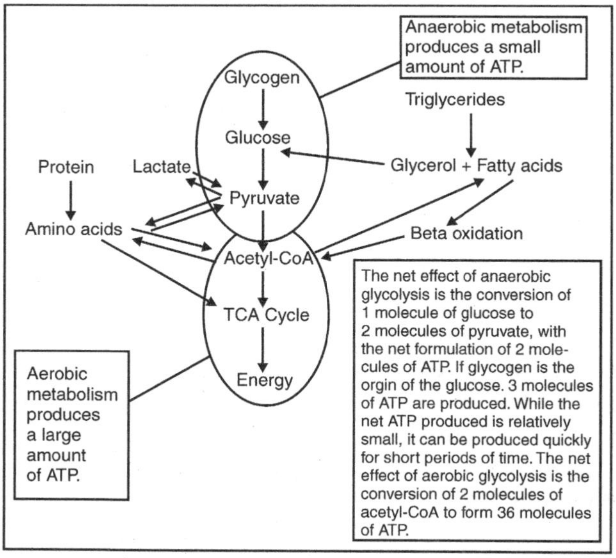

Of the energy stores available to us, fat is the most efficiently stored and provides the greatest mass from which we can derive ATP energy. Glycogen requires approximately 3 grams of water for storage, while fat storage is essentially anhydrous, making fat a more efficient form of energy storage. Muscle and liver glycogen stores represent a small fraction of the energy in fat stores but have the advantage of being able to be metabolized either anaerobically or aerobically, while fat can only be metabolized aerobically. Protein stores are from functional tissues that, under ideal conditions, would never be catabolized as a source of energy. Nevertheless, a small amount of protein (approximately 5 percent of total energy needs) does appear to be metabolized to meet energy requirements in most activities.

In the absence of carbohydrate, protein stores are catabolized at a faster rate to provide a source of glucose (the amino acid alanine can be converted to glucose by the liver) and a source of acetyl-CoA and oxidative metabolism. However, this protein catabolism is not desirable and can be avoided with a regular supply of carbohydrates and adequate total energy consumption. At the initiation of exercise, the majority of ATP is derived anaerobically. For highly intense, maximal-effort activities, the requirement for a high volume of energy mandates a continuous dependence on anaerobic processes. However, for lower-intensity activities the majority of ATP is initially provided anaerobically, but then the activity switches to aerobic metabolism to meet most ATP needs.

Anaerobic and aerobic metabolic processes should be thought of as proceeding simultaneously, with the intensity of the activity determining the predominant metabolic pathway for the supply of ATP. High-intensity, maximal-effort activities rely more on anaerobic metabolism, while lower-intensity activities rely more on aerobic metabolism. Because far more energy is available to us aerobically (fat can only be metabolized aerobically), the high energy needs of endurance athletes force them to train muscles to be more aerobically competent. Cells of well-trained aerobic athletes have more mitochondria and more aerobic enzymes in the mitochondria, resulting in a higher capacity to derive energy aerobically.

Considerations for Endurance Sports

Endurance sports are those in which the predominant form of energy metabolism is oxidative (it occurs in the presence of oxygen). Aerobic athletes must be capable of acquiring and delivering enough oxygen to the working muscles to support the physical work that is being done. Endurance work occurs at intensities below a person's maximal work capacity. By definition, working at a higher or maximal capacity exceeds the athlete's capacity to meet oxygen needs. Sprinters, who are working at maximal capacity, can run very fast but only for relatively short distances. Endurance athletes can't run as fast as sprinters but can go much longer distances because they are metabolizing energy using a much more clean-burning and energy-efficient oxidative system. To maintain the efficiency of their systems, endurance athletes should consider factors that can influence their aerobic capacity, including overtraining, overuse injuries, and dietary adequacy.

OVERTRAINING

A world-class runner wrote the following in an e-mail communication: I've just come off three weeks of particularly intense training. I went"hard"four days a week. I have plateaued and am now resting to allow my body to absorb all the good work I did but am still feeling a little bit lethargic. I sleep well but don't feel rested. My coach is concerned that I may have become anemic. As you know, my nutrition analysis has always come up good. Should I get some blood tests? Take iron? Take something else? I'm worried.

These signs are typical of an"overtrained"athlete who is suffering the consequences. Overtraining has some well-established warning signs, including increased muscle soreness, delay of muscular recovery, inability to perform at the previous training load, poor-quality sleep, decreased vigor, swelling of lymph nodes, high illness frequency, and loss of appetite. Many of these signs are a result of working at a level harder than the body's capacity to adequately recover. Overtraining rarely leads to an improvement in performance and, in fact, commonly reduces performance because it ultimately increases the likelihood that the athlete will become sick or injured.

A 26-year-old athlete who transferred to a more competitive team increased his training volume by 200 percent and after 2 months experienced continuous fatigue, tinnitus, palpitations, and insomnia. Nevertheless, he continued to play for 3 months but then became totally unfit, with sleepless nights and severe mental depression.5 This is precisely what should never happen.

Overtraining is a problem for many athletes (10 to 20 percent of those who train intensively) and appears to be relatively common in endurance athletes. Among other factors that may increase the risk of developing this condition, a poor intake of carbohydrates and fluids is known to be a problem. Overtraining syndrome is an untreated excessive training overload with inadequate rest, resulting in chronic decreases in performance and in the ability to train. Other problems may result and may require medical attention.

Factors associated with the development of overtraining syndrome include:

- Frequent competition, particularly if it involves quality efforts;
- Monotonous training with insufficient rest;
- Preexisting medical conditions (e.g., colds or allergies);
- Poor diet, particularly inadequate intake of carbohydrates, or dehydration;
- Environmental stress (e.g., altitude, high temperatures, and humidity); and
- Psychosocial stressors (e.g., work or school conflicts).

According to the American College of Sports Medicine, overtraining syndrome can be effectively eliminated through a logical training program that allows for adequate rest and recovery with proper nutrition and hydration. Studies of marathon runners suggest that even athletes who consume a high-carbohydrate diet require 7 days after a marathon to return muscle glycogen to prerace levels. A continuation of regular training before full muscle glycogen resynthesis will inevitably lead to performance degradation. Athletes must therefore understand that rest is a useful and necessary part of training, particularly after a hard and intensive training session.

Athletes fearing that a reduction in training may diminish competitiveness may resist getting enough rest. Therefore, everyone in the athlete's training circle (family, coach, athletic trainer, and so on) should support the concept that overtraining is associated with reduced performance. Put simply, rest and recovery should be an integral part of the training plan.

OVERUSE INJURIES

Overuse injuries occur when an athlete chronically repeats the same physical task; they may be particularly problematic in adolescent athletes experiencing rapid growth. In its simplest and most benign form, a heel blister caused by the rubbing of an ill-fitting running shoe is an overuse injury.

A more serious form would be, for instance, the constant pounding of legs on hard pavement that causes sufficient vibrational bone stress to induce a stress fracture. This is analogous to taking a wire clothes hanger and bending it repeatedly in the same place. After a while, the hanger develops a crack and eventually breaks. Because endurance athletes spend so many hours training, overuse injuries are a real concern. A study of triathletes found that some developed skeletal injury early in the competition, which became worse as the competition progressed. These injuries to muscles may alter the use of energy substrates as the triathlon progresses and as the body heals after the competition.

Although protein breakdown and muscular damage occur during a race, well-trained athletes should experience no alteration in fitness provided nutrition status is maintained. Adequately nourished athletes have a better capacity to heal the minor tissue damage that occurs during training and competition.

Additionally, athletes who can maintain carbohydrate and fluid levels during exercise are likely to have better brain function, which translates into a smoother running style that is less prone to injury development. Loss of mental capacity, which can easily occur with either a carbohydrate or fluid storage deficit, causes a breakdown in coordination that can increase structural stresses that lead to injury.

DIETARY ADEQUACY

Low glycogen stores reduce the time an athlete is capable of exercising, a fact that mandates the regular consumption of carbohydrates to maintain or replace limited glycogen stores. This requires, ideally, a carbohydrate intake of between 7 and 10 grams per kilogram of body weight per day. The timing of carbohydrate ingestion is also important and may influence glycogen storage and resynthesis. A study of highly trained male cross country runners found that food intake was generally adequate and well timed except for the period after competition.

Although it is recommended that endurance athletes consume carbohydrates immediately after competition to encourage restoration of glycogen stores, these athletes delayed eating carbohydrate foods until, on average, 2.5 hours after competition. A delay of this magnitude leads to poor glycogen replacement, and subsequent days of exercise showed the negative effect of reduced endurance. A study of marathon runners found that a significant proportion of total energy intake occurs after 4:00 P.M. rather than earlier in the day when it's needed the most. Delayed eating represents a missed opportunity for trained endurance athletes to maximize muscle glycogen storage after exercise.

There is no substitute for consuming sufficient energy and carbohydrates for endurance events. Supplements and ergogenic aids appear to be ineffective. The only strategy that works effectively is to eat enough and to eat at a time that is most useful for delivering energy to needy muscles or for maximizing glycogen storage. There is also evidence that endurance athletes must focus on glycogen storage activities about 1 week before a major competition. A failure to do so can have a negative effect on the athlete's endurance capacity. Endurance athletes, because of the time spent training and competing, may need to develop their own strategies for obtaining energy and fluids during the activity. This is not easy because consuming the wrong foods or fluids or consuming them at the wrong time will have a performance-decreasing effect. The "nervous stomach" many athletes experience just before a race makes the consumption of the right energy or fluid package even more difficult.

Athletes should undergo a purposeful experimentation to discover what sports beverages, carbohydrate gels, and other sources of energy and nutrients are well tolerated so they can use this information to optimize training sessions and achieve peak performances during competitions. The reliance on recommendations for the general athlete population may be a good starting point, but it is not enough to generate winning performances in tough competitive fields. Each type of race alters slightly the proportion of carbohydrate and fat that is burned as fuel (more intensity equals proportionately more carbohydrate; lower intensity equals proportionately more fat), but it's the carbohydrate level that ultimately determines if the athlete will "hit the wall." That is, when the glycogen stores are depleted, the athlete will no longer be capable of maintaining a sufficiently strong pace. Since endurance events are long, every available opportunity must be capitalized on to ensure that the athlete has enough food energy to continue the race and to store enough energy (glycogen) to do well during the next day of racing.

Athletes should do whatever is necessary to take in sufficient energy and nutrients (bringing baggies filled with food to meetings, eating while walking to class, snacking while going to the car), or the benefits derived from training will be wasted. Athletes who do not eat effectively become more easily fatigued and injured and are more likely to try unproven products touted as having ergogenic properties. There is no doubt whatsoever that much of the attention given to ergogenic aids and nutrient supplements by athletes, whether they've cyclists, runners, or swimmers, is to overcome a failure in planning to eat enough and to eat on time.

Maintaining hydration status is important for operating at optimal physiological levels of efficiency. Endurance athletes should practice consuming fluids frequently, even in the absence of thirst, to reduce the chance of dehydration. Consumption of a carbohydrate-containing beverage with small amounts of sodium is useful for fluid absorption and for maintaining the drive to drink.

Nutrient Supplementation

Athletes who fail to consume sufficient carbohydrate or enough total energy are likely to be at increased risk of inadequate vitamin C, thiamin, riboflavin, niacin, calcium, magnesium, and iron intakes. A study of marathon runners found that supplement usage (especially vitamins C and E, calcium, and zinc) was common. Forty-eight percent of the runners questioned reported using at least one supplement within the 3-day period surrounding the Los Angeles Marathon.

Other studies confirm that nonsupplemented marathon runners, soccer players, wrestlers, and basketball players have adequate serum concentrations of vitamin C and vitamin B_6, so supplementation of these vitamins does not appear to be warranted. In a study evaluating the effectiveness of magnesium supplementation on marathon runners, the supplementation did not improve resistance to muscle damage during the race, did not enhance muscle recovery after the race, and did not improve running performance.

Male marathon runners were evaluated to determine if the consumption of a commercial ergogenic aid supplement containing vitamins, minerals, amino acids, and unsaturated fatty acids was useful in improving performance. Results indicate that the ergogenic aid had no effect on oxygen consumption or any other important metabolic or physiological parameter that might be useful to endurance athletes.

Nutritional Concerns for Female Endurance Athletes

Female endurance athletes must consume sufficient energy and nutrients to avoid amenorrhea (cessation of regular menstrual periods). Amenorrhea occurs for many reasons, including high physical stress, high psychological stress, inadequate energy intake, poor iron status, high cortisol levels, and low body fat levels. It is conceivable that female endurance athletes have all of these factors working against them.Although some of these factors are clearly out of a woman's control, food intake is not one of those.

Female athletes should do whatever is within their means to consume sufficient total energy with a balanced nutrient intake to support good health. Amenorrhea is strongly associated with a loss in bone density and an increase in stress fracture risk. Additionally, low bone density during the running years places the athlete at higher osteoporosis risk later in life.

A Sampling of Sports Relying on Aerobic Metabolism

Aerobic sports involve submaximal effort over long periods of time. The aerobic nature of these activities results in a steady creation of ATP energy in working muscles, allowing athletes to continue working for as long as there is sufficient fuel and for as long as the body does not become overheated (hydration state is important). Athletes who can rely more efficiently on fats and less on carbohydrate (glycogen) typically have better endurance because they have more fat fuel than glycogen fuel, which has limited stores. Because fats require a high amount of oxygen for efficient metabolism, the endurance athlete with the best oxygen system (e.g., most oxidative enzymes, best ability to carry oxygen to working cells) has the best endurance. What follows is a sample of endurance sports and their special nutrition considerations.

NUTRITION TACTICS FOR SPORTS REQUIRING POWER AND ENDURANCE

This chapter provides nutrition information for sports where athletes have intermittent bouts of high-intensity activity followed by periods of lower-intensity activity. Team sports such as basketball, volleyball, rugby, team handball, and soccer all have combinations of high-intensity and lower-intensity activity interspersed throughout the competition. The mixed intensity of certain individual sports, such as figure skating and tennis, may also fall into this category. This differs from other sports, where the focus is either predominantly endurance or predominantly power or speed. There is nothing in artistic gymnastics training or competition that requires, for instance, a great deal of aerobic endurance; and marathon runners rarely require the explosive power exhibited by gymnasts. Team sport athletes must focus on speed, power, *and* endurance.

Soccer players must run the field at a controlled pace until a sudden opening requires a quick burst of speed. Basketball players may jog back and forth in a steady aerobic pace, but each player must have the capacity for a powerful jump to grab a rebound or for a quick sprint to make a defensive play. The intermittent high and low intensity of team sports creates a requirement for energy that is derived from a combination of aerobic and anaerobic means. Although the anaerobic metabolic processes are solely reliant on existing stores of ATP, phosphocreatine (PCr), and muscle glycogen, the aerobic processes derive energy from muscle glycogen, blood glucose, fat, and to a lesser extent, protein.

There is a heavy reliance on muscle glycogen for the majority of muscle energy during team sport activity, with nearly an equal reliance on fat and blood glucose for most of the remaining source of energy. Fat is almost never in short supply. However, the amount of glucose energy in blood is small and requires constant vigilance by the athlete to ensure a continuing source of glucose during the competition. The heavy reliance on muscle glycogen and blood glucose to fulfill the energy needs of working muscles demands a high level of consumed carbohydrates before exercise and carbohydrate-containing sports beverages during exercise.

In a study of the performance outcomes of a moderate (39 percent of total calories) versus high (65 percent of total calories) carbohydrate intake, the higher-carbohydrate diet significantly improved intermittent exercise performance. Repeated sprint work is enhanced by consumption of a arbohydrate-electrolyte beverage. Although it has been long established that consumption of a carbohydrate-electrolyte beverage enhances submaximal endurance performance, only recent studies have clearly shown the benefits of these beverages during high-intensity, short-duration efforts such as those found in football and basketball.

Subjects performed seven additional 1-minute cycling sprints at 120 to 130 percent of peak $VO_2$when they consumed a 6 percent carbohydrate-electrolyte beverage, as compared with a water placebo. This finding suggests that a dramatic improvement in sprint capability during the last 5 to 10 minutes of a basketball game is possible if players methodically consume a sports beverage. A similar study determined that sports drinks (i.e., carbohydrate-electrolyte beverages) can help maintain high-intensity efforts during high-intensity activities consisting of intermittent sprinting, running, and jogging. Again, these findings have strong positive implications for sustaining high-intensity activity over the course of a typical basketball or soccer game.

The effects of consumption of the individual components of a sports beverage (electrolytes, water, or carbohydrate) and of the combination of all components have also been assessed. Compared with the electrolyte-only trial, performance during the water-only and carbohydrate-only trials was approximately 6 percent faster. However, the combination of carbohydrate and water caused a performance enhancement that was approximately 12 percent faster than the electrolyte trial and 5 to 6 percent faster than when water only or carbohydrate only were consumed.

These findings support the thesis that carbohydrate enhances water absorption and that the limited carbohydrate storage mandates consumption of carbohydrate during exercise. The high demands on circulating and stored carbohydrate during high-intensity work require a constant vigilance to ensure proper and speedy replacement. This study was based on earlier work that showed exercise performance improved significantly with carbohydrate feedings. It has also been found that the optimal level of carbohydrate concentration during exercise is 6 to 7 percent. This concentration is best for fluid absorption and also helps to efficiently deliver carbohydrate. An 8 percent carbohydrate solution causes a slower fluid absorption. A basketball player leaping for the ball and a soccer player sprinting toward the ball and jumping high to kick it are activities comparable to certain forms of strength training. A study of resistance-trained athletes found that athletes tended to perform more repetitions of the same weight when carbohydrate was consumed versus a water placebo.

Blood glucose and lactate concentrations were higher with the carbohydrate trial, suggesting that more carbohydrate was available and used to sustain the high-intensity exercise. A study of head-to-head comparisons of Gatorade, Powerade, and All Sport found that Gatorade stimulates fluid absorption faster than either Powerade or All Sport.

This difference can be attributed to both the type of carbohydrate and the concentration of carbohydrate in the beverages. Gatorade has a carbohydrate concentration level that is consistent with the positive findings in virtually all the studies (6 percent) and contains an equal mixture of sucrose and glucose. Powerade and All Sport have higher carbohydrate concentrations, mainly from fructose. Fructose has been shown to cause gastrointestinal (GI) distress; it is also less efficient in sustaining blood glucose because it requires a secondary conversion in the liver after absorption.

Table. ACSM Fluid-Consumption Guidelines.

Timing	Amount	Adaptation
2 hr before exercise	Drink 500 ml (.5 L or approximately 17 oz).	None
During exercise	Drink 600 to 1,200 ml (.6 to 1.2 L. or approximately 20 to 40 oz)	Drink 150 to 300 ml (about 5 to 10 oz) every 15 to 20 min. per hour.
After exercise weight changes, compensates for urine	Based on pre- and postexercise body weight. This amount	Drink 150% of the amount needed to restore body drink enough fluid to restore body weight losses, which may induce (16 oz fluid – 1 lb of hypohydration when only body weight).100% of the weight-replacement fluid is consumed.

Athletes who perform repeated or sustained high-power efforts experience a reduction in performance when they are dehydrated. A 6 percent carbohydrate solution aids in fluid delivery, a fact that should be considered when team sport athletes select a rehydration beverage. Several general nutrition guidelines-covering what to do before, during, and after exercise and competition-are important for virtually all athletes involved in sports that have intermittent periods of maximal intensity.

The two keys to these guidelines are fluids and carbohydrates in the context of a generally varied diet. Athletes should explore workable strategies to consume both fluids and carbohydrates at every opportunity. Recent findings tend to contradict the traditional and commonly followed belief that carbohydrate-containing beverages are useful only for endurance (aerobic) activities lasting longer than 60 minutes. The best predictors of athletic performance are maintenance of blood volume and maintenance of glycogen and glucose. What follows are some strategies that might be useful for achieving both enhanced hydration and improved maintenance of system carbohydrate in different sports.

A Sampling of Sports Relying on a Combination of Anaerobic and Aerobic Metabolism

Certain sports require a combination of aerobic and anaerobic metabolic processes. At times, these athletes work at maximal intensity (anaerobic), while at other times they work at submaximal intensity (aerobic). Imagine a soccer striker who is jogging to cover her area of the field, working to get an open position. At the moment the ball is passed to her, she sprints to the ball with maximal effort. This combination of anaerobic and aerobic effort has special nutrition considerations that are presented as examples in the following sports.

NUTRITION TACTICS FOR POWER ATHLETES

Different activities place special metabolic requirements on muscle systems, and these differences alter the nutrition requirements among athletes involved in various types of sports. Sports that require a high level of power and speed over short distances have a high anaerobic component. Athletes in these sports are not interested in their ability to move efficiently over long distances for long periods of time; they want to be there first in short distances. When a baseball player steals a base, there is virtually nothing about that 4- to 5-second experience that requires aerobic efficiency. The sprint to the next base is entirely dependent on anaerobic metabolism, which is almost entirely dependent on phosphocreatine and glycogen as fuels. Bodybuilders need explosive power to train but almost never place continuous stress on the muscles for longer than 1.5 minutes, which is the approximate time limit for anaerobic activities.

There has been an evolution in the way athletes eat to support top athletic performance. Around AD 200, Diogenes Laertius wrote that the training diet of Greek athletes of the time consisted of dried figs, moist cheeses, and wheat products.1 American Olympians at the Berlin Games of 1936 had a daily intake that included beefsteak, lots of butter, three eggs, custard, 1.5 liters of milk, and as much as they could consume of white bread, dinner rolls, fresh vegetables, and salads. With each successive Olympic Games, athletes have consumed certain foods and avoided other foods depending on the state of nutrition knowledge. Since the 1960s, however, there has been a purposeful scientific effort to learn what athletes need and why they need it. This scientific endeavor has led to a much-improved understanding of how muscles work for power and how they work. for speed. The science of sports nutrition has also helped us understand the different nutrition demands associated with different types of activities. A failure to consider the nutrition implications of the activity will most certainly lead to problems in training and to performance outcomes that are below the capabilities of the athlete.

Our current knowledge of nutrition requirements for anaerobic activity is substantial, with the clear understanding of what it is that muscles use in this type of activity: phosphocreatine and glycogen. Of course, there is also a question of how to obtain and sustain the larger muscle mass typically needed by athletes involved in anaerobic activity, and the answer to this question is also known: more calories. However, despite the well-established nature of these facts, anaerobic athletes place an unyielding focus on protein intake to satisfy the phosphocreatine, glycogen, and muscle requirements of their activities.

ANAEROBIC METABOLIC PATHWAYS

Athletes have the capacity to obtain a limited amount of energy quickly without oxygen. Events that are predominantly anaerobic (i.e., they require maximal power and energy over a limited period of time) rarely last longer than 90 seconds because the anaerobic energy supply would be exhausted. In some anaerobic events, such as boxing, each round is followed by a period of rest to allow the cells to prepare for the next bout of intense work. What follows is a description of the anaerobic metabolic pathways.

PHOSPHOCREATINE (PHOSPHAGEN) SYSTEM

Anaerobic metabolic processes supply ATP from phosphocreatine (PCr) and glycolysis without oxygen. The in-muscle concentration of preformed ATP is 25 to 33 percent of the concentration of PCr. The enzyme creatine kinase can break apart PCr into inorganic phosphate and creatine, with a resultant release of energy. The free inorganic phosphate is united with ADP to reform ATP. The breakdown of PCr is not reversible until energy is obtained from other sources (mainly through oxidative metabolic processes). The volume of energy that can be supplied by the breakdown of PCr is vast, and it can be produced instantaneously. However, the length of time that this high volume of energy can be supplied is never greater than 10 seconds because of the limited amount of PCr stored in tissue.

The formation of energy from PCr breakdown is directly linked to the intensity of exercise: The higher the exercise intensity, the greater the reliance on PCr breakdown as a source of energy. Athletes performing maximal exercise for 8 to 10 seconds (sprint, vault, jumps) must take a break of 2 to 4 minutes to allow for the regeneration of PCr before undertaking another maximal bout of exercise. Creatine monohydrate supplementation is popular because athletes want to increase the storage of PCr, with the hope of increasing both capacity and power.

GLYCOLYSIS (GLYCOLYTIC SYSTEM)

Glycolysis refers to the anaerobic breakdown of glucose or glycogen for energy. There is a delay of 5 to 10 seconds after the initiation of activity before the glycolytic system can supply energy to working tissues. The six-carbon glucose molecule is phosphorylated and broken into two three-carbon molecules (glyceraldehydes-3-phosphate, or G3P). Each molecule of G3P is converted into pyruvate with the formation of ATP.

The glycolytic reaction creates two molecules of ATP for each molecule of glucose; it creates three molecules of ATP for each molecule of glucose if glycogen is the initial substrate. The pyruvate may be converted to acetyl-CoA for later storage as fat or conversion to lactate. In either case, the fat or lactate created from pyruvate can be oxidative sources of energy. Glycolysis has half the power to create energy as the PCr system but has three times the capacity. The combination of PCr and glycolysis can support predominantly anaerobic maximal work for approximately 90 seconds, often referred to as the anaerobic maximum.

A Sampling of Sports Relying on Anerobic Metabolism

Anaerobic sports require maximal effort over relatively short periods of time. Imagine a baseball player swinging the bat or running to first base; imagine a gymnast sprinting down the runway to perform a vault. In both of these cases the athlete is using mainly existing energy stores that are both limited and easily depleted.

You could never imagine a baseball player running to first base, back to home, and back to first base over and over again for 30 minutes with the same power and speed shown while running to first base a single time because it is physiologically impossible to do so. Nor could you ever ask a gymnast to repeatedly perform the floor routine over and over again without a break because it would be impossible to do. What follows is a sampling of sports that have these special characteristics: They involve extreme intensity performed with breaks between each bout of intense effort.

BASEBALL

Baseball is a wonderful sport that requires an almost equal combi-nation of teamwork and individual effort. It's also a highly mental game, requiring that the athlete stay constantly alert to make split-second judgments for the right play. It's safe to say that physically tired baseball players are also likely to be mentally tired (glucose is the fuel for both the brain and muscles) and prone to bad judgment and poor physical performance.

David Halberstam, in his book *Summer of '49*, describes the 1949 pennant race between the Red Sox and Yankees. A central theme of this book is how players get worn out during the long baseball season, with the outcome of the pennant race determined, to a degree, by the number of players who remain relatively fresh by the end of the year. Clearly, many factors contribute to wear players down during a long season, including frequent travel, hard-fought games, and constant time-zone changes.2 Nutrition factors also come into play: The foods and fluids consumed over the long summer and fall season do make a difference in the outcome.

When steak and beer are constantly on the menu, as was common for many baseball players in the past, it is predictable that physical and mental fatigue will eventually take its toll. Alcohol interferes with B-vitamin metabolism (and therefore energy metabolism) and also increases dehydration risk. Red meat is a useful means of supplying good-quality protein, iron, and zinc, but it should not be the focus of a baseball player's diet. What baseball players really need is plenty of bread, cereal, fruits, and vegetables to constantly replace the glycogen used up in the quick and powerful actions of the game, as well as enough calories to support their muscle mass. Keeping this in mind, baseball players must consider the following nutritionally relevant factors for their sport.

As a primarily summer sport, baseball is often played in a hot and humid environment. Optimally hydrated muscles are composed of more than 70 percent water, and it should be the athlete's goal to maintain this optimal hydration state. A failure to do so will lead to a progressive reduction in total body water with a concomitant reduction in athletic performance.

Evidence suggests that poor hydration makes an athlete more prone to injury by reducing mental function (poor hydration is associated with higher core temperatures that can reduce coordination) and by making muscles less resilient (thus increasing the risk of muscle tears and pulls).

Baseball players (particularly pitchers) are known to experience a reduction in peak torque arm strength between pre- and postseason measurements, with some of this power reduction caused by overuse injury to the pitching arm. Reduced throwing power could also be due to reduced leg strength, which could negatively alter the throwing motion and exacerbate the risk of injury. It seems likely that the degree to which the progressive reduction in power occurs could be reduced with a regular program of optimized hydration and energy intake. A study of baseball players strongly suggests that conditioning plays an important role in the ability of the athletes to maintain an optimal hydration state. At fixed exercise intensities, the better-conditioned baseball players were able to maintain body temperature with a lower sweat rate than that of players who were less fit.

Another study found that blood flow to the pitching arm (in pitchers) increased up to 40 pitches but steadily declined after that. By the 100th pitch, blood flow to the pitching arm was 30 percent below baseline. The decrease in blood flow to the pitching arm matches a decrease in the general hydration state of the pitchers. Since it is established that blood volume is a key factor in the maintenance of athletic performance, the performance of pitchers may be strongly influenced by their ability to stay hydrated. Given the possibility of frequent exposure to hot and humid environments, baseball players should consider the following strategy for maintaining their hydration state:

- During the preseason, weigh yourself before and after each game and practice to learn how much weight has been lost. Then determine how much fluid should be consumed during a typical game (1 pint of fluid equals 1 pound of body weight). The goal should be to drink enough fluids to maintain body weight (±1 pound) during the game. Different people have different sweat rates, so your drinking schedule is likely to be different from those of your teammates.
- Drink plenty of fluids before each game (a minimum of 16 ounces 1 hour before the game, followed by a constant sipping of fluids).
- Use each opportunity between innings to consume fluids. Since baseball involves multiple bouts of high-intensity movements, the drain on stored glycogen is high. Therefore, fluids consumed should be carbohydrate containing. Fluids containing a 6 to 7 percent carbohydrate solution are best both for replacing carbohydrate and for encouraging a fast fluid absorption.
- Immediately after the game, eat and drink enough carbohydrate to reconstitute your glycogen stores and body water.
- Avoid or limit the consumption of alcohol and caffeinated beverages. With sufficient intake, both alcohol and caffeine have a diuretic effect that could place an athlete in a negative water balance.

Baseball requires a combination of power and speed, both of which place a high reliance on phosphocreatine and carbohydrate (primarily glycogen) for muscular fuel. Phosphocreatine is synthesized from three amino acids (from protein), so an adequate intake of protein is necessary to ensure that sufficient phosphocreatine can be manufactured.

However, the protein consumption must be in the context of an adequate total intake of energy (calories). It is easier for fully grown baseball players to obtain sufficient energy (baseball players reach their peak at around age 28).

However, younger players must supply enough energy to support the activity plus enough energy to support growth. Inadequate energy intake causes the consumed protein to be burned as a fuel, making it unavailable for use as a substrate for the manufacture of other substances, such as creatine. With sufficient energy, even a modest protein intake of between 1.2 and 1.7

grams per kilogram is sufficient to support the synthesis of creatine and support a stable or growing muscle mass. In the context of an adequate energy intake, baseball players should consume a diet that derives 60 to 65 percent of its energy from carbohydrate, 20 to 25 percent of its energy from fat, and 15 percent of its energy from protein.

Many games are played during the season, with several games played each week. This frequency of games and practices can easily lead to overtraining, with the associated problems of fatigue, weakness, and increased risk of illness.

A key to limiting the impact of overtraining is adequate rest and the consumption of a high-carbohydrate diet to maintain muscle glycogen levels. It has been shown that daily practices or competitions will lead to a progressive reduction in muscle glycogen storage, with a related reduction in endurance and performance. Since baseball players are highly dependent on muscle glycogen as a fuel, a reduction in glycogen from inadequate carbo hydrate intake would lead to a discernable lowering of performance with time.

Games typically last 2 to 3 hours. Normal blood glucose flux is approximately 3 hours. That is, from the time you finish eating a meal, blood glucose stays in the normal range for about 3 hours. After this, blood glucose falls below the normal range (80 to 120 milligrams per deciliter), a physiological event normally associated with hunger. In the exercising person, blood glucose is likely to fall below the normal range more quickly. Because blood glucose is an important factor in maintaining normal mental function, and is also important for delivering fuel to muscles that have exhausted stored carbohydrate, baseball players should take steps to maintain blood glucose during the entire game by consuming a carbohydrate-rich beverage at every opportunity.

Pitchers work harder when pitching than other players on the team, which is why they are able to pitch effectively only every three to five games. Pitchers are better able to maintain muscular power (both in the legs and arms) by maximizing glycogen storage and hydration state before and during each game.

The principle of glycogen loading (a general tapering of activity coupled with a high carbohydrate and fluid intake information) can be followed with starting pitchers because they typically have several days between starts.

The weight and insulating effect of the equipment worn by catchers adds to their energy and fluid requirements. Catchers are constantly in motion, working with the pitcher, and since they tend to play on a more frequent rotation than starting pitchers, it's safe to say that, pound for pound, catchers have the highest energy and fluid requirements of any of the baseball positions.

Index

A

Absorption 106, 107, 109, 110, 111, 112, 113, 114, 115, 116, 117, 118, 123, 125, 126, 127, 128, 129, 130, 131, 132, 133, 134, 216, 223
Acetylation 160
ACYCLOVIR 11
Acyclovir 11, 12
Adaptation 255, 277
Adsorption 51, 52, 53, 54, 56, 57, 59, 107, 108, 109
Adverse Effects 5, 6, 7, 9, 11, 12, 13, 15, 17
Aerobic Metabolism 277, 278
Agonists 38
Allergens 236
Azole Antifungal Drugs 8

B

Berberine 30, 31
Bile Salts 117, 129
Biotransformation 30, 32, 143, 144, 145, 147, 148, 149, 151, 152, 153, 155, 156, 157, 158, 159, 160, 161, 162
Buccal 111, 125

C

Caffeine 245, 251, 252, 253, 256, 282
Carbohydrates 165, 243, 247, 258, 259, 260, 261, 264, 268, 270, 271, 273, 274, 275, 276, 278
Carbon 146, 147, 148, 149, 150, 151, 152, 159
Carcinogenicity 227
Cardinolides 31, 32
Cardiovascular 158, 160
Cotransporters 133, 134
Cycling 276

D

Dehydration 273, 277
Demographic 251
Dietary Adequacy 272
Dietary Supplements 238, 239
Dosage 38, 41
Drugs 233, 236, 240, 263, 266, 267
Dry Powder Inhalation 107

E

Effects of Adsorption 51
Effects of Moisture 58
Elimination 37, 38, 39, 40, 41, 42
Embraces 236
Endocytosis 134
Endurance 255, 267, 268, 270, 271, 272, 273, 274, 275, 276, 278, 283
Enterococcus 67, 79

Epidemiology 66
Ergogenic Aids 273, 274
Esophagus 262, 264
Excipient 51, 52, 53, 54, 56, 57, 58, 59, 108, 109
Extrapolation 220

F

Facilitated Diffusion 133
Farmaceuticals 239, 240
Fatigue 254, 255, 256, 281
Fermentation 72, 74, 75, 76, 141
Formulation 51, 56, 57, 222, 228
Functional Foods 239

G

Garlic 240
Gastric Emptying 261
Gastric Juice 116, 117
Glucuronidation 158
Glycerol 260
Granulation 51, 53, 54, 56, 108
Griseofulvin 6, 7
Gymnastics 275

H

Harmonization 227, 233
Health Concerns 252
Hepatic 40, 42
Herb 233, 235, 236, 238, 240, 248, 250
Hyponatremia 247

I

Idoxuridine 14
Imidazoles 8, 9
Immunoglobulins 15
Indications 6, 9, 13, 14
Inhalation 107, 108, 111, 112, 125, 126
Inhibitors 70, 79, 105
Interactions 7, 8, 13, 14

K

Ketoconazole 8

L

Lactose 74, 75
Lipid Structure 261
Living Organisms 60

M

Magnesium 262, 274
Materia Medica 235, 236
Meaning Medicinal 235
Mental Agility 245
Mental Disorders 246
Metabolic Syndrome 246, 247
Mucosa 110, 113, 115, 117, 125, 127, 128, 129, 130, 131

N

Nanoparticles 128, 130, 131
Nasal 109, 110, 111, 123, 124, 125
Nomenclature 36
Nutraceutical 238, 240, 241

O

Occurrence 225
Opium 235, 236
Overtraining 270, 271, 272, 283
Overuse Injuries 272

P

Parenteral Routes 109, 123
Particle Size 51, 52, 53, 59, 108
Pediatrics 160
Permeability Glycoprotein 134
Placebo Effect 241
Polysaccharides 103
Pulmonary 112, 126

R

Rationale 233
Receptors 38
Reduction 5, 12, 13
Regularities 58
Rejuvenation 250